Also in the Variorum Collected Studies Series

PAUL KUNITZSCH
Stars and Numbers: Astronomy and Mathematics in the Medieval Arab and Western Worlds

PAUL KUNITZSCH
The Arabs and the Stars: Texts and Traditions on the Fixed Stars and Their Influence in Medieval Europe

MARTIN J.S. RUDWICK
The New Science of Geology: Studies in the Earth Sciences in the Age of Revolution

WILLIAM A. WALLACE
Domingo de Soto and the Early Galileo: Essays on Intellectual History

DONALD CARDWELL, edited by RICHARD L. HILLS
The Development of Science and Technology in Nineteenth-Century Britain: The Importance of Manchester

MENSO FOLKERTS
Essays on Early Medieval Mathematics: The Latin Tradition

GERARD L'E. TURNER
Renaissance Astrolabes and their Makers

DAVID ELLISTON ALLEN
Naturalists and Society: The Culture of Natural History in Britain, 1700–1900

WILLIAM H. BROCK
Science for All: Studies in the History of Victorian Science and Education

JOHN HEDLEY BROOKE
Thinking about Matter: Studies in the History of Chemical Philosophy

DAVID L. COWEN
Pharmacopoeias and Related Literature in Britain and America, 1618–1847

BRUCE S. EASTWOOD
The Revival of Planetary Astronomy in Carolingian and Post-Carolingian Europe

ROGER FRENCH
Ancients and Moderns in the Medical Sciences: From Hippocrates to Harvey

VARIORUM COLLECTED STUDIES SERIES

Studies on the Transmission of Medieval Mathematical Astronomy

Dr Raymond Mercier

Raymond Mercier

Studies on the Transmission of Medieval Mathematical Astronomy

Routledge
Taylor & Francis Group
LONDON AND NEW YORK

First published 2004 by Ashgate Publishing

Published 2016 by Routledge
2 Park Square, Milton Park, Abingdon, Oxon OX14 4RN
52 Vanderbilt Avenue, New York, NY 10017

Routledge is an imprint of the Taylor & Francis Group, an informa business

ISBN 13: 978-0-86078-949-9 (hbk)

British Library Cataloguing-in-Publication Data
Mercier, Raymond
Studies on the Transmission of Medieval Mathematical Astronomy. (Variorum Collected Studies Series: CS787).
1. Astronomy – Mathematics – History – To 1500. 2. Astronomy, Medieval.
3. Astronomy, Arab. 4. Astronomy, Renaissance. 5. Indian astronomy.
6. Astronomy, Greek. I. Title.
520.9'02

US Library of Congress Cataloging-in-Publication Data
Mercier, Raymond
Studies on the Transmission of Medieval Mathematical Astronomy / Raymond Mercier.
p. cm. – (Variorum Collected Studies Series: CS787).
Includes bibliographical references and index.
1. Astronomy, Medieval. 2. Astronomy – Mathematics. 3. Science – Greece.
4. Science, Ancient. I. Title. II. Collected Studies.
QB23.M525 2003
520'.9'02–dc22 2003063730

VARIORUM COLLECTED STUDIES SERIES CS787

CONTENTS

This volumes consists of x + 322 pages

PREFACE

The individual articles are placed in context in an introductory chapter, and there is no need to anticipate them here.

However, by way of acknowledging personal debts, I must single out a mentor of my undergraduate days in Montreal, Andrew O'Connor, Arabist and erstwhile chemist, who taught me so much of medieval philosophy and science in classes at the Université de Montréal, classes which I attended in the evening after normal undergraduate life at McGill University. So although my undergraduate studies were grounded in Physics and Applied Mathematics, my interest in the history of science, especially its medieval and oriental aspects, also goes back to those times. The Islamic Institute of McGill University gave me an opportunity not only to begin Arabic, but to discover the pleasure of personal relations with people from world-wide Islam. Those were formative years. Afterwards three years at Cambridge University led to a Doctorate in Physics, but that was a time also of happy acquaintance with Joseph Needham, just when he completed volume three of his Science and Civilisation in China, the volume concerned with Mathematics and Astronomy.

By the early seventies I had given up Physics, or perhaps it would be truer to say that Physics had given me up. In any case I then turned wholly to historical research. Although many people in the community of historians drew me into the currents of research, notably John North and the late Olaf Pedersen (Aarhus), one person who left a profound mark on my work was the late French Sanskritist Roger Billard. He wrote the book which I wish I had written, detailing the results of his rigorous approach to the problems of Indian astronomy, marked by his bold application of exact astronomical calculation to the problem of establishing the scientific content of the Sanskrit texts. The next figure who has strongly influenced me is the Belgian philologist Anne Tihon, whose professional competence with Greek manuscripts is without equal. She is professional also in her enviable determination to get research published, and not to let it gather dust, a fate that still afflicts much of my work.

RAYMOND MERCIER
Spring 2004

PUBLISHER'S NOTE

The articles in this volume, as in all others in the Collected Studies Series, have not been given a new, continuous pagination. In order to avoid confusion, and to facilitate their use where these same studies have been referred to elsewhere, the original pagination has been maintained wherever possible.

Each article has been given a Roman numeral in order of appearance, as listed in the Contents. This number is repeated on each page and quoted in the index entries.

ACKNOWLEDGEMENTS

For permission to reprint these essays I am grateful to the following publishers, institutions and individuals: R. Halleux, *Archives Internationales d'Histoire des Sciences* (IIa+B, VI, IX) ; Julio Samsó, Barcelona (III); Harrassowitz Verlag, Wiesbaden (IV); Cambridge University Press (V); Warburg Institute, London (VII); Dār al-Fikr, Beirut (VIII); M. Hoskin, *Journal for the History of Astronomy*, Cambridge (X); R.C. Gupta, *Gaṇita Bhāratī, Bulletin of the Indian Society for the History of Mathematics*, New Delhi (XI).

ACKNOWLEDGEMENTS

[illegible]

I

The Transmission of the Hellenistic Legacy of Astronomy: an Introduction

The diversity of topics covered in this volume is more apparent than real. The mathematical astronomy of this broad multicultural area is of a piece. Greek, Arabic, Sanskrit, Hebrew, Latin texts all present technical astronomy along the same lines, which is only to be expected since the common thread running through this activity was Hellenistic science. There was also relatively easy communication across this vast region, initially within the Roman and Sasanid empires, and then within the world of Islam. So the term 'medieval' in this context covers a period from late Alexandrian up to, let us say, the time of Copernicus. The geographical extent does not really extend so far as China, even though there was some transmission through Sanskrit, and then Persian into Chinese.

European astronomy, like European science in general, comes into its own from around the time of Copernicus, from which time the scientific thought of the old medieval world ceases to be of much interest. In contrast to what might be termed the European period, medieval astronomy had an astonishing breadth, spanning virtually the whole Eurasian world.

This chapter is divided into two parts dealing respectively with the technical and historical aspects of these articles.

Technical procedures

Faced with a volume of astronomical tables there is clearly a need to understand the various underlying procedures. The author assumed certain primary parameters in its construction, and may or may not have made these explicit in some text that accompanies the tables. The majority of medieval astronomical handbooks can be traced to either the Handy Tables of Ptolemy, or to his *Almagest*. In the *Almagest* the derivations are virtually always quite explicitly given by Ptolemy, and one can easily confirm that the tables conform to the parameters and methods that he gives. In the later handbooks matters are not always so straightforward.

The tables of mean motion tabulate the values of some linear function of time T, of the form

$$L = L_0 + (T - T_0)M.$$

The constant M is the rate of motion, T_0 the epoch, and L_0, which is the value of L at the time of the epoch is sometimes called the *radix* or root of the motion, although it has no particular astronomical significance, being just the value of L at a certain time. The handbooks normally have the full set of nine mean motions, for the Sun, Moon, lunar apogee, lunar node, and the mean longitudes of the five planets.

If the basic constants L_0, T_0, M are not expressly given in the text one can usually fix them with little difficulty by an examination of the tabulation.

In order to pass from these mean motions to the observed, or 'apparent' values of the longitudes, one must add the so-called equation to the mean. This has to be established for the seven objects, the Sun, Moon, and planets. The equations of the Moon and planets depend on the position of the mean Sun calculated at that moment, but are otherwise independent of each other. These equations depended on apogees that were subject to mean motions. Although the equations are grounded more or less in geometrical arguments, these are represented by a variety of functions which are made available through a collection of tables.

Deviations

With T_0, along with each pair L_0, M, 19 parameters in all, we have a complete specification of the mean motions in the system. Once you believe that you have fixed these values there are two obvious questions : are these the correct values for this system, and how do they compare with such a collection of parameters from another system ? Both questions can be answered efficiently by plotting the deviations, a technique established by Roger Billard in his studies of Indian astronomical systems.[1] The deviation is the difference between the value of L found for the system in question, and its value found from modern astronomical parameters. In present day dynamical astronomy there is still a valid distinction between mean and equation, although the mean now depends on non-linear but slowly varying terms as well as a linear expression such as that given for L. The bundle of deviation curves provides a sort of X-ray of the system of mean parameters.

If a deviation curve hardly varies at all in time, and is nearly equal to zero, then we can be sure that the values of L_0, T_0, M have been reasonably well established. For experience shows that generally speaking the medieval handbook provides positions of the Sun, Moon, and planets that are close to those calculated by modern astronomy for those years, at least for the time of the composition of the system. The major exception throughout the period is Mercury, the planet for which seriously accurate parameters were never really obtained.

[1] Billard (1971).

If the deviation curve varies slowly in time, this shows that the motion M in the medieval system differs by a small amount from its modern counterpart. If the slope is negative, this might indicate that the M represents motion in a sidereal reference system (as in the Indian systems), or it might just indicate that in that system the year is too long even though the reference system is tropical (as in the *Almagest*). If the system is sidereal, then the slope represents the rate of precession. There is nothing inherently incorrect in the use of sidereal longitudes, indeed it is the normal practice in the Indian systems.

If the deviation curve is nicely horizontal, but at some distance above or below the axis, then it may mean that a certain bias was introduced deliberately in the system of mean longitude and equation. This was done in a number of medieval Arabic systems, by al-Bīrūnī, al-Ṭūsī, and others, for the sole purpose of avoiding subtraction in the application of the equation. For while negative numbers do not trouble us, in the Arabic context (as in the Greek) it was necessary to lay down rules as to when the equation had to be subtracted rather than added. The use of such a bias, added to the equation, and subtracted from the mean, resulted in a system in which the equation was always additive. Such arrangements are usually clear from the tables and associated text.

Beyond these possibilities one may simply find that a 'poor' set of deviations just indicates that the values L_0, T_0, M have been badly determined, and then there is a need to investigate the source further. It may happen, for example, that the initial value chosen for T_0 is in error by one day, which is quite enough to disturb the whole set of curves, introducing, for example, an error of some 13 degrees in the lunar deviation. It may also happen that one has assumed a Noon epoch in error for a Midnight epoch, which is also quite enough to disturb everything by large amounts.

Clearly access to such a set of deviation curves provides a valuable 'reality check', not only to show the quality of the mean parameters within the system, but to assist in controlling our own determination of those parameters.

Now given the sets of mean parameters from two systems one would naturally like to compare them. If the two systems have different values of T_0, which is more than likely, then the values L_0 will also differ. Then, even if the values of M appear to be similar, a mere inspection of the numerical values of L_0 will reveal little. However from the visual inspection of the two bundles of deviations it will be apparent at once whether the two systems are closely similar or differ substantially.

In Roger Billard's original researches, in which deviation curves play such a central role, he was concerned primarily to find the year for which the bundle of curves forms a tight node. For some systems, such as that of Āryabhaṭa, this is especially striking, the curves forming a tight node around AD 500; the precise year depends on which subset of curves is selected.

Associated with the determination of T_0 is the question of the meridian of reference, which is effectively a fractional part of T_0. If the meridian of reference is varied then the bundle of curves is altered, and if they form a good coherent set there will be a value of that meridian for which the bundle is tightest, and that should correspond to the meridian for which the observations underlying the system were made. Carrying this approach further, one can look for a best fit for both the year and the meridian simultaneously. The quantity to be examined is a variance based on the deviations, as explained in the article V 'The meridians of reference of Indian astronomical canons'. There the variance denoted by Q reaches a minimum for certain values of time and meridian t_0, ϕ_0, and may be approximated near that point by the expression

$$Q = h_{11}(t - t_0)^2 + 2h_{12}(t - t_0)(\phi - \phi_0) + h_{22}(\phi - \phi_0)^2 + h_{00}.$$

In Billard's studies only the variation in year was considered, for he always assumed that the meridian of reference was Ujjain, whose Greenwich longitude is 76;46, and in that case the quantity Q would be $Q = h_{11}(t - t_0)^2 + h_{00}$. When (t, ϕ) are both allowed to vary, (t_0, ϕ_0) is the minimum point of an ellipsoidal surface, and in order to visualise this, it is helpful to draw the level curves. An example is shown in Fig.1, calculated from the canon of Āryabhaṭa, taking the Sun and all the other parameters save Mercury (a selection indicated by the index[2] (01111011111)). In this case the optimum year is 502.1±4.6 and the meridian is 81.0±1.6.[3] The basis of the modern procedures differs from that used by Billard. For while he took the mean longitudes from the old formulae of Brown, Newcomb and Leverrier, I now use VSOP87, the procedure established by Pierre Bretagnon and colleagues at the Paris Observatory, and where he assumed the formula for ΔT that was established by Spencer Jones, I now use a version established by Morrison and Stephenson. In truth neither revision makes a great deal of difference. It is also important to realize that while VSOP87 provides an improvement in accuracy compared with the formulae of Brown, Newcomb and Leverrier, this improvement is slight compared with the uncertainty in the value of ΔT. As to Āryabhaṭa the important conclusion is that not only were the observations carried out in his lifetime but also within India.

---

[2] This notation is explained in article V, 'The meridians of reference of Indian astronomical canons'. The confidence limit is set at 68%.

[3] This argument can be turned around: if the meridian of reference is assumed to be Ujjain then the value of ΔT is implied, and such values agree well with those found by Morrison and Stephenson (1997).

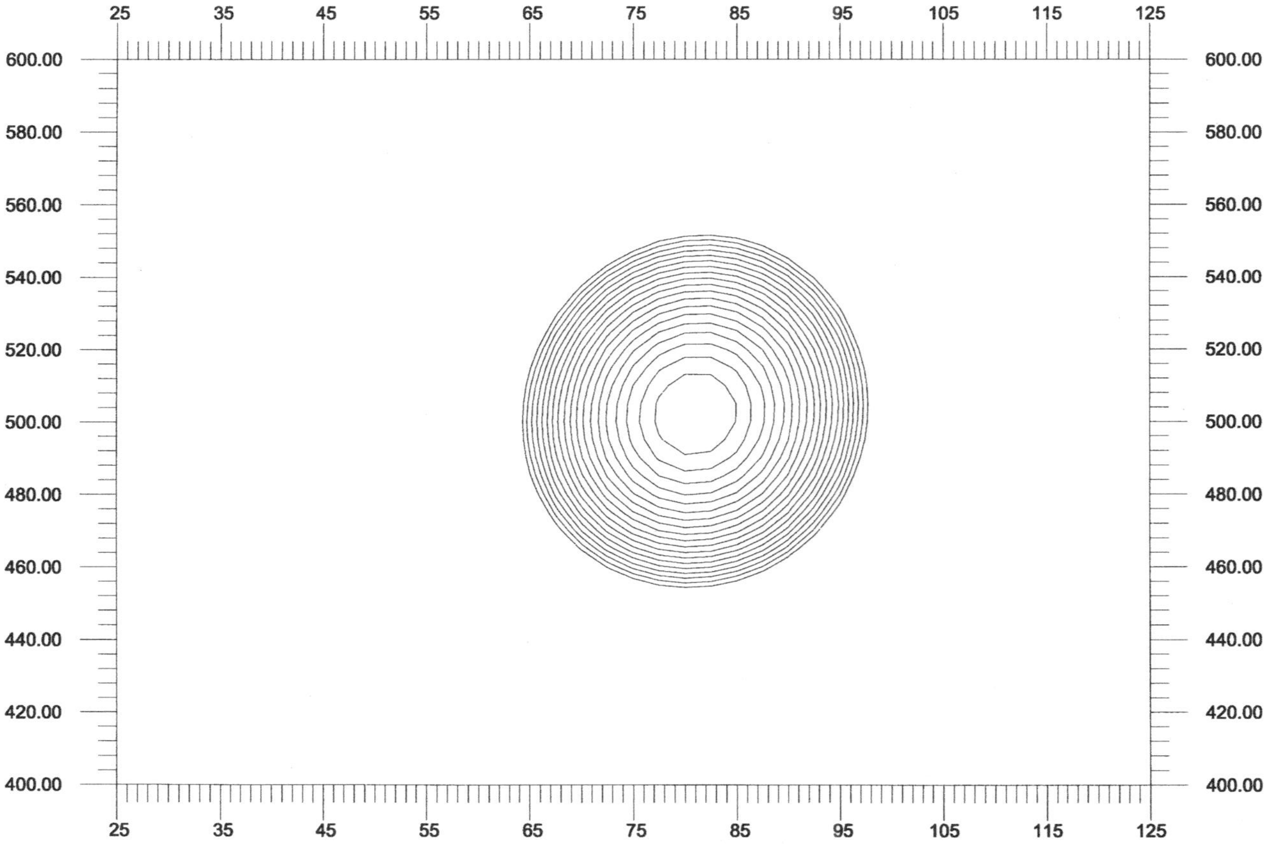

ARYABHATA (mean) statistic 0111101111
Optimum: Year 502.1±5.1, Meridian 81.0±1.7

Fig. 1

The deviation curves so far considered represent only the difference between mean longitudes, modern and ancient. However the study of the differences between true longitudes is also important, but for a different reason. First of all such true deviations, depending on the difference between ancient and modern equations, will reveal very rapid fluctuations, some with periods as short as a month. If such a bundle of curves were plotted over a range of hundreds of years they would appear only as an unintelligible blur, and the first question to be asked would be, what are the mean lines through the rapid oscillations ? The mean lines are of course just the mean deviations already considered, from which essentially statistical conclusions are to be drawn. However the true deviations may be usefully plotted over range of, say, two years, and from such curves one may observe where the true longitudes computed from the system are in exact agreement with the true position at that precise date. In other words this is about individual observations, not about broad statistics.

History

Introduction

In the centuries following Ptolemy the level of scientific thought in the Hellenistic world certainly appears to decline, even in Alexandria. There is some degree of continuity, but one can see in the works of Theon, Pappus, Proclus and Simplicius, and in the specifically astronomical texts of those centuries, a gradual loss of vigour of scientific and mathematical investigation. This is even more evident in the Syrian and Persian context. Syriac literature, in as far as it has any scientific content, is but a pale reflection of Greek sources. The Persian side is difficult to evaluate, because so much of the Pahlavi material has disappeared altogether, although something of its content can be inferred from Arabic versions. In these circumstances the well known rapid revival of Greek science under early Islamic patronage is perhaps difficult to understand. Such a rapid recovery depended on contact with a living tradition of teaching in astronomy, mathematics and geography, not to mention other areas, such as medicine. If Syriac literature as we know it contains relatively few examples of activity in the mathematical sciences, we can be sure nevertheless that only by means of learning from people who were seriously involved, is it possible to understand the extremely rapid maturing of the sciences during the early ʿAbbāsid period. It is impossible to learn these things simply by reading books. Books were indeed read, and beyond that people carried out observations or other relevant practices in order to understand what they read in the books, but a living contact with teachers was essential. It is as if ancient science was already part of the soul of the non-Greek community, and awaited liberation from the heavy hand of New Rome. Scientific activity in the early Islamic period is a true continuation of ancient science.

If the stages of conservation from the Hellenistic world of Ptolemy to that of Baghdad are still rather mysterious, the transmission to India is much more so. There is nothing new in the observation that the Sanskrit astronomical works employed Greek loan words and epicyclic models, and even adopted a canon entitled the Romaka Siddhānta[4] whose meridian of reference was Alexandria, and yet the real paths of transmission cannot yet be documented.

The Latin world of the twelfth century was not as privileged as that of early Islam. Here the transmission was almost entirely through books, even when the Latin translations were made in Toledo, or elsewhere in al-Andalus. There must have been only very limited contact with the living mathematical practitioners in the Arabic or Hebrew speaking world. An exception appears to be provided by the thirteenth century mathematician Leonardo of Pisa (Fibonacci), who as we understand, had direct access to the mathematical community in Islamic North Africa, at Bejaïa (Algeria). The brilliant creative work which he produced shows well what could be achieved in the Latin world when living teachers were involved. The history of Latin science from the twelfth to the early sixteenth centuries is largely one of a struggle to transcend book learning. Only at the end of that long period do we observe Europeans as true masters of scientific subjects.

Sanskrit

For our present purposes we date the origin of Indian scientific astronomy to the *Romaka Siddhānta*, in the early fifth century A.D. This is known from the *Pañcasiddhāntikā* of Varāhamihira, who flourished in the latter part of the next century. He was a prolific author in the category of *jyotiṣa*, which includes not only what we regard as scientific astronomy, but also astrology. The *Pañcasiddhāntikā* is also our only source for the *Sūryasiddhānta* in its early form, a work indeed which is considered by Roger Billard to be by Āryabhaṭa, of the early sixth century.

There is a great number of authors of Sanskrit astronomical canons, constituting a tradition that extended almost to our own day. The individual works are almost always composed in verse, while the commentaries are always in prose. The commentaries are quite essential for a comprehension of the verses, which are concise and compressed in the extreme.

---

[4] So named from Ῥωμαῖος, Roman, but applied, after the rise of New Rome (Constantinople), to Greeks as well as to Romans, and to the provinces of the Eastern Roman Empire. In Arabic *bilād al-rūm* referred to Asia Minor.

For a long period, at least since the authoritative writings of H.T. Colebrooke, people have been of the opinion that the multitude of numerical data that constitute the parameters of any canon were arrived at in a way that owed little or nothing to observation, at least not to observations by the Indian authors themselves. Colebrooke did allow in principle 'that the date of a set of astronomical tables, or of a system for the computation of the places of the planets, is deducible from the ascertainment of a time when that system or set of tables gave results nearest to the truth', but went on to say that 'The Hindus ... cultivated astronomy for the sake of astrology, and for the regulation of religious feasts. They have been content with a very inaccurate practice of it which was, however, sufficient for the purposes of divination and a festal calendar.'[5] In fact from these canons quite respectable calendars had been derived, calendars that were on the whole compliant with at least luni-solar phenomena. If it were not for the courageous efforts of Roger Billard people would even now be of Colebrooke's opinion. While his approach is founded in the common sense process of comparing calculations from the canon with retrospective calculations from modern astronomy, already recognised in principle by Colebrooke, Billard's results nevertheless conflicted with long-held prejudices, and met with an almost totally negative response. This is partly because so many scholars are in thrall to the opinions of David Pingree, and partly because they do not have sufficient confidence in their own capacity for scientific investigation. This in spite of the fact that Billard's account of his work is clear and self-contained, and anyone with any programming skill can readily confirm his results.

In the first section of this chapter this technical development was explained in some detail. As remarked there, the new method of investigation shows that the Indian canons from the time of Āryabhaṭa were based on observations made in India, not on parameters derived from an earlier Greek source. In the concluding section, *Unity in diversity*, of this introduction, I give some details of the method used in Sanskrit sources to calculate the equations of motion.

Arabic

From about the year AD 800 until AD 1500 one finds in the Islamic culture area, and mainly in the Middle East, a series of astronomical activities, based usually on an observatory, and centred in each case on one distinguished figure. These figures were patronised by the rulers of the time. Observatories seemed to have flourished only during the lifetime of the director, or perhaps only as long as one particular ruler was inclined to support it. This contrasts with the situation in modern Europe where for example the observatories at Paris and Greenwich

[5] Colebrooke (1837), pp. 343, 374.

have continued for centuries, long after their original patrons died.[6] In any case the cumulative effect was a more or less continuous tradition of observation and composition of texts which presented the new parameters reduced from the observations.

The astronomical handbooks, invariably referred to as *zījes*[7], consisted of a set of tables by which one determined the positions of the Sun, Moon, and planets, together with tables by which one could calculate the details of eclipses, both solar and lunar. The latter require also tables of conjunctions and tables of parallax. Refraction was ignored, and in any case was not sufficiently understood. The use of the tables was explained in a series of short chapters. Normally these covered only the use of the tables, and said nothing about the derivation of the parameters from observations. In more comprehensive treatises, however, such as the works of al-Battānī, Ibn Yūnis, or al-Bīrūnī, details of some observations were provided, and perhaps an actual derivation of the parameters from them.

In the beginning Islamic astronomy had access to non-Greek materials as well as Greek, and among these the most notable was the *Zīj* of Khwārezmī, an unusual amalgam of Indian, Persian and Greek sources. This is one of the first scientific works to appear in Arabic at this early stage of the Islamic interest in science, that is at the beginning of the ninth century. Al-Khwārezmī, i.e. Muḥammed ibn Mūsa al-Khwārezmī, was not the primary author, but essentially one who adapted the work to Arabic and Persian usage. It was, to begin with, the Sanskrit *Brāhmasphuṭasiddhānta* of Brahmagupta, composed in the early seventh century. The original was a wide-ranging treatise in which the astronomical parameters were presented in the way customary in Indian treatises, that is as numbers written out in verses, without any tabular display. This led by some route to a Persian account of the astronomical section, in a version associated with al-Fazārī. We have no reason to believe that the whole work (with all its mathematical chapters) was translated into either Persian or Arabic. In any case it was referred to as the *Sindhind* by Arabic writers. In form, the *Zīj* of Khwārezmī had a full display of tables, with short chapters giving instructions in the use of the tables. Such an arrangement seems to owe something to the *Handy Tables* of Ptolemy. Since in fact our knowledge of the *Zīj* is based on Latin versions, further remarks will be given in the section on Latin astronomy.

The foundations of astronomical activity in the Islamic world, patronised by the ʿAbbāsid Caliphate, were laid at an observatory set in a suburb of Baghdad, the Shammāsīya. These included primary parameters such as the lengths of the year and the month, and the obliquity of the ecliptic. This was the beginning of

---

[6] Note however that the Greenwich Observatory ceased to exist in October 1998.

[7] The origin of the Arabic word *zīj* is discussed in article IV 'From tantra to zīj',

the observatory in the Islamic world, institutionalised here to a degree hitherto unknown.The underlying theoretical model in later handbooks was that defined in Ptolemy's *Almagest*, but the parameters were rarely identical to those given by Ptolemy.[8] When that work was assimilated by the earliest Arabic scholars in Baghdad, observations were carried out and new parameters substituted.

All the astronomical handbooks included tables of geographical coordinates of various towns, in order that the calculations might be adjusted to sites away from the assumed prime meridian. A very extensive Geography was composed by al-Khwārezmī, effectively a revision of Ptolemy's Geography, and both were drawn upon by the composers of the handbooks. In the early ʿAbbāsid period efforts were made to establish a clear basis for geodesy, by establishing the relation between the Arabic mile and the degree of latitude.[9] This was partly a matter of deciding what was meant by the Roman mile, which had to be distinguished from the Arabic mile. In this geodetic work we see illustrated the tendency to make use of fresh observations in order to clarify the translation of works into Arabic. In this sense their science is to a certain extent 'book-centred', in a way that makes it different from modern science. We do not now carry out scientific activity simply to throw light on the received literature.

AL-BATTĀNĪ (Abu Abdu'llāh Muḥ. ibn Sinān ibn Jābir al-Ḥarrānī). He was active in Raqqa in northern Iraq *ca* 880. He carried out a number of observations there, sufficient to reform all the principal parameters of the motion of Sun, Moon and planets. Much more than a mere observer and technician, he composed the *Zīj al-Ṣābiʾ*, which not only presented a new set of tables, based on the observations, but also a critical review of the problems, with historical notes and details of some of his observations. Manuscripts of his work are not as numerous as its importance would suggest. There is only one copy of the Arabic text, that preserved in the Escorial Library, near Toledo. In Fig. 2 the deviation curves for this zīj are shown. The deviations are smallest near the time of his own observations, but in the second century A.D. a number of the deviations are close to those of Ptolemy's *Almagest*, corresponding to the fact that al-Battānī made use of the observations given in the *Almagest*.

IBN YŪNIS (Abu l-Ḥasan ʿAlī ibn ʿAbdu'l-Raḥmān Aḥmad). He was active at an observatory in Cairo *ca* 1000. Although his work does not survive in a complete copy, there is quite enough to show that he carried out observations with a precision and control rare in any period of the history of astronomy.

---

[8] Perhaps the outstanding exception was the length of the synodic month, a value inherited from Babylonian astronomy, and already so precise that revision was not really needed.

[9] Mercier (1992).

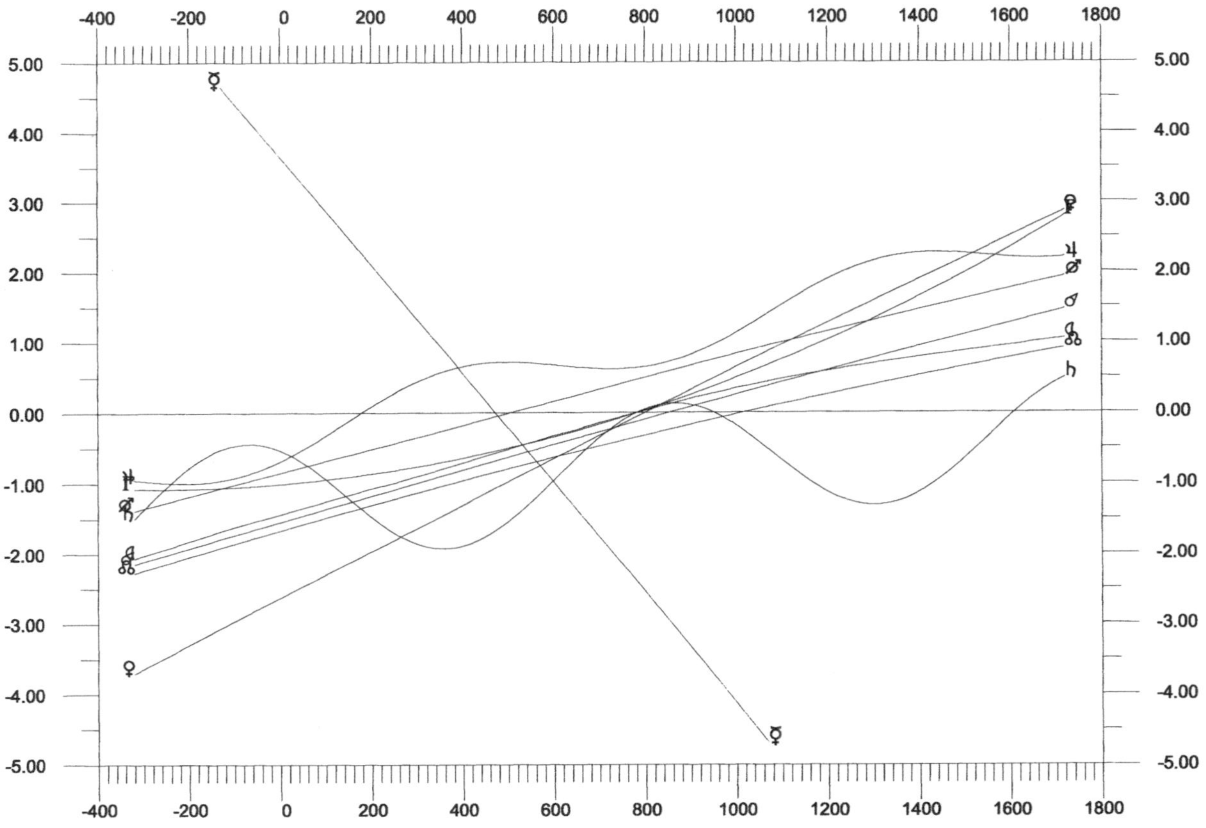

AL-BATTANI (mean)
Meridian: 39.000
Deviations

Fig. 2

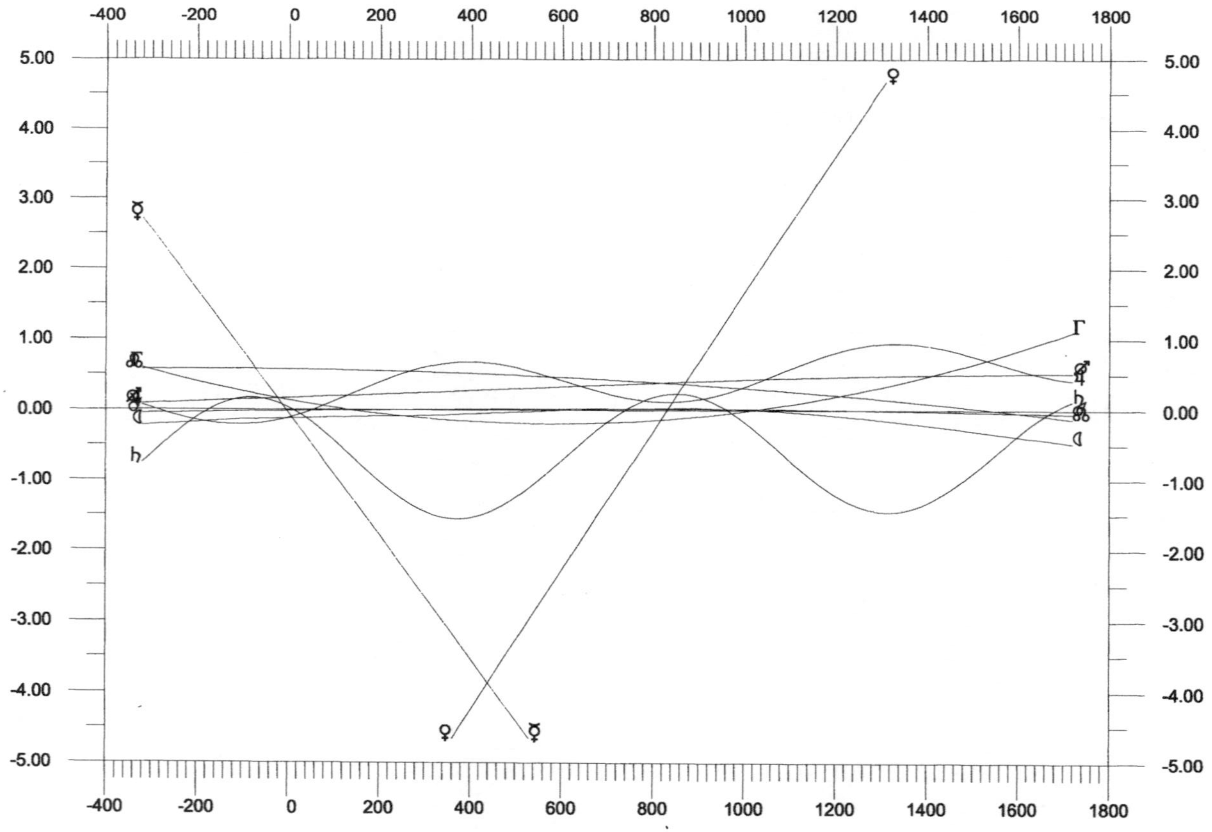

IBN YUNIS (mean)
Meridian: 31.250
Deviations

Fig. 3

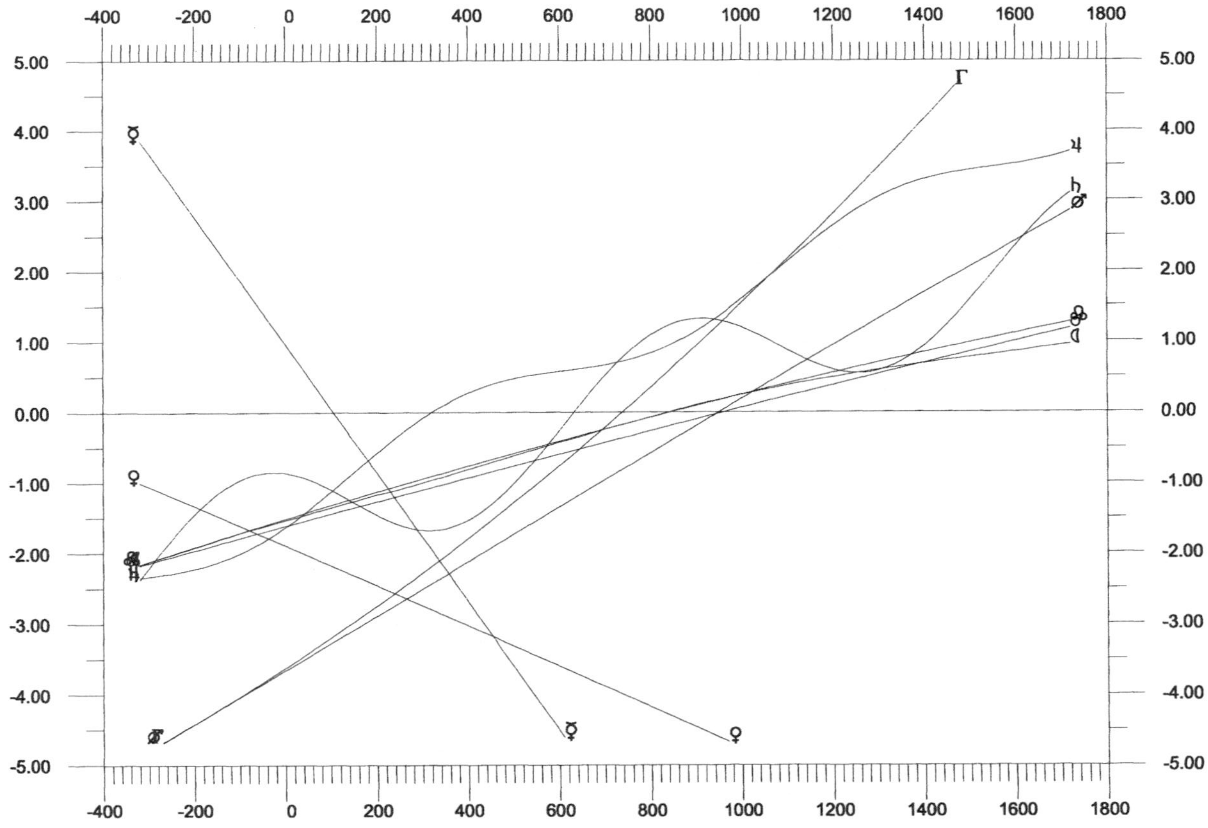

QANUN AL-MAS'UDI (mean)
Meridian: 68.433
Deviations

Fig. 4

The incomplete Arabic text is preserved in the University Library in Leiden. His account, the Zīj al-Kabīr al-Ḥākimī, is of immense historical importance because he lists numerous observations by his predecessors in Baghdad and elsewhere. From this we learn of many works now lost.

When the deviation curves of the Zīj al-Ḥakimī of Ibn Yūnis are examined selecting only the Sun and Moon, with the statistical index (111100000), the optimum year and meridian are found to AD 1006.1 ± 28.0, and 31.1 ± 0.7 degrees East of Greenwich, consistent with observations at Cairo (Greenwich Longitude 31;15), the site of the observations of Ibn Yūnis. Clearly such an approach reveals the serious scientific quality, and the genuine historicity of these famous systems.

In Fig. 3 the deviation curves for this zīj are shown. Apart from Mercury and Venus there is a great improvement over earlier systems, primarily owing to the great improvement in the parameters of the Sun. This was as much the result of his intelligent use of earlier observations as of his own. This improvement in the Sun directly controls the improvement in all the planetary motions. He was notably sceptical of Ptolemy's solar 'observations', which we now know to have been merely calculated from the work of Hipparchus. This is evident from the deviations at the time of Ptolemy.

AL-BĪRŪNĪ (Abu'l-Raiḥān Muḥammed ibn Aḥmad), died 1048. He was active in his native Khwarezm as well as in Iran and India. He was a scientist and scholar in the widest sense, but not a specialist astronomer. His astronomical treatise, the Qanūn al-Mas'ūdī, is of great importance for the historian of the subject, although as a contribution to the astronomical parameters it does not stand comparison with the work of Ibn Yūnis, or even al-Battānī. The deviation curves for the system defined in the Qanūn are shown in Fig. 4, and the contrast with the curves for the system of Ibn Yūnis is painfully evident. Al-Bīrūnī's influence in the Islamic world does not actually seem to have been very great, as if people understood well enough that his work was not really of professional quality, in contrast to Ibn Yūnis, whose astronomical parameters were with good reason widely adopted.

AL-ANDALUS

Islamic astronomical activities in the West, that is in al-Andalus, as the Arabs referred to the Iberian peninsula, seem to have been centred at first on Toledo. The most outstanding of several figures was Zarqāla, *ca* 1080. In spite of his importance it has to be understood that even his work distinctly lagged behind that in the East. Scientific activity began two centuries later than in the East, and was always somewhat provincial in character. The Zījes of al-Battānī and of

Khwārezmī, works that had been truly superseded in the East two centuries earlier, were the basis of research in al-Andalus in the beginning of the eleventh century. The works of Ibn Yūnis and al-Bīrūnī seem to have been unknown in al-Andalus, although the Zīj of al-Ṣūfī, com`posed in the late tenth century, was known, as well as his star catalogue.

ZARQĀLA (Ibrāhīm b Yaḥya), active in Toledo and Cordoba in the second half of the twelfth century. His contributions were centred in two areas: fresh and precise observations of the Sun *ca* 1080, and new types of astrolabe.

Ptolemy quoted a number of solar observations in the *Almagest*, both for the time of Hipparchus, *ca* BC 130, and also for his own time, *ca* 140 A.D.. It is clear not only to us, but to many Islamic astronomers that while the observations attributed to Hipparchus are genuine, those claimed by Ptolemy for his own time were simply calculated from the parameters derived from the Hipparchian observations. These parameters were erroneous to the extent that the implied times of solstice and equinox for the time of Ptolemy were about one day in error. However, in contrast to Ibn Yūnis and others, who simply ignored the pseudo observations of the equinoctial times given by Ptolemy, Zarqāla sought to create a model of the Sun which was consistent with all the solar observations quoted in the *Almagest* and with his own (genuine) observations. This was achieved by means of a non-linear model of precession, represented by an epicyclic motion of the equinoctial points around equinoctial points fixed in relation to the stars. This is the famous 'trepidation', more correctly referred to as 'accession and recession'.[10] The commonly used term 'trepidation' is something of a misnomer, insofar as the Latin *trepidatio* occurs first of all in the Alfonsine tables and its derivatives. Precession in the medieval tradition is the subject of the articles IIA, IIB 'Studies in the medieval conception of precession'. Originally I accepted the common view that the model known through the Latin tract *de motu octave spere* was composed by Thābit ibn Qurra.

However, this is the occasion for a more subtle application of the study of deviations. The Toledan Tables form a system in which the longitudes are sidereal, so that the deviation curves have a downward slope. If however the model of accession and recession is added to the Sun, then we get a deviation which vanishes around 1075, at the time of Zarqāla. This is part of the argument that leads to the conclusion that this model of accession and recession was not due to Thābit ibn Qurra, but to Zarqāla. Certainly without the use of the deviations in this way I could never had discovered this. Duhem, many years ago, had strongly suspected that the model was due to Zarqāla, but the proof awaited this new technique. This was summarised in article VII 'Astronomical

---

[10] In Sanskrit typically *prāk-paścāt*, in Arabic *iqbāl wa idbār*, in Latin *accessio et recessio*.

Tables in the Twelfth Century', and developed fully in II 'Accession and Recession: Reconstruction of the Parameters'. Apart from this innovation, Zarqāla followed al-Battānī closely. These results were incorporated in new sets of tables, one version of which came to be known in Latin as the Toledan Tables, which secured the dominant position in Latin astronomy in the twelfth and thirteenth centuries.

Hebrew

The account of non-Christian activities in Spain and would scarcely be complete without references to work by Jews, who were active in Provence too. Although earlier in the Middle East Jews did not play a strong role in the mathematical sciences, they evidently devoted much effort to reforming and consolidating their own calendar. This calendar, luni-solar in character, and derived from the ancient Babylonian calendar, came to be fixed in essentially its present form in the late Talmudic period. Jewish scholars before the twelfth century were not very forthcoming about it, and the oldest descriptions of it were recorded by Islamic mathematicians including al-Khwārezmī and al-Bīrūnī. In Spain Abraham bar Ḥiyya wrote in 1123 in Hebrew the first detailed account of the Jewish calendar, and also composed other influential works on arithmetic, geometry, and astronomy. Most of these scientific works, but not the work on the calendar, were translated into Latin.

ABRAHAM IBN EZRA While Adelard of Bath represents the exploration of a Christian into the Spanish Islamic world, Abraham ibn Ezra represents in a way the movement in the opposite direction. For this reason it worthwhile to review his life and work in some detail. He lived a long and prolific life, writing on a wide range of topics, not least Biblical exegesis. His scientific work included a number of tracts on astronomical tables. He generally wrote in Hebrew, but it is possible that some at least of his scientific work was composed in Latin. Born *ca* 1090 in Spain, in the town of Lucena in al-Andalus[11], he left Spain for Italy *ca* 1140, going first to Rome, and then to towns in northern Italy: Venice, Verona, Lucca, Pisa, Mantua. After that he resided in southern France for some time, and is associated with Narbonne and Béziers, and other places not identified with certainty. Late in life, shortly before 1160, he visited England, certainly London. He died *ca* 1165, aged about 75. The evidence of these complex travels has been studied by Graetz, and more recently by Ochs, Fleischer and Ben-Menachem.[12] They accept that he travelled beyond Europe, through Morocco, Egypt, the Holy

---

[11] According to Fleischer (1929d), 333; others give his place of birth as Toledo.

[12] Graetz (1861), Vol. 6, 198 seq., and Note 8; articles by Ochs, Fleischer and Ben-Menachem are listed in the References. These relatively obscure writings are gathered here to encourage further study of Ibn Ezra's life and works.

Land, and Mesopotamia, although here the few details must be gleaned from remarks made in passing in his Biblical commentaries. Travels in those regions appear to have taken place before he left Spain for Italy, in 1140, but much is uncertain. His interest in the Zīj of al-Ṣūfī encourages at least the speculation that he was responsible for bringing his works to al-Andalus.[13]

His interest in astronomy is known from two major examples, his Hebrew version of Ibn al-Muthannā's Commentary on the Astronomical tables of al-Khwārezmī, and a work known in Latin which may be seen as an introduction to his tables for Pisa.[14] See articles VII (pp. 108 112) and VIII. The tables of al-Khwārezmī have been noted above. They were the subject of a commentary by Ibn al-Muthannā (Andalusian, perhaps tenth century), a work whose Arabic original is now lost, but which is preserved, at least in part, in Hebrew and Latin. Fleischer showed that this was made by Ibn Ezra in 1160, when he was in London.[15]

Byzantine Greek

The proximity of Byzantium to the Islamic world naturally leads one to expect movement in both directions across the frontier. Indeed some of the earlier transmission to Arabic depended not only on translations via Syriac to Arabic from Greek materials found in Syria, but also from Greek to Arabic directly using, as we know, materials collected in *Bilād al-Rūm* (Byzantine Asia Minor) and Alexandria by various scholars, notably the renowned translator Ḥunain ibn Isḥaq (d. 873).[16]

Translations in the other direction were much more limited in extent, and yet do occur. In this collection article VI, 'The parameters of the *Zīj* of Ibn al-A'lam', presents an analysis of the parameters of an Arabic zīj of the late tenth century that found its way into Greek, where the author was given as Αλιμ or Αληµ The analysis of the Greek material was presented in a companion paper by Anne Tihon.[17] There had been some questions about the true Arabic identity of Αλιμ but this was settled by the investigations in these two articles. In the manuscripts Oxford Bodley Selden supra 16 and Naples Bibilioteca Nazionale II

---

[13] There is no reason to suppose that Adelard brought any scientific material from the Middle East.

[14] Millás Vallicrosa (1947).

[15] Fleischer (1929e), 189. In fact there are two Hebrew versions which have both been edited by Goldstein (1967). While it had been generally assumed that both versions were due to Ibn Ezra, Goldstein argues that only the version that is preserved in an incomplete manuscript in Parma is due to him.

[16] Anawati (1980), p. 230.

[17] Tihon (1989), reproduced in Tihon (1994).

C 33, both Greek MSS of the fifteenth century, there are two horoscopes for the years 1153 and 1162, calculated according to Αλιμ, while in another part of the former MS, a short text gives the mean parameters of Αλιμ. In Vat gr. 1594 the radices for 1032 March 1 are given. On the other hand the Arabic sources, first cited by Kennedy, include Paris Bibliothèque Nationale Supp. pers. 1488 (Zīj-i Ashrāfī), and Bibliothèque Nationale arabe 5968.

Towards the end of the thirteenth century there was a fresh source of material from the Islamic world, or to be more precise the Il-Khanid part of the Mongol Empire, centred on Tabrīz, and beginning with Hülegü Khan (1256 1265). At Marāgheh, not far from Tabrīz, an observatory was established, led by Naṣīr al-Dīn al-Ṭūsī. His principal astronomical production, the *Zīj-i Īlkhānī*, included not only tables of Sun, Moon and planets, but calendrical tables of Chinese and Uighur origin. On the Greek side there appears about this time the *Persian Syntaxis*, associated first of all with Chioniades, and then in the early fourteenth century with Chrysococces, who composed a lengthy tract to accompany the *Syntaxis*. This work turns out to be a direct rendering into Greek of the tables (but not the text) of the *Zīj-i Īlkhānī*, and the account of this transmission is the subject of article IX 'The Greek 'Persian Syntaxis' and the Zīj-i Īlkhānī'.

Nearly contemporary with the *Persian Syntaxis* there appeared an anonymous almanac for Trebizond for the year 1336, based essentially on the works translated by Chioniades, the *Zīj-i Īlkhānī* and the *Zīj al-'Alā'ī* of al-Fahhād.[18] It is argued in my edition that its author was the priest Manuel, who had learned from Chioniades and who had taught Chrysococces.

On the eve of the final collapse of the Byzantine Empire, when it was reduced to a tiny region around Constantinople, and to a small region of the Peloponnese, there appears the philosopher and polymath George Gemistus, known as Plethon, who died in 1455. His great ambition was to reinvigorate ancient Hellenism, and to this end he created a new calendar, with true lunar months and an intercalation, similar in style to that of Hellenic Greece. This was founded on a complex set of tables, not only for the Sun and Moon, but for the planets as well. This work was published by Anne Tihon and myself.[19] This study is summarised in article X 'The Astronomical Tables of George Gemistus Plethon'. The work is remarkable for its dependence on Hebrew sources, such as a Hebrew version of the tables of al-Battānī. But then Plethon no doubt appreciated that among the calendars current in his own day only that of the Jews resembled that of ancient Greece, in that it used both true lunar months and intercalation.

---

[18] The edition of the almanac is Mercier (1994a).

[19] Tihon (1998).

Latin

TWELFTH CENTURY

In the early twelfth century a number of Christian scholars began to take an active interest in scientific activity in al-Andalus. To be sure the interest was not confined to astronomy, but ranged widely over the sciences and philosophy. Some of these scholars were permanently situated in Spain, while others travelled there from a great distance, from points in Northern Europe.

It is important to understand that these people arrived as *students*, prepared to learn not only new materials and texts, but whole new subjects and disciplines, some of which were totally foreign to Latin learning, which was innocent of the mathematical rigour of Greek science.

Apart from the question of language, of the need to fetch material from Arabic, there was evidently the problem of finding people who would act as teachers. It was as true then as now that one cannot really learn mathematics by reading a book. In the East there was evidently a living tradition in the transmission through Greek, Syriac and Persian learning, so that from the outset the new Islamic society could benefit from teachers relatively near to the scientific disciplines of the ancient world. Moreover in the earliest phase of Islam there was evidently an effort to promote peaceful relationships with Jews and Syriac Christians, and the result was a real enough liberation of latent talent.

In the West it was all very different. It does not seem to be the case that the mathematical disciplines were actively taught to Christian scholars in Spain, but only that assistance was given in the translations from Arabic to Latin. Christian scholars of the twelfth century started from the lowest possible level, and one can understand the slowness in the development of an active mastery of the new disciplines. They were very isolated, an isolation that is characteristic of Latin science in almost all its aspects from this time forward. During the long interval up to the time of Copernicus people learned how to manipulate the techniques prescribed in the astronomical texts, and understood a little of the way by which observations can be used to control the numerous parameters. Yet observations were rare and unsystematic, never more than when the Alfonsine tables became officially adopted in the early fourteenth century. Instead people pursued techniques such as the construction of multiple entry tables extended over countless pages in the manuscripts, even to the end of the fifteenth century; that was the only way that astronomy was understood. This is a measure of the profound isolation of the Latin Christian world. With Copernicus there was at last not only a clear grasp of the way in which systematic observations can be used as the foundation of all the parameters, something understood from the very beginning of the Islamic phase, but beyond that Copernicus introduced the wholly original concept 'solar system', on which everything depended in the subsequent development of astronomy. It is only with this concept that one can actually *interpret* the complex calculations of astronomy.

A detailed survey is given in article VII 'Astronomical Tables in the Twelfth Century'. Among the many important works which were translated into Latin we may single out the following, in chronological order of original composition.

Zīj of Khwārezmī. It is curious that the first astronomical handbook to come to the notice of Latin scholars was also the one which was the first to be studied in the Eastern Islamic period, although by the time we know of it in al-Andalus the work had undergone further revision at the hands of al-Majrīṭī (i.e. from Madrid, late tenth century). It was also commented on by Ibn al-Muthannā. In the early 12th century, it was translated into Latin in several stages by Petrus Alfonsi and Adelard of Bath, followed by a revision by Robert of Chester. Further the commentary by Ibn al-Muthannā was translated into Hebrew by Ibn Ezra, and from there (or otherwise) into Latin. This commentary provided a useful introduction to the trigonometrical notions employed in astronomy. It ought be a matter for surprise that the work continued to be of interest in the al-Andalus as late as the 11th century, long after it had been superseded in the East, but that is just a measure of the provincial character of science in al-Andalus. Indeed Christian scholars had no way to know that it was out of date.

Petrus Alfonsi was a converted Jew, formerly Moshe Sefardi, who passed some time in England, apparently as physician to the English King Henry I. His version was not a translation, but a rather unsatisfactory attempt to adapt the work to the Latin calendar, an attempt showing many deficiencies in calculations, and curious intrusions from the Zīj of al-Battānī. All three manuscripts of his work are found in English collections.

The first full translation was undertaken by Adelard of Bath, prompted, one would suppose, by the work of Petrus Alfonsi. Although it is a translation several steps away from the Arabic text known in the ninth century, it is an immensely important work, because all the earlier versions are lost. We see here for example procedures for the determination of the true positions of Sun, Moon and planets which are entirely Indian in character, indeed the same set of rules as prescribed by Āryabhaṭa, and differing from those of the *Brāhmasphuṭasiddhānta.* We know therefore that the work as it passed into Persian and Arabic hands was already the subject of some experimentation within India, or by people who knew a variety of Indian sources and procedures. The problem with this translation is that many of the Arabic technical terms are merely transcribed, without a proper Latin equivalent. If Adelard really visited the Middle East, there is no reason to suppose that he brought back any astronomical material.

The next translation, by Robert of Chester, showed a distinct improvement in the matter of technical vocabulary, for we see here new Latin terms, notably *sinus* for the sine function. There was also a shift in the reference meridian, for while Adelard's version still referred to the Indian meridian through Ujjain, Robert's is shifted 55° west, as if for Toledo.

Zīj of al-Battānī. This work, important and self-contained as it was, nevertheless had been superseded in the Middle East by the *Zīj* of Ibn Yūnis, and others. As already mentioned it was central in the work of Zarqāla. Certainly the work was known to Jewish and Christian scholars in the twelfth century, who not only referred to it but made calculations which demonstrably depended on its parameters. It is a little surprising that no copy of the tables as such survives in Latin, although there are extant copies of the long explanatory text, translated into Latin by Plato of Tivoli. The tables are known from a unique Arabic manuscript. This text contains an account of al-Battānī's determination of the parameters, including for this purpose his citation of eclipse observations. One can just manage to determine from this account the luni-solar parameters which are required for the missing tables. These parameters, as established by Edmund Halley (died 1742) from this text, were employed by him in his tentative proposal of a correction to the lunar motion since referred to as the 'secular acceleration of the Moon'.[20]

Toledan Tables. This set of tables was of the greatest importance in the Christian world of the twelfth and thirteenth centuries. The work was superseded only in the early fourteenth century when the Alfonsine tables entered upon the scene. Most extant Latin manuscripts present a version produced by Gerard of Cremona *ca* 1175. However earlier versions date from *ca* 1120, although they were not always called the Toledan Tables. The tables are the work of Zarqāla, although in the form in which they came to be used in Latin they do not correspond precisely to extant Arabic material. We have in both Arabic and Spanish an 'almanac' of Zarqāla, the parameters of which may be regarded as a revision of those of the Toledan tables. The tables of Zarqāla's pupil Ibn Kammād, known only in Latin, represent another revision.

The first Latin version was prepared of Raymond of Marseille *ca* 1140, adapted to the meridian through Marseille and the Latin calendar. It is clear that he knew well how to use the tables, and he calculated a stationary point for Mars in mid-November 1140. Somewhat later we find the tables adapted by Roger of Hereford, *ca* 1175, to the meridian of Hereford, working as he says from the Marseille version. Around this time there appeared a Latin version of these tables much closer to the Arabic original, and tabulated against the Arabic calendar. This was attributed to Gerard of Cremona. The tables were now

[20] Mercier (1994b), p.192.

accompanied by a translation of the instructions, which indeed went beyond the mere use of the tables, and included a good amount of the trigonometrical material.

During the next century there are further examples of adaptations to other meridians, Novara, Toulouse, etc. There is for example a four-year daily almanac of the Sun, found widely distributed in the manuscripts of the period, demonstrably calculated from the tables for Toulouse.

Tables of London and Pisa. This material which consists of tables and texts is much less well known than the Toledan Tables and their derivatives, but is mentioned in a number of Latin sources associated with Abraham ibn Ezra and Robert of Chester. The true source, according to Abraham, lies in a *zīj* created by late tenth-century Persian astronomer Abdu al-Raḥmān al-Ṣūfī, best known for his star catalogue with its elaborate constellation figures. Since the *zīj* is lost in Arabic this Latin material is of real interest. There is an account of the sources of these tables in article VIII, 'The lost Zīj of al-Ṣūfī in the twelfth-century tables for London and Pisa'.

The tables for Pisa were composed in 1143, as Ibn Ezra says, and we know from Fleischer's extensive researches that Ibn Ezra was in northern Italy at that time. Among his other writings at that time (AD 1146) we have his tract on the Jewish calendar, *Sefer ha-Ibbur*, written in Verona. In another later tract on the Pisan tables he says that he is writing in 1154. The Latin manuscripts and the Arabic references to al-Ṣūfī are explored in article VIII.

THIRTEENTH CENTURY

In the unique attempt by a Latin astronomer of the time to control parameters by observation, William of St Cloud in Paris observed in late March of 1290 the retrograde motion of Mars which in its retrogradation looped around the star β Scorpii; he noted its distance above and below the star on the nights when its longitude was the same as that of the star. Comparing these observations with the calculations from the Tables for Toulouse (based very closely on the Toledan Tables), he inferred the need to make a certain adjustment in the planet's parameters. After making this, and other adjustments, to the tables of Toulouse, according to his own observations, he prepared a full ephemeris for a period of 20 years from 1290. This ephemeris survives in a unique acephalous copy in MS Paris Bibliothèque Nationale latin 16210. It is sad to note that his fine effort, far from marking the start of observational work in the Latin world, was disregarded in favour of the Alfonsine tables which arrived in Paris within 20 years, and which were followed blindly until the time of Copernicus.

Unity in diversity

By way of illustrating the variety of mathematical techniques within the astronomical handbooks, I give here illustrative passages from Sanskrit, Arabic and Latin texts.

From the time of Āryabhaṭa the underlying theoretical framework in all the Sanskrit texts was largely derived from Greek origins, and we find the familiar elements of mean motion, with an equation defined by epicycles.[21] Take for example the statement of the equations of the Sun and Moon from the early *Sūrya Siddhānta,* presumed to be contemporary with the *Āryabhaṭīya*. The Sanskrit text is taken from the part of the *Pañcasiddhāntikā* of Varāhamihira that deals with that work.[22] This is among the oldest of such Sanskrit materials, and although it is not as secure in its manuscript tradition as one would like, it will serve to illustrate the simplest of the equations.

Pañcasiddhāntikā IX.7-8 :

अंशऽशीत्या हीनोऽर्कः केन्द्रं स्वोच्चवर्जितश्चन्द्रः ।
तज्जयाऽर्कसय मनुघ्नी रूपाऽग्नी गुणाशसाङ्कस्य ॥७॥
व्योमरसाऽनल भक्ते तच्चापं द्विस्थितं स्वकेन्द्रवशात ।
प्रथमे चक्रस्यार्धे क्षयश्चयः पश्चिमे भागे ॥८॥

aṃśā'śītyā hīno'rkaḥ kendraṃ svoccavarjitaścandraḥ
tajjyā'rkasya manughnī rūpā'gni guṇā śaśāṅkasya 7
vyomarasā'nala bhakte taccāpaṃ dvisthitaṃ svakendravaśāt[23]
prathame cakrasyārdhe kṣayaścayaḥ paścime bhāge 8

Subtract 80 from the Sun to get its centre (kendra), subtract the lunar apsis from the Moon [to get its centre]. Multiply the sine of the Sun's centre by 14, that of the Moon by 31. Divide by 360, put down twice. According to the centre, in the first half-circle subtract, add in the latter, multiply by the daily motion of the Sun, divide by 21600 as was done with the Sun, likewise with the Moon.

---

[21] Of the various canons described in the *Pañcasiddhāntikā*, however, the *Pauliśasiddhānta* owes something to Babylonian procedures.

[22] There are three published editions of the *Pañcasiddhāntikā*, Thibault (1889), Neugebauer (1970) and Sarma (1993).

[23] In Neugebauer (1970) *svakendravaśāt* is 'corrected' to *śaśāṅkaravau*, Moon and Sun.

Expressed in terms familiar to us, the rule for the equation is evidently $\arcsin((r/360)\sin(\text{mean} - \text{apogee}))$, where $r = 14$ for the Sun, and 31 for the Moon. For both the Sun and Moon the rule is distinctly simpler than that of the Greek models, but one sees the broad unity with Greek astronomy, and not merely because kendra is a loan word from the Greek κέντρον. However there is already the use of the sine function, unavailable to Greek authors. The rules in the early *Sūrya Siddhānta* for the planets similarly follow steps that, while they do not correspond exactly with the Ptolemaic analysis of the deferent-epicycle models, are clearly anchored in that concept.

outer planets	inner planets
$\alpha = \Gamma - \frac{1}{2}\sigma(\lambda_s - \lambda)$	$\alpha = \Gamma - \frac{1}{2}\sigma(\lambda - \lambda_s)$
$\beta = \alpha + \frac{1}{2}\mu(\lambda - \alpha)$	$\beta = \alpha + \frac{1}{2}\mu(\lambda_s - \alpha)$
$\gamma = \lambda - \mu(\lambda - \beta)$	$\gamma = \lambda - \mu(\lambda_s - \beta)$
$L = \gamma + \sigma(\lambda_s - \gamma)$	$L = \gamma + \sigma(\lambda - \gamma)$

where λ and λ_s are respectively the mean planet and the mean Sun, Γ is the planet's apogee, L is the true longitude, and finally, the functions μ and σ, the *manda* (slow) and *śighra* (fast) equations, are defined by these expressions

$$\mu(\theta) = \arcsin(\sin \mu_m \sin \theta) \quad \sigma(\theta) = \arctan\left\{\frac{\sin \sigma_m \sin \theta}{1 + \sin \sigma_m \cos \theta}\right\}.$$ [24]

The intermediate quantities α, β, γ receive names in the Sanskrit texts, not always in the clearest way, although γ at least is named *sphuṭamadhya*, 'true-mean'.

Very similar planetary procedures appear in the Latin version of the *Zīj* of Khwārezmī, where the tables are based on a revision of the parameters of the *Brāhmasphuṭasiddhānta* of. Brahmagupta (seventh century). In the *Zīj* of al-Khwārezmī the steps of the procedure may be recast into a form that is directly comparable to those in Indian texts.[25] The rules in the *Zīj* for the outer planet are as follows :

---

[24] These formulae take advantage of the modern definitions of sin, cos, and tan, and are rather simpler than the Indian formulae.

[25] This is stated rather summarily in article VII 'Astronomical Tables in the Twelfth Century', p. 93.

$$\alpha_1 = \lambda_s - \lambda,\ s = \Gamma - \tfrac{1}{2}\sigma(\alpha_1)$$
$$\kappa_2 = \lambda - s,\ \kappa_3 = \kappa_2 - \mu(\kappa_2)$$
$$\alpha_2 = \alpha_1 + \mu(\kappa_2),\ \kappa_4 = \kappa_3 + \sigma(\alpha_2)$$
$$L = s + \kappa_4$$

Here the notation follows that employed above, but with the addition of various intermediate quantities, $\alpha_1, s, \kappa_2, \kappa_3, \alpha_2, \kappa_4, L$. For the inner planets it is only a matter of interchanging λ_s and λ.

These steps can be combined as follows

$$L = s + \kappa_4 = s + \kappa_3 + \sigma(\alpha_2) = s + \kappa_2 + \mu(\kappa_2) + \sigma(\alpha_2) = \lambda - \mu(\lambda - s) + \sigma(\alpha_2)$$

If we define $\gamma = \lambda - \mu(\lambda - s)$, then $\alpha_2 = \lambda_s - \lambda + \mu(\lambda - s) = \lambda_s - \gamma$.
With this expression for α_2, we have

$$L = \lambda - \mu(\lambda - s) + \sigma(\lambda_s - \gamma) = \gamma + \sigma(\lambda_s - \gamma).$$

Renaming s as β we recover the sequence of steps as they would be given in the corresponding Sanskrit text :

outer planets	inner planets
$\beta = \Gamma - \tfrac{1}{2}\sigma(\lambda_s - \lambda)$	$\beta = \Gamma - \tfrac{1}{2}\sigma(\lambda - \lambda_s)$
$\gamma = \lambda - \mu(\lambda - \beta)$	$\gamma = \lambda_s - \mu(\lambda_s - \beta)$
$L = \gamma + \sigma(\lambda_s - \gamma)$	$L = \gamma + \sigma(\lambda - \gamma)$

This should be compared with the procedures of the *Āryabhaṭīya* (where in this canon both μ and σ depend on variable amplitudes):

outer planets	inner planets
$\alpha = \lambda - \tfrac{1}{2}\mu(\lambda - \Gamma)$	$\beta = \Gamma - \tfrac{1}{2}\sigma(\lambda - \lambda_s)$
$\beta = \alpha + \tfrac{1}{2}\sigma(\lambda_s - \alpha)$	$\gamma = \lambda_s - \mu(\lambda_s - \beta)$
$\gamma = \lambda - \mu(\beta - \Gamma)$	$L = \gamma + \sigma(\lambda - \gamma)$
$L = \gamma + \sigma(\lambda_s - \gamma)$	

Thus the sequence of formal steps of the *Āryabhaṭīya* for the inner planets is that used in the *Zīj* of al-Khwārezmī for both inner and outer planets.[26] From the *Brāhmasphuṭasiddhānta* itself, which was the source of the mean parameters of the *Zīj* of al-Khwārezmī, we have something significantly different :

[26] Note that there is an error in the first line of Billard's formulae for the Āryabhaṭīya, paragraph 'Planète inférieure', where he has M+½, in place of Γ-½; Billard (1971), p. 79.

$$
\begin{array}{lll}
\text{Saturn, Jupiter:} & \text{Mars} & \text{inner planets:} \\
\alpha = \lambda - \mu(\lambda - \Gamma) & \alpha = \lambda - \tfrac{1}{2}\mu(\lambda - \Gamma) & \alpha = \lambda_s - \mu(\lambda_s - \Gamma) \\
\beta = \alpha + \sigma(\lambda_s - \alpha) & \beta = \alpha + \tfrac{1}{2}\sigma(\lambda_s - \alpha) & \beta = \alpha + \sigma(\lambda - \alpha) \\
\gamma = \lambda - \mu(\beta - \Gamma) & \gamma = \lambda - \mu(\beta - \Gamma) & \gamma = \lambda_s - \mu(\beta - \Gamma) \\
L = \gamma + \sigma(\lambda_s - \gamma) & L = \gamma + \sigma(\lambda_s - \gamma) & L = \gamma + \sigma(\lambda - \gamma)
\end{array}
$$

Of the outer planets, only Mars has a variable amplitude, in fact both its apogee and amplitude vary in a very complex way with $\lambda_s - \lambda$. Apart from that, its procedure is the same as that used by Āryabhaṭa for the outer planets. Of the inner planets, Venus has variable amplitude, and apart from the interchange of λ and λ_s, the steps of the procedure are the same as for Saturn and Jupiter.

Turning to an Arabic example, al-Battānī in his *Zīj* gives the procedure for finding the equation of the Sun. This now represents precisely the Ptolemaic deferent-epicycle model. After establishing the model and its parameters, al-Battānī states the actual steps of calculation as follows.[27]

وكلما اردت ان تعمل ذلك حساباً فانظر الى الاجزاء التي يقطع الكوكب او الشمس او القمر من فلك التّدوير من نقطة البعد الابعد وهي الحاصة المذكورة للشمس والقمر وسائرالكواكب فان كانت هذه الحاصّة اقل من قف فاعمل بها وان كانت اكثر من قف فانقصها من شس واعمل بما يبقى. ووجه العمل ان تأخذ الاجزاء التي تحصل لك من احدى الجهتين فان كان اقل ص من فخذ وتره ووتر ما يبقى لتمام ذلك الى ص فاضرب كل واحد من الوترين في نصف قطر فلك التّدوير الكوكب الذي هو وتر جملة التّعديل المنصَّف فما بلغ فاقسمه على نصف القطر فما حصل لوتر تمام الاجزاء فزده على الستّين التي هي نصف القطر فما بلغ فاضربه في مثله وزد عليه ما يحصل لوترفي مثله وخذ جذر ما اجتمع فاحفظه. ثم عد الى ما حصل لتلك الاجزاء مضروبا الاجزاء ايضاً فاضربه في نصفالقطر واقسمه على الجذر الذي حفظت

When you want to do this by calculation, note the degrees from the apogee of the epicycle reached by the Sun, Moon, or planet, this is the anomaly of the Sun, Moon, or planet. If this anomaly is less than 180, use this; if greater, subtract from 360 and use the remainder. Take the resulting degrees in either case. If the degrees are less than 90 take the sine and the

---

27 Nallino (1899), Arabic text, p. 72-3.

sine of the complement (cosine), and multiply each one by the radius (r) of the epicycle of the 'star', divide by the radius (R), increase the cosine by 60 (which is the radius R), square it, add the square of the sine, take the square root of the sum, keep it. Then revert to what resulted for these degrees[28] [the sine times r/R], and multiply by the radius and divide by the square root that you found.

The Arabic passage continues to cover the case when the 'degrees' exceed 90, and finally directs to convert the final ratio to an arc, which constitutes the equation.

This procedure for the equation of the Sun may be written in modern notation as follows; here A is the anomaly, r is the radius of the epicycle, and R is the radius of the deferent,

$$\text{equation} = \arcsin\left(\frac{(r/R)\text{Sin A}}{\sqrt{(60+(r/R)\text{Cos A})^2+((r/R)\text{Sin A})^2}}\right).$$

The more common Arabic term for sine, *jaib*, is not used here, but instead *watar*, literally chord, although the sine is actually the half-chord. In this modern formula the Arabic sine and cosine are distinguished as Sin and Cos, which differ by the factor R, the radius of the defining circle. Al-Battānī draws on a tabulation of the sine (again called *watar*) at intervals of ½ degree, while the values of the equation are tabulated in steps of one degree.

These steps owe nothing to Greek works, yet might have been taken from any one of a number of Indian works, for it is the same as the procedure for $\sigma(\theta)$, the *śīghra* part of the equation defined above, and yet it is here employed by an Arabic writer for the purpose of Greek astronomy. In a typical Sanskrit text, in the course of commenting on the *śīghra*, there occurs the instruction [29]

कोटीफलेन रहिता सहिता त्रिभज्या कार्या कुलीरमकरादिगते स्वकेन्द्रे ॥५(२)॥
तद्वर्गबाहुफलवर्गसमासमूलं कर्णो भवेद् भुजफलं गुणितं त्रिमौर्वया
कर्णोद्धृतं .. ॥६॥

[28] A rather obscure phrase, but the intention is clear.

[29] Quoted from the 9th-century work of Lalla, *Śiṣyadhīvṛddhidatantra*, iii.5-6, Chatterjee (1981), Vol. 1, p. 43. Much the same expressions are to be found in texts and commentaries from the time of Bhāskara I (sixth century).

koṭīphalena rahitā sahitā tribhajyā kāryā kulīramakarādigate svakendre 5b
tadvargabāhuphalavargasamāsamūlaṃ karṇo bhaved. bhujaphalaṃ guṇitaṃ trimaurvyā 6a
karṇoddhṛtaṃ 6b
Karṇa is the square root of the square of the sine plus the square of (cos+radius). The karṇa is divided into the product of the sine and the radius.

Here *karṇa*, 'diagonal', is just the square root expression seen in the modern formula shown above. This clear and concise instruction for $\sigma(\theta)$ is virtually identical to the steps given by al-Battānī. Such an approach seems to us perfectly natural, but that is only because we inherit Indian trigonometry.

Ptolemy's approach is very different, in that he always solves a triangle by considering the circumscribed circle. Thus in any triangle, if the angle at a vertex is A, then the opposing side subtends the angle 2A at the centre of the circumscribed circle; Euclid, *Elem.* iii.20. Ptolemy regards the side of the triangle as a chord of the circumscribed circle, and accordingly draws up a table of such chords as a function of the angle subtended by the chord at the centre of the circle. Ptolemy's procedure, found repeatedly in his demonstrations of eccentric and epicyclic models, seems very awkward and tedious compared with those of the Indian texts, and it is not to wondered at if the Arabic authors followed Indian trigonometry even when developing Greek astronomical demonstrations. As far as I know Ptolemy's use of the circumscribed circle was never used by Arabic authors. The historical problem remains, however: what led the Indian writers to develop their methods ? Although Indian mathematicians receive due praise for their development of an arithmetic based on place value notation, the influence of their trigonometry is even greater, especially in the domain of astronomy.

References

Anawati (1980). 'Ḥunayn ibn Isḥaq al-'Ibādī, Abū Zayd' by G.C. Anawati, *Dictionary of Scientific Biography,* Vol. 15, ed. C.C. Gillispie, New York: Charles Scribner's Sons, 1978.

Billard (1971). *L'Astronomie indienne. Investigation des textes Sanskrits et des données numériques,* Roger Billard, Publications de l'École française d'Extrême-Orient, Volume LXXXIII, Paris, 1971.

Chatterjee (1981). *Śiṣyadhīvṛddhida Tantra of Lalla, with the commentary of Mallikārjuna Sūri,* Critical edition with Introduction, English Translation, Mathematical Notes and Indices, by Bina Chatterjee, New Delhi: Indian National Science Science Academy, 1981.

Colebrooke (1837). *Miscellaneous Essays,* H.T. Colebrooke, Vol. 1, London, 1837.

Fleischer (1929a). 'In what year did Abraham ibn Ezra die ?', Y. L. Fleischer, *Mizraḥ u-Ma'arav* 2 (5689 = 1929-1931) 245-256; [in Hebrew].

Fleischer (1929b). 'Abraham ibn Ezra in Africa', Y. L. Fleischer, *Mizraḥ u-Ma'arav* 3 (5689 = 1929-1930) 81-91; [in Hebrew].

Fleischer (1929c). 'Abraham ibn Ezra in France', Y. L. Fleischer, *Mizraḥ u-Ma'arav* 4 (5690 = 1930-1) 352-360; 5 (5690 = 1930-1) 38-46; 217-224; 289-300; [in Hebrew].

Fleischer (1929d). 'Why did Abraham ibn Ezra leave Spain ?', Y. L. Fleischer, *Mizraḥ u-Ma'arav* 3 (5689 = 1929-1930), 325 - 335; [in Hebrew].

Fleischer (1929e). 'Ibn Ezra and his literary activity in England', Y. L. Fleischer, *Otsar ha-Ḥayyim* 7 (5689 = 1929-1930) 189 - 203; [in Hebrew].

Fleischer (1932). 'Ibn Ezra and his literary activity in Rome', Y. L. Fleischer, *Otsar ha-Ḥayyim* 8 (5692 = 1932-3) 97-100; [in Hebrew].

Fleischer (1934). 'Abraham ibn Ezra and his literary activity in Lucca in Italy', Y. L. Fleischer, *ha-Soqer* 2 (5694 = 1934-5) 77-85; [in Hebrew].

Goldstein (1967). *Ibn al-Muthannā's Commentary on the Astronomical tables of al-Khwārizmī. Two Hebrew versions, edited and translated with an astronomical commentary,* Bernard R. Goldstein, Yale University Press, 1967.

Graetz (1861), *Geschichte der Juden,* H. Graetz, Leipzig, 1861.

Ben-Menachem (1941). 'A study of Abraham ibn Ezra', N. Ben Menachem, *Sinai* 10 (1941) 276-287; [in Hebrew].

Mercier (1992). 'Geodesy', in *History of Cartography,* Raymond Mercier, Volume Two, Book One, *Cartography in the Traditional Islamic and South Asian Societies,* ed. J.B. Harley and David Woodward, Chicago: University of Chicago, 1992, pp.175-188

Mercier (1994a). *An Almanac for Trebizond for the Year 1336,* Raymond Mercier [Corpus des Astronomes Byzantins, VII], Louvain-la-Neuve: Institut Orientaliste de Louvain, 1994.

Mercier (1994b). 'English orientalists and mathematical astronomy', Raymond Mercier, in *The 'Arabick' interest of the natural philosophers in seventeenth-century England*, edited by G. A. Russell, Leiden, 1994, pp. 158-214.

Millás Vallicrosa (1947). *El libro de los fundamentos de las Tablas astronómicas de R. Abraham ibn 'Ezra.* Edición crítica, con introducción y notas por José Millás Vallicrosa. Madrid-Barcelona, 1947.

Nallino (1899). *Al-Battānī sive Albatenii Opus Astronomicu. Pars tertia textum arabicum continens,* Carlo Alfonso Nallino, Milan, 1899.

Neugebauer (1970). *The Pañcasiddhāntikā of Varāhamihira,* O. Neugebauer and D. Pingree, Det Kongelige Danske Videnskabernes Selskab Historisk-filosofiske Skrifter 6,1, two parts, København 1970.

Ochs (1916), 'Die Wiederherstellung der Kommentare Ibn Ezras zu den Büchern Jeremias, Ezechiel, Sprüchen Salomos, Esra, Nehemia und Chronik.', S. Ochs, *Monatschrift für Geschichte und Wissenschaft des Judenthums,* Neue Folge, 24 Jahrgang, Breslau, 1916, 41-58, 118-134, 193-212.

Pedersen (2002). *The Toledan Tables,* Fritz S. Pedersen. Det Kongelige Danske Videnskabernes Selskab, Historisk-filosofiske Skrifter 24, Copenhagen, 2002.

Sarma (1993). *Pañcasiddhāntikā of Varāhamihirā, with translation and notes by T.S. Kuppanna Sastry, critically edited with introduction and appendices by K.V. Sarma*, P.P.S.T. Foundation, Adyar, Madras, 1993.

Stephenson (1997). *Historical Eclipses and Earth's Rotation*, F. Richard Stephenson, Cambridge : University Press, 1997

Thibault (1889). *The Pañchasiddhāntikā, the astronomical work of Varāha Mihira. The text edited with an original comentary in Sanskrit and an English translation and introduction by G. Thibault and Mahāmahopādhyāya Sudhākara Dvivedī*, Benares, 1889.

Tihon (1989). 'Sur l'identité de l'astronome Alim', Anne Tihon, *Archives Internationales d'Histoire des Sciences*, 39 (No. 122, 1989) pp. 3-12.

Tihon (1994). *Études d'astronomie byzantine*, Anne Tihon, Variorum Series, Aldershot: Ashgate, 1994.

Tihon (1998). *Georges Gémiste Pléthon Manuel d'Astronomie* par Anne Tihon et Raymond Mercier, [Corpus des Astronomes Byzantins, IX], Louvain-la-Neuve: Institut Orientaliste de Louvain, 1998.

IIa

STUDIES IN THE MEDIEVAL CONCEPTION OF PRECESSION

(Part I)

1. Introduction

The phenomenon of precession presents the astronomer with a problem of relative motion. Pre-Copernican astronomers however had at best only an intuitive grasp of relative motion, and certainly did not possess sufficient maturity of understanding of kinematics to be able to analyse such a problem in terms of the addition of velocities. Instead they would discover some geometric locus entailed by the constraints of the model representing the physical situation.

The ancient value of the rate of precession, 1° per 100 years, which we owe to Ptolemy, and which was probably used by Hipparchos, was derived by them apparently by a direct study of stellar longitudes. In fact however, both Hipparchos and Babylonian astronomers were in possession of luni-solar data, partly at least of Babylonian origin as Kugler showed, which entailed a much more accurate value for the rate of precession, but which could only have been exploited by an analysis involving the addition of velocities, or something very much like it. From the Almagest (iv. 2), the 'exeligmos'[1] of 126007 $\frac{1}{24}$ days contained 4267 synodic months, and 4612 sidereal lunar revolutions less 7.5°. The synodic and sidereal velocities are therefore 12.1907472763 and 13.1763469568 degrees per day, the difference of which is the sidereal velocity of the Sun, 0.9855996805. When compared with the Hipparchan tropical year, which gives a tropical velocity $360/365.24\dot{6} = 0.9856352784$, we obtain the rate of precession, the velocity of the equinoctial point against the stars, of 0.0000355979, or 46.807″ p.a. All this was pointed out by Biot and Sédillot (12, p. 83; 52, p. 437), and seems obvious enough, and yet Delambre missed it in spite of his extensive study of Hipparchos's discovery, and Manitius in a footnote to his translation of the Almagest (72, vol. 2, p. 196 n.b) clearly makes the mistake of thinking that the exeligmos entails a value of the tropical year. These same considerations led Kugler and Schnabel to feel certain that the Babylonians had discovered precession, since values of the sidereal and tropical years are implicit in Babylonian records, although it became

[1] This long period, which is approximately the product of the 19-year seasonal period with the 18 year eclipse period, must have been discovered in Babylonian eclipse records, and of course the Babylonians must have known of it. I have calculated its mean value, taking the dates of lunar eclipses from Oppolzer (67) and the more exact times from Goldstine (37), and with 44 pairs of lunar eclipses from (A.D. –519, A.D. –174) to (A.D. –471, A.D. –126), the mean value is 126007 days, 75.7 minutes; the number of surplus minutes ranges from 33 to 116. The interval 7.5° is the least accurate element in this description, for it should be nearer to 6.5°.

clear eventually that they did not (58). It is therefore clear that ancient astronomers, both Babylonian and Greek, were not able to make these calculations. A value for the sidereal year (365.2568126) is often presented as 'Ptolemy's', but we must understand that he never quoted any such figure, nor does it seem likely that he could have calculated it exactly in terms of the tropical year and the rate of precession[2]. This is not due to any lack of mathematical skill as such, but to an ignorance of kinematics.

He did not even calculate the sidereal month, nor make any use of the interval 7.5°. The sidereal month occurs only in Babylonian and Indian astronomy, the Babylonian value agreeing exactly with that derived from the exeligmos using 7.5°, as Kugler showed (45, p. 46). The Indian values agree nearly with that derived if one uses 6.5°.

An intuitive approach to the problem is found first of all in Medieval astronomy. Thābit ibn Qurra, for example, calculates the tropical year from values of the sidereal year and a rate of precession (62, p. 284), and it is interesting to translate his argument into a formula expressed in terms of velocities. He writes,

> If you, however, wish to know the time of the solar year, defined by the return to a point of the zodiac, by means of the known motion of the fixed stars, then you can easily find this, provided you know the point from which the sun begins to move in the zodiac; and you should know how much belongs to the sun from the motion of the fixed stars during the year from its mean motion at the point from where it began its motion. And then let us know the length of time in which the sun traverses this amount with its mean motion. Then you subtract this time from the solar year, whose beginning lies at a point of its circle, until it returns to the same.

If v_p, v_s, v_t are the rates of change of stellar longitude, sidereal solar longitude and tropical solar longitude (so that we ought to have $v_t = v_s + v_p$) then Thābit says that we calculate the movement of precession in one sidereal year, which is $v_p(360/v_s)$, calculate the time occupied by the sun in moving through this angle, $360v_p/v_s^2$, and subtract from the sidereal year, $360/v_s$, to obtain the tropical year $360/v_t$. This implies the relation

$$v_t = \frac{v_s^2}{v_s - v_p} = v_s + v_p$$

which only agrees well with the correct relation, because $v_p \ll v_s$.

Abraham ibn Ezra (12th C) (89, p. 82–3) gives an argument to relate the two year lengths, an argument which is repeated uncritically by Roger Bacon (8, p. 16). According to them, one associates with the tropical year a rotation in proportion to the excess of 365 days, so that with Ptolemy's value $365.24\dot{6}$ we have $0.24\dot{6} \times 360 = 88;48$. From the rate of precession 1° per 100 years one obtains another rotation $360/100 = 3;36$, and upon adding these we have 92;24 as the rotation of the

[2] Aaboe (1) has presented calculations corresponding to this discussion of the exeligmos, but directly in terms of the Babylonian parameters, which entail a sidereal year equal to 12;22,8 synodic months (365.2506 days). Babylonian tables of equinoctial dates depend however on a tropical year equal to 12;22,6,10 synodic months. See § 9.2.

sidereal year, 92;24/360 = 0.25$\dot{6}$. This argument, which is much weaker than Thābit's, is both dimensional and kinematic nonsense, and illustrates well the desperate weakness of this vital aspect of medieval astronomy.

Copernicus calculates the rate of change of the tropical solar longitude quite correctly by adding velocities (de Revolutionibus iii. 14), and it is quite clear that by his time there had been achieved, presumably by his immediate predecessors, a clear operational grasp of velocity. So far, it would seem, historians have documented little if any of this development.

There is a curious feature of the Alfonsine tables which may offer a clue to the growing awareness of the addition of velocities. For, the mean tropical rate of the sun's motion as used in these Tables is given quite precisely[3] by the formula

$$\frac{360}{365.25}\left(1+\frac{1}{49000}\right)=0;59,8,19,37,19,13,56,13,\ldots$$

which allows one to believe that this rate was seen as the sum of two elementary rates, both, unfortunately, rather artificial. In any case there is a strict adherence within the Alfonsine tables to a standardized daily rate of motion and to a standardized measure of time measured in natural days and sexagesimal fractions of a day. This represents surely a distinct conceptual advance, and not just a saving of paper!

In calculating the sidereal year itself medieval tables in the Babylonian tradition come closer to the notion of the addition of velocities. Indeed the mean motion of the Sun against the stars was calculated correctly from the lunar rates in the Indian works such as the Sūrya Siddhānta, and in the Khwārizmian and Toledan Tables[4]. In the Sūrya Siddhānta (i. 35; 19, p. 25) we are given the number of lunar sidereal and synodic revolutions in a mahāyuga of 4320000 years as 57753336 and 53433336 = 57753336 — 4320000, respectively, and this relation entails the same result as if one subtracted velocities. Nevertheless it is important to recognize the conceptual difference from the kinematic approach.

In the Khwārizmian Tables the mean sidereal motion of the Sun is 0.985603528464 (calculated from Table 4 in Suter's edition according to which the displacement is 304;15,25 in 720 Arab years; 82, p. 115), while the sidereal motion of the moon is 13.176355401234 (61, p. 92). This entails a synodic rate 12.190751872770, which is equivalent to the larger of the two synodic months, 29;31,50,5,45, used in the table of lunar conjunctions (61, p. 109). The same feature is found in the Toledan Tables (87, p. 80).

It is perhaps strange therefore, but nevertheless true, that precession was not treated in this way, although again the Indian texts seem to verge upon this approach. For example, in the Siddhānta Śiromaṇi (§ 7 of this paper), which takes precession as a continuous rotation, not as trepidation, we are told that the

[3] The full result quoted is found in a version of the tables in Br. Mus. MS Royal 12. D. VI fol. 7r (c. 1400). The printed tables give the first seven figures only.

[4] It must be understood that the sidereal year can never be the result of direct solar observation, but is purely derivative from the lunar periods: the question is whether this derivation is correct or not.

equinoctial point makes 199669 revolutions in a kalpa of 4320000000 years, which implies that 4320199669 tropical years = 4320000000 sidereal years. For some reason however this clear inference is never made, nor is the tropical year ever mentioned.

The notion of velocity clearly requires in turn some notion of a universal time as a dimension of the universe independent of spatial coordinates. The tendency in pre-Copernican astronomy had generally been to regard time as a complex patchwork of incommensurable periods, rather than as a formally continuous mathematical quantity, and the periods in question were expected to have some celestial significance. At first sight nothing would seem more 'natural' than the natural day as a unit of time, but in the Babylonian tradition this seems to have been regarded as a terrestrial rather than a celestial unit. They preferred instead to use complicated calendars geared to celestial phenomena, in particular phenomena plotted against the stars in the night sky[5].

The Babylonian tables present one with a self-consistent account of the night sky, not as unchanging, to be sure, but one in which the changes of planetary position are correlated amongst themselves, rather than with the Sun. The sense of time is therefore rather weak, since one experiences time primarily through solar phenomena. Greek astronomy shows all the opposite characteristics, and seems to be primarily concerned to record the precise solar time of events. It followed that positions of the planets, Sun and Moon were measured from the equinox of date, and that their motion was expressed in terms of tropical and synodic periods.

Clearly a knowledge of precession was necessary in order to relate these two astronomical traditions, and this presents apparently a greater problem to the Babylonian tradition, for there all results calculated from the table need to be corrected for precession in order to obtain a prediction in terms of local time. In the Greek tradition on the other hand, because planetary positions are given directly in terms of the equinox of date, and so in local time, it would seem that an understanding of precession is less important. However, the problem in this tradition is then that of making basic observations in the night sky, making use of the stars for primary measurements, and so the stellar position must be given in relation to the equinox. The Babylonian approach makes more sense on the whole from the point of view of experimental method, and the results were in fact more accurate in this tradition in the Middle Ages. In particular the sidereal year was known with greater accuracy than the tropical year.

In order to find a satisfactory articulation between these two astronomical traditions people had to develop a notion of relative motion, which in turn required a notion of time which embraced phenomena in the nightly as well as the

[5] Even in modern astronomy one has to distinguish between 'ephemeris time' the continuous variable appropriate to Newtonian dynamical theory, and 'universal time', which is a mere counting of days, the difference arising from the fact that the rate of spin of the earth on its axis is slowing down. In modern physics and cosmology the notion of time is once again in the melting pot, largely as a result of Einstein's theories, and it is easier for us, than for Kant for example, to appreciate that time as used in Copernican and Newtonian theory is a specific scientific hypothesis.

daily sky. From this deep problem of selfconsistency there emerged gradually the notions of absolute universal time, and kinematics, which form the backbone of the Copernican synthesis.

We begin the discussion in this paper with an example of a medieval criticism of one of the most influential models of precession, that of Thābit ibn Qurra.

2. *The Toledan Tables in A.D. 1300*

When Guillaume de St. Cloud, a leading figure in Parisian astronomy towards the end of the thirteenth century, wished to compare and corroborate astronomical tables known to him at that time, he proceeded to determine the precise time of the Spring equinox in the year 1290, and he deduced from this the 'movement of the eighth sphere' to be 10;13. This is recounted by Guillaume in his work 'Almanach Guillelmi de Sancto Clodoaldo ad annos 20 incipit anno 1292', BN MS 7281 fol. 141a–144b. Duhem (31, vol. 4, 16–17) and Littré (51) give full accounts of Guillaume's observations. To be more exact, Duhem and Littré repeat clearly what Guillaume reports, but do not explain just how it is that one could ever determine a movement of the eighth sphere, that is a movement of precession, from the date of the equinox. The explanation, however, turns out to be simple enough. The tables which Guillaume had with him included those of Ptolemy, Alexandria, Tolosa and Toledo. The Toledan tables, unlike those of Ptolemy, were based on the sidereal year (taken as 6,5;15,23,29,24 . . . days), and have this feature in common with the Khwārizmian and Indian tables which precede them in a tradition associated, one might say, with astrology and with Babylonian methods.

The calculation of the Sun's position then, according to the Toledan tables, would give its position with respect to a point in the ecliptic which is fixed in relation to the stars, and this would be the correct longitude of the Sun with respect to the equinoctial point only at one particular date. For all tables in this 'astrological' tradition, that date is located somewhere in the range A.D. 450–650, and the point of the ecliptic which serves as a zero of longitude agreed in right ascension with the end point of the lunar mansion known as Baṭn al-ḥūt in Arabic, Revatī in Sanskrit, and K'uei (奎) in Chinese, determined by the junction stars, β Andromedae, ζ Piscium and η Andromedae, respectively.

We may verify Guillaume's figure of 10;13 by making a calculation of the Sun's position for the point in time which he determines as that of the equinox, 16^h after noon of Sunday 12th March 1290 (Julian date 2192301.$\dot{6}$, ignoring the correction due to the difference in longitude between Greenwich and Paris). We simply substitute this date into the Toledan Tables and we ought to get—10;13, so that the movement of the eighth sphere in question is the precessional movement from that date when the equinox was located at a point near ζ Piscium.

The Toledan Tables have been surveyed by Zinner (100) and Toomer (87), but they do not give quite enough information to enable us to carry out the complete calculation of the Sun's true position, for in particular, they do not give the complete table of the equation of the Sun. Moreover in calculating the mean motion of the Sun, one might try to work with a reduction of the table to a modern formula, but this would only obscure certain irregularities which are inherent in the structure of the tables, due for example, to the irregular distribution of the eleven Arab leap years in each cycle of thirty.

Therefore we have made the calculation using one particular copy of these tables, included in Cam. Univ. Lib. MS Ii. 3.3 fol. 117–216 (A.D. 1276). We first calculate the number of days elapsed from 31st Dec. 1 B.C., given by the table on fol. 153a,

1288 years	2, 10, 40, 42 days
1	6, 5
Jan.	31
Feb.	28
12 days of Mar.	12
total =	2, 10, 47, 58

From this one subtracts 1, 3, 3, 35 which is the number of days elapsed from 31st Dec. 1 B.C. up to 14th July A.D. 622, leaving 1, 7, 44, 23 days. Making use of the next table, fol. 153b, this interval is distributed over Arab years and months,

660 years	1, 4, 58, 2 days
28	2, 45, 22
2 months	59
total =	1, 7, 44, 23

On fol. 181a, 181b we have the table of the longitude of the mean Sun in terms of the Arab calendar

660 years	7^s 19;48,54
28	1^s 29;11,33
2 months	1^s 28;9,3
16 hours	0;39,25
total =	11^s 17;48,55

Note that $1^s = 30°$, so the longitude of the mean Sun is 347°48′55″ in the more familiar notation.

In order to calculate the equation of the Sun, which must be added to the longitude of the mean Sun, we first find the anomaly, that is the distance of the mean Sun from the apogee, which is given in the Canons of the tables, fol. 126a, as 2^s 18;50. The anomaly is therefore 8^s 29;58,55. The tables of the equation are given following those of the mean motion and we must make a linear interpolation from the entries.

8^s 29 equation = 1;59,10
9^s 0 ,, = 1;59,4.

Interpolation gives 1;59,4 and so the true Sun has the longitude 11^s 19;47,59. This falls short of the full circle by 10; 12, 1, which is to be compared with Guillaume's figure of 10;13. These are close enough to permit us to claim that he was calculating his result by using the Toledan tables.

Guillaume goes on to refer to that particular theory of precession, Thābit ibn Qurra's model of trepidation, which seems to have been generally associated with the Toledan tables. If this model gave a correct account, and if Guillaume's observation of the date of the equinox were correct, then the precession could be calculated as 10;13 (or 10;12), whereas Guillaume's calculation from Thābit's tables gave him 9;23. Because of this discrepancy he rejected, it seems, not only the theory of trepidation, but the Toledan tables as well. These tables were in fact based on an exceptionally accurate value of the sidereal year, 6,5;15,23,29, 24, the modern value being 6,5;15,22,54. After this time Latin astronomers generally used the Alfonsine tables which were based on the tropical year, and it was not until Copernicus succeeded in formulating a new synthesis between this 'astrological' tradition and the Ptolomaic tradition, that astronomers returned to the use of tables based on the sidereal year.

Turning now to Guillaume's calculation of the figure 9;23, we find that the tables in Thābit's work *de motu octauae sphaerae*, which are given also in the MS mentioned above (fol. 88b–89a), require us to calculate first the *motus octauae sphaerae*, which is given by the rule,

1;34,2 + (2;34,58) (Arab year/30).

The tables give the values of this expression for intervals of 30 years and 1 year, so that we have

660 years	*motus*	=	1^s 28;23,20
28			2;24,38
1/6			0; 0,51
	total		2^s 0;48,49

One now turns to the second table for the precession itself, here called *equatio diversitatis longitudinis capitis arietis*, whose values are given very roughly by the rule 10;45 sin *(motus)*. The entries are in steps of 5°,

motus	*equatio*
60	9;17,44
65	9;43,53

A linear interpolation gives 9;21,59 corresponding to the *motus* 60;48,49. While this differs slightly from Guillaume's result, it is near enough, like the former calculation of 10;13, to permit us to claim to have followed his own calculation.

The table of values of the *equatio* is given in full later (Table III).

The precise time of the Spring Equinox can be determined in accordance with modern parameters by using Tuckerman's Tables (88), from which we find 2192301.73[6], i.e., 17;50 hours after noon of 12th March. Guillaume's error, then, is about as small as one could expect with non-telescopic observations.

[6] These Tables give the solar longitude at 10 day intervals to an accuracy of 0.01°. At each of the years 1288–1292 a four point inverse interpolation is made using dates either side of the zero in longitude, and then a linear regression curve is fitted to these five interpolated equinoctial

Although Guillaume's observation of the time of the equinox is very accurate, it is useful to calculate the deviation he would have found had he employed the correct time of 0.73 day after noon of 12th March. In the additional 0.0613 day the movement of the Sun is 0;3,38, and this would reduce the deviation to about 10;9.

3. The Toledan and Khwārizmian Tables in the sixth century

The deviation in the calculation of the Sun's position according to the Toledan Tables arises, as explained above, from the difference between the true tropical year and the value of the sidereal year implied in the Toledan Solar Tables, and in view of this we can work out the rate of change of the error. Rough calculations indicate that the deviation is zero at some point in the sixth century, and we make use of this in arranging a more precise calculation.

The Toledan Tables of the mean Sun are all based on the assumption that the motion in 30 Arab years, 10631 days, is 10478;0,21, and this entails a daily rate 0.9856 0867 5885. On the other hand, making use of the modern formula for the length of the tropical year (7, p. 537)

$$365.24219878 - 0.00000614\ T \text{ days},$$

where T is the time elapsed since A.D. 1900 in units of 36525 days, we find that the mean rate of motion of the Sun, in tropical longitude, for the year A.D. 900, situated midway between A.D. 1300 and A.D. 500, is 0.9856 4716 9778. The rate of change of the error is therefore 0.00003849 3893, equivalent to a "precession" of 50.616″ p.a. The deviation in 1290 (2192302) is about 10;9, and so it is zero in A.D. 564.

One ought to consider a small correction due to the increase in the equation of the Sun at the time of the true equinox, for this increases from 1;56,36 to 1;59,4 (as the anomaly decreases from 280° to 270°). This decreases very slightly the rate of change of the error to 0.00003838, which however puts back the date of the zero deviation by less than one year.

This may be confirmed by calculating directly the movement of the eighth sphere in the years A.D. 560–4. The equinoctial dates are calculated by a four-point inverse interpolation from Tuckerman's Tables, to which is fitted a linear regression as for the years 1288–1292 in § 2.

It is interesting to compare corresponding results for the Khwārizmian Tables, which have been edited by Suter (82), and translated with a commentary by Neugebauer (61). These also are based on the sidereal year, although all the parameters are slightly different as seen in the table below. In these Tables the Sun is referred to the meridian of Arin, which is the ancient site of Ujjain, situated 75;46 East of Greenwich (11, p. 53). However Neugebauer gives two illustrations in his commentary (pp. 94, 111) which indicate that in this Spanish recension at least, the real

dates. This procedure is expected to give dates reliable in the second decimal place, perhaps with an accuracy ± 0.002.

meridian for the Solar Tables is around Baghdad, 45° East of Greenwich. The date of the Spring Equinox referred to Baghdad time would be 0.125 days less than for Toledo.

The relevant parameters for the Toledan and Khwārizmian Tables are next given, followed by the dates of Spring Equinox, and the movement of the eighth sphere in these years.

SUN	Khwārizmian	Toledan
mean rate	0.9856 0352 8464	0.9856 0867 5885
radix	113;25,47	113;41,11
apogee	77;55	77;50
equation at equinox	2;11,46	1;56,32

MOVEMENT OF EIGHTH SPHERE

A.D.	Equinox	Khwārizmian	Toledan
560	1925674.82_{40}	—0;2,10	—0;2,41
561	1926040.06_{71}	—0;1,15	—0;1,54
562	1926405.31_{03}	—0;0,21	—0;1,7
563	1926770.55_{35}	0;0,33	—0;0,19
564	1927135.79_{66}	0;1,27	0;0,28

It is seen that the movement changes sign between 562–3 for the Khwārizmian Tables, and between 563–4 for the Toledan Tables. It is very remarkable that these two tables, which differ in every one of the four solar parameters, point nevertheless to the same year when the sidereal longitude of the Sun equals its tropical longitude.

Of course these calculations which we have made to find the deviation at the true equinox, were not available to Medieval astronomers, and there is no way in which they might have inferred from the internal structure of these solar tables that they were correct in A.D. 563. It is also unreasonable to suppose that the date of the equinox for that year was known with sufficient accuracy to enable the solar tables to have been adjusted then, or at a later time, so that they were without error in A.D. 563. In any case it is not clear in what way medieval users were concerned to interpret what we have called the deviation.

In fact one can only make a simpler inference, viz., that as part of some convention of measurement, the solar longitudes were measured from some point fixed in relation to the stars, since after all, the tables are based on the sidereal year. While we can easily find the position of the equinoctial point for A.D. 563, it is again not the case that the medieval astronomer could have calculated the same position. The modern determination is made using the formula (4, p. 530)

$$p = 50.2564'' + 0.0222''\ T \text{ p.a.}$$

where T is the time elapsed since A.D. 1900, expressed in units of 36525 days. Between A.D. 563 and A.D. 1900 the mean rate is therefore 50.108″ p.a., and the longitude of the equinoctial point is 18;36,34 (A.D. 1900). It is of historical interest to compare this with the longitude of ζ Piscium, which is found by the use of P. V. Neugebauer's Table (38) to be 18;26 (A.D. 1900). Thus the equinoctial point, to which longitudes were referred in the Toledan and Khwārizmian tables, is about 0;10 East of ζ Piscium. This, it must be emphasized, is not necessarily where the astronomer who used these Tables believed the origin of longitude to be situated. Nevertheless a model of precession described in the Sūrya Siddhānta does indeed take the point in question to be just there. This has been known since the researches of Colebrooke into the Indian accounts of the lunar mansions (nakṣatra), (22, Vol. 2, pp. 283–303).

3.1 Indian and Sassanian sources behind al-Khwārizmī

This question of the zero point of sidereal longitude will occupy much of this paper, and in § 5 we explain the historical origin of the zero point as a product of the technique of star mapping in early Hellenistic astronomy.

Although the Ancient and Medieval astronomers knew where to locate the zero point in relation to the stars, there were considerable variations in the choice of date on which the equinoctial point would coincide with it. I know of only one explicit medieval statement of this date, and otherwise the historian is left to infer it from models of precession and trepidation, apart from the study of the Tables such as we have just given. This solitary reference is provided by Isaac Israeli in his elaborate treatise on the Jewish calendar, Yĕsōd ha ʻŌlām (43, p. ix of the first part of the translation), who says that the moving and fixed zodiacs coincided around 750 years before the time of composition of his work, which can be firmly placed in 1310; therefore coincidence occurred in around 560. This is not only the only explicit statement of the date of coincidence, it is also surprisingly accurate. In none of the models of precession to be surveyed in the following pages is the implicit date of coincidence so accurately given.

There is an interesting problem here regarding the date 562/3. The authors of both the Khwārizmian and Toledan Tables must have had it explicitly in mind, for in no other way could it have been arranged that the two solar Tables could agree. One has to decide between a calculation of this date found by making use of some model of precession related to the Greek star catalogue, and actual observations made very close to A.D. 560. No model of precession known to us would give the equinoctial point 0;10 East of ζ Pisc at exactly this date, and certainly none of the pre-Islamic models come near to so doing. This leaves one to suppose there were observations of Indian or Sassanian origin made sufficiently carefully around A.D. 560.

In his recent important study of Indian material, Billard (11) has studied the deviations of the mean longitudes of Sun, Moon and planets compared with the corresponding modern formulae. He actually calculates not the exact date at which any particular deviation vanishes, but rather the date at which the sum of squares of selected deviations is minimum. In the case of the Brāhmasphuṭasiddhānta (A.D. 628) these dates for various selections of deviations range over the sixth century, while in particular the date of minimum value for the sum of squared deviations for Sun, Moon, lunar apse and node, and all the planets save Mercury, is A.D. 562.8 (11, p. 117). From Billard's point of view this date might therefore be regarded as the date which represents overall the origin of the sidereal coordinates. This date is virtually the same as that found above in the analysis of the Toledan and Khwārizmian Tables, and reinforces the conclusion already argued by Suter, Burckhardt and Neugebauer, that the Khwārizmian Tables descend in part from the Brāhmasphuṭasiddhānta, taken to be that Indian system known to the Arabs as the Sindhind. Indeed the parameters relating to the mean motion are virtually the same in these two works. (See: 61, pp. 93, 131 and the references to Burckhardt on p. 235; 82, p. 32; 71, pp. 16, 28.)

However one ought to calculate the date on which the true, not the mean, Sun is exact at the Spring Equinox. The procedure for calculating the equation of the Sun is explained (if that is the right word) in the Brāhmasphuṭasiddhānta (17, ii. 20 seq.), where the basic sinusoidal formulation is inflated by empirical modifications. These do not seem to be of importance in the present calculation, so we take the equation to be given by the rule,

$$\sin \eta = -\frac{41}{1080} \sin (\mathcal{L} - \tilde{\omega})$$

where $\mathcal{L}$ = longitude of the mean Sun and $\tilde{\omega}$ = longitude of the apogee. The date sought may be shown to be A.D. 580, when at the Spring Equinox 1932979.76_5,

$$\begin{aligned} \tilde{\omega} &= 77.9072 \\ \left.\begin{aligned} \mathcal{L} &= 357.8418 \\ \eta &= 2.1415 \end{aligned}\right\} & \ \mathcal{L} + \eta = -0.0167 \end{aligned}$$

The value of η is obtained by a self consistent solution of the equation for η, when $\mathcal{L} = 360 - \eta$, and when the values of sin given in the text are interpolated; one is given the sin for multiples of 3;45:

$$\sin 3;45 = \frac{214}{3270} \quad \sin 78;45 = \frac{3207}{3270} \quad \sin 82;30 = \frac{3242}{3270}$$

(15, ii. 5). The procedure of obtaining $\mathcal{L}$ and $\tilde{\omega}$ is straightforward and one may refer to Billard's useful summary of the formulae.

It is now clear that for the true position of the Sun the Khwārizmian Tables do not derive from the Brāhmasphuṭasiddhānta. Another source must be sought for the explanation of the year 562/3, and this turns out to be the Sassanian Tables known as the Zīj al-Shāh, associated with the reforms of Khusrau Anūshirwān (reigned 431–579). These Tables are no longer extant, and our information concerning them derives from various Arabic sources: the Fihrist of ibn al-Nadīm (33, p. 578), the Qānūn'l-Mas'ūdī of al-Bīrūnī (14, p. 1473) and the summary of Māshā' allāh's astrological history by ibn Hibintā (71, passim). Modern studies are due mainly to Taqīzādeh (84), Kennedy (44), Pingree (71) and van der Waerden (97). For our purposes we need the passage in al-Bīrūnī's work to which attention was first drawn by Taqīzādeh (84, p. 321). This passage gives the date A.D. 556 when Khusrau convened Persian astronomers, and is the principal source associating the Zīj al-Shāh, or at least its reform, with Khusrau specifically. However a closer analysis of the passage enables us to draw much deeper conclusions. The passage is as follows, taken from al-Bīrūnī's account of the astrological system known as the 'Thousands' *(hazarāt)*, (13, pp. 1473–4)[7].

> In their opinion the seven planets and the two nodes (Head and Tail) follow one another with a constant number of years for each; we call this number *fardāriyāt*[1]. They agree that in the twenty-fifth year of Anūshirwān[2], four years of the *fardāriyāt* of Jupiter had elapsed, and there remained eight[3]. Then the *fardāriyāt* of Mercury, 13 years; then Saturn, 11; then the Tail, 2; then Mars, 7; Venus, 8 and the Sun for 10 and the Moon 9 and the Head 3[4], then again Jupiter around the Zodiac in relation to its exaltations, during the space of 75 years.
> We mentioned the above date (twenty-fifth year of Anūshirwān) because at that time the astronomers of Persia met to correct the Zīj al-Shahrīyārān known as al-Shāh, and they recorded the measures of times and stations (al-naubat wa'l-mīnāhā). They knew from the position of Jupiter at the end of the month Ābān[5] that three *hazar* had passed and of the fourth, 851 years. Thus if we discard 75-year cycles 51 times, there remain 26 years[6], which begins with the *fardāriyāt* of the Sun and ends with the first four years of the *fardāriyāt* of Jupiter. From this point to the beginning of the reign of Yazdagird there are 76 years: 23 of Anūshirwān, 12 of Hormuz, 37 of Parvīz, 4 of Shīruya[7]. Thus at the beginning of the reign of Yazdagird 5 years had elapsed of the *fardāriyāt* of Jupiter.

Notes. 1) *fardāriyāt*: for this other related technical terms of astrological history, see Pingree. 2) twenty-fifth year of Anūshirwān: A.D. 556. 3) the *fardāriyāt* of Jupiter ran from 552 to 564. 4) of all these *fardāriyāt*, only that of Jupiter has any astronomical significance, since we have then the sidereal period.5) end of Ābān = early March[8]. 6) 3851 = 51×75 + 26. 7) Khusrau Anūshirwān, 531–579; Hormuz IV, 579–591; Khusrau Parvīz, 591–628; Shīruya (and others), 628–632; Yezdagird III, 632–651.

The planet Jupiter was manifestly the key to this astrological system of *fardāriyāt*, as also to the reform of the Zīj al-Shāh. When the planet was observed in Ābān 556 it was visible during much of the night, but at the termini of the *fardāriyāt*, in 552 and 564, it was in conjunction with the Sun almost exactly at the moment of the Spring Equinox. Nowhere else in the sequence of *fardāriyāt* of Jupiter or the other 'stars' is there any clear astronomical phenomenon at the termini. A fuller discussion of the behaviour of Jupiter will be given later, after Thābit's model of trepidation has been studied. Here we conclude that the true historical explanation of the use of the year A.D. 564 as a 'canonical' point stems from this Sassanian reform, which happened to take place just when the equinoctial point coincided with the sidereal zero point (see § 5), although in the Brāhmasphuṭasiddhānta the same year is relevant in an average sense. These considerations show that the Khwārizmian Tables used the mean motions of the Brāhmasphuṭasiddhānta, but

[7] I am most indebted to Mr H. R. Farran, a research student of this department for the translation.

[8] The New Year, Nauroz, 1st Farvardin, occurred on 16/6/632 at the accession of Yazdagird III, and it is believed that al-Bīrūnī used the vague year, not the intercalated Zoroastrian year; Ābān is the eighth month counting from Farvardīn.

adopted an equation (and perhaps a method of calculating an equation) exactly in line with the Sassanian reform. Suter (82, p. 33) has quoted ibnu'l-Qifṭī who in fact reports just this, 'al-Khwārizmī in his Tables relied on the mean positions of the Sindhind, but departed from it regarding the equations and the declination (of the ecliptic); he established his equations according to the method of the Persians, and the declination of the Sun according to Ptolemy's method'. (See also van der Waerden, 96, 97.)

4. *Thābit ibn Qurra's model of precession*

The theory of trepidation may derive from the desire in various culture areas, to allow for the phenomenon of precession by way of reconstructing in post-Hipparchan times, the appearance of the night sky as it was in 'antiquity', so that ancient religious or astrological interpretations could be conserved. To this extent we are not dealing with pure Astronomy, but with 'astrological history', a recurrent feature in the history of Astronomy. Apart from the well known remarks of Rhaeticus in the Narratio Prima, who gives a historical interpretation of Copernican trepidation, one may also cite recent studies of Abū Ma'shar and Māshā'allāh (71), and Longomontanus (55). We find sometimes that the Eastern limit of the movement of trepidation is a position in the ecliptic of some phenomenon which had marked the beginning of the year; for example, in this way in the Indian models one appears to encounter the heliacal rising of α Arietis and the heliacal setting of the Pleiades.

Thābit ibn Qurra's[9] contribution to the theory of precession is available only in a Latin version entitled *de Motu Octavae Sphaerae* which has been edited by Carmody (83), and translated with a commentary by Neugebauer (62). The geometry of the model has been the subject also of recent papers by Goldstein (36), Dobrzycki (29) and North (64). In spite of these studies there is still much that needs to be established about the model in order to clarify its position in the history of pre-Copernican astronomy. Thābit's contribution was dismissed by Delambre with the famous judgement 'Son malheureaux système de la trepidation infecta les tables astronomiques jusqu'a Tycho.' (26, p. 73), a remark echoed by many later historians. This is harsh and unjustified, for Thābit's model was no more improbable, or lacking in self-consistency than any other feature of pre-Copernican astronomy, and provided moreover better agreement with the stellar longitudes known to Hipparchos, Ptolemy and Arab astronomers than did any other model earlier than that developed by Copernicus.

[9] Thābit is a product of the extremely important culture of Ḥarrān in Northern Mesopotamia (lat. N 36;51, long E 39;1) where one had not only a very conservative maintainance of pagan worship, in the form of a Moon-cult, but also a centre for the diffusion of Greek and Syriac learning. There is an extensive article on the subject in the Fihrist, and Dodge (33, p. 745) in his translation of this work gives further references, particularly to the monumental study of Chwolson, Die Ssabier und der Ssabismus (21), where one will find a biography of Thābit, and a list of his very numerous writings. Ḥarrān was situated 100 km due north of al-Raqqa, the site of al-Battānī's observations.

4.1 *The geometry*

The familiar construction of great circles representing the equator and ecliptic is supplemented for present purposes by a second ecliptic circle. These are called the fixed and moving ecliptics respectively (see Fig. 1), the moving ecliptic being

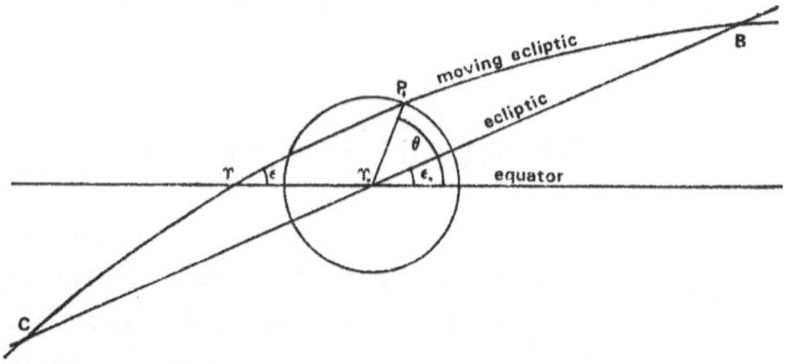

Fig. 1. Thābit's model of trepidation. View from the outdie of the sphere showing the small circle at the Spring Equinox, and the three great circles.

in fact fixed in relation to the fixed stars in the eighth sphere. To complete the construction one proceeds as follows. Taking as centres the two points of intersection of equator and fixed ecliptic, one describes two small circles lying in the sphere, and having a radius r in angular measure. The moving ecliptic intersects the small circles in diametrically opposed points P_1 and P_2, P_1 (Caput Arietis) being on the circle centred on the spring equinox. The moving ecliptic is determined finally, for each position of P_1, by constraining the intersections B and C, where the two ecliptics meet, so that $P_1B = 90°$. An alternative constraint has been proposed by North (64) and Goldstein (36) in which the point B is kept a distance 90° from the centre of the small circle Υ_0, but for reasons which will be given later the present choice is to be preferred. The position of P_1 on the circle is located by the angle θ, called the motus, measured from the equator, as shown in the diagram. Thābit's treatise leaves some uncertainty perhaps as to whether the motus should be measured from this point, or from the point where the small circle intersects the ecliptic. We follow Neugebauer's reading (62, p. 296)[10], measuring θ from the equator. P_1 moves at a constant rate around the circle in the direction of increasing θ, which is indeed clockwise as seen by an earthbound observer (counter-clockwise in the case of P_2). This rotation causes a seesaw motion of the moving ecliptic, and in turn a motion of the eighth sphere, the sphere of the fixed stars, for the moving ecliptic, and also the two points P_1 and P_2 are fixed in that sphere. The point of intersection of the equator and the moving

[10] Thābit remarks (83, p. 103, line 17) that the movement of accession or recession is slowest when $\theta = 90°$, but this does not directly settle the question because we have to decide also whether he measures this movement by the arc $P_1\Upsilon$ along the ecliptic, or by the arc $\Upsilon_0\Upsilon$ along the equator, for these choices are consistent respectively with a measurement of θ from the equator or the ecliptic. Naturally the movement $P_1\Upsilon$ along the ecliptic corresponds best with our notion of precession, and is the choice made here. However Carmody takes the other view. See also the remarks made later, on the early analysis by Pedro Nuñez.

ecliptic denoted ♈ in Fig. 1 is the spring equinoctial point, which moves to-and-fro in relation to P_1, the distance between them reducing to zero when $\theta = 0°$, or $\theta = 180°$. The distance P_1♈ is measured as an angle $E(\theta)$, a certain function of θ, the angle E being positive when $0 \leqslant \theta \leqslant 180°$, negative otherwise. In the Latin text the angle E is called *equatio diversitatis longitudinis capitis arietis ab equatore diei*, while the angle θ is called *motus accessionis et recessionis 8e sphere*. When the *motus* increases from —90° to 90°, ♈ moves Westward relative to the stars, as it is actually observed to do, and we have *accessio* (Arabic 'iqbāl, from the root qabala, to come, arrive), and the end point $\theta = 90°$ is the limit of accession (Arabic al-nihāyat al-'iqbāl). In the next two quadrants $90° < \theta < 270°$, moves Eastward, in disagreement with observation, and we have *recessio* (Arabic 'idbār, from root dabara, to turn away). *Accessio* and *recessio* therefore describe the movement of ♈ when it is respectively with and against the daily motion of the stars.

Thābit, like most writers, tends to use the terms *accessio* and *recessio* together without distinguishing which is which. However he does say (83, p. 104, line 17) 'Huic uero motui predicto accidit ut inuenitur tarde diuersitatis et uelocis diuersitatis, et illud est quoniam cum caput Arietis fuerit super 90 ab equatore diei in duabus partibus septentrionis et meridiei, accessio erit tunc tarde diuersitatis; et cum caput Arietis fuerit propinquum sectioni circuli parui cum equatore diei erit accessio tunc uelocis motus; et similiter inuenerunt consideratores.' In the following line he applies these observations to the work of Ptolemy and others, and so clearly intends that *accessio* is the observed motion of the equinoctial point. Another example helps to fill in the interpretation. The diagram of a model of trepidation reproduced by Hartner from a work of Quṭb al-Dīn al-Shīrāzī (40, p. 628) marks the Western limit of the movement of ♈ as al-nihāyat al-'iqbāl. Many other references to trepidation found in the widely scattered literature bear out this interpretation of *accessio*.

We shall see later that, in Thābit's model, the angle E vanishes near the end of the year A.D. 604. The fixed and moving ecliptics do not coincide then, but at a later date during Thābit's lifetime[11], A.D. 870, at which time $E = r$.

There emerges a complex picture in which the 'moving ecliptic' with its points P_1 and P_2 are fixed in relation to the stars, while the equator, the fixed ecliptic, and the small circles are continually moving. However for the medieval observer it is the equinoctial point ♈ and the equator which are fixed, while the stars and the ecliptic move.

In Thābit's model the moving ecliptic, which is the real ecliptic of observational astronomy, makes an angle ε with the equator, and this angle of obliquity varies between narrow limits as P_1 traverses the circle. The angle between the fixed ecliptic and the equator, ε_0, lies between these limits, and when $\theta = \varepsilon_0$ we have $\varepsilon(\varepsilon_0) = \varepsilon_0$, so that the two ecliptics coincide, and this happens, as noted above, during A.D. 870.

[11] It has been argued by Duhem (31, vol. 2, 257) that this model of trepidation is not actually due to Thābit, but rather to Zarqalu. If that were the case if is not likely that the date A.D. 870 would figure so prominently in the working of the model. Millás Vallicrosa (90) has attempted to refute Duhem's views.

This model of trepidation is worked out in detail in the appendix, where the functions $E(\theta)$ and $\varepsilon(\theta)$ are found to be given by the expressions,

$$\sin E_{\mathrm{I}} = \sin\theta \tan r \left\{\frac{1 + \tan^2 r \cos^2(\theta - \varepsilon_0)}{\tan^2 r \sin^2(\theta - \varepsilon_0) + [\sin\varepsilon_0 + \tan^2 r \cos(\theta - \varepsilon_0)\sin\theta]^2}\right\}^{1/2} \quad (1\text{a})$$

$$\cos\varepsilon_{\mathrm{I}} = \frac{\cos r \cos\varepsilon_0 + \tan r \sin r \cos(\theta - \varepsilon_0)\cos\theta}{[1 + \tan^2 r \cos^2(\theta - \varepsilon_0)]^{1/2}} \quad (1\text{b})$$

When the moving ecliptic is constrained in the alternative way so that $B\Upsilon_0 = 90°$ then E and ε are equal to the functions $E_{\mathrm{II}}(\theta)$ and $\varepsilon_{\mathrm{II}}(\theta)$ where

$$\sin E_{\mathrm{II}} = \sin\theta \sin r \left\{\frac{1 + \tan^2 r \sin^2(\theta - \varepsilon_0)}{\sin^2\varepsilon_0 + \tan^2 r \sin^2(\theta - \varepsilon_0)}\right\}^{1/2} \quad (2\text{a})$$

$$\cos\varepsilon_{\mathrm{II}} = \frac{\cos\varepsilon_0}{[1 + \tan^2 r \sin^2(\theta - \varepsilon_0)]^{1/2}} \quad (2\text{b})$$

These four expressions have been calculated for the range of θ, —90(10)90, and the results are given in Table I, and the graph of E_{I} is shown in Fig. 6. The values of the basic parameters are $r = 4;18,43$ and $\varepsilon_0 = 23;33$, both of which are taken from Thābit's text. Values of E and ε outside this range may be found by using the relations $E(\theta + \pi) = -E(\theta)$ and $\varepsilon(\theta + \pi) = \varepsilon(\theta)$ which apply to both the equations (1) and (2). Note that the actual maximum of E occurs not at $\theta = 90°$, but at $\theta \doteqdot 88;7,46$, when $E \doteqdot 10;40,25$. The entries in Table I for E_{I} and ε_{I} agree with Dobrzycki's results (29, p. 14).

In the Latin translations of Thābit's work the quantity E is tabulated for the range 0(5)90, and one is also given values of a quantity called *equatio diametri dimidii circuli parvi*, or in one MS at least (Cam. Univ. Lib. Mm. 3.11 fol. 105a) *distantia caput arietis ab equatore diei latitudo*. Its interpretation has been discussed by Neugebauer (62, p. 298) and following him we take this quantity, here denoted by δ, to be the distance from P_1 to the equator (see Fig. 1). Therefore

$$\sin\delta = \sin E \sin\varepsilon. \quad (3)$$

In Table II we give Thābit's values of E, and ε. In the third column are given the corresponding values of the obliquity, which are calculated from equation (3), taking values of E and ε from the first two columns.

The point P_1, *caput arietis*, as seen from the point of view of the equator and the instantaneous equinox Υ has polar coordinates E, ε. The locus of this point is nearly a straight arc of length about 21° set at an angle to the equator of about 23°. Looked at closely however this locus is seen to be a very narrow closed curve intersecting itself once. This structure is shown enlarged in the graph in Fig. 2. The curve for the alternative model is drawn there also, and this is seen to be a more open figure of eight.

The coordinates E and ε taken from Thābit's data are also plotted in Fig. 2. These values lie very scattered about the exact curves, and only fairly negative conclusions might be drawn from a comparison. One could not really claim that these points lie particularly close to either curve, although there is at least a tendency for ε to increase with E, as if some crude attempt had been made to allow for the small change in obliquity.

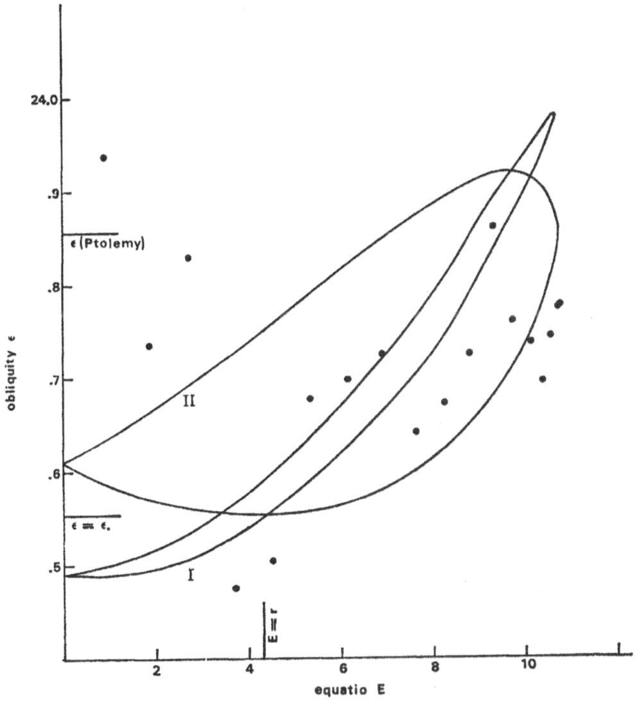

Fig. 2. Graph showing the movement of caput Arietis, with E, ε as Cartesian coordinates. Legend: I, exact calculation with preferred constraint ($P_1B = 90°$); II, exact calculation with alternative constraint ($\Upsilon_0 B = 90°$). The points are based on Thābit's tables. As θ increases each curve is traced counterclockwise beginning at $E = 0$ when $\theta = 0$. Only the range . . . ordinate.

The maximum value of E is 10;45, (according to Thābit) and people have sometimes been inclined to argue that the points might have been derived from a Medieval table of sines, so that E would be written 10;45 sin θ. There is surely no question of this interpretation of Thābit's date, although the analogous equation of the eighth sphere in the Alfonsine model is given with fair accuracy by 9 sin θ. Equations (1a) and (2a) yield the approximation

$$\sin E \doteqdot \frac{\tan r}{\sin \varepsilon_0} \sin \theta, \ \varepsilon \doteqdot \varepsilon_0$$

but there would be no historical justification in attempting to relate ($\tan r/\sin \varepsilon_0$) to the figure 10;45. Nor is there any good reason to regard this maximum value of E as a number of fundamental importance in the genesis of the theory, for this value is not related in any particular way to the other values of E.

The arc joining ♈ to $♈_0$, denoted by *c*, is calculated also in the appendix, where the preferred constraint is assumed. The formula obtained there is identical to that derived for this quantity by both Delambre (26, p. 278–281) when he develops Pedro Nuñez' clear argument (66, p. 304–7), and by Carmody (83, p. 95), all of whom use the preferred constraint. They assume that the *motus* is measured in relation to the ecliptic rather than the equator, by an angle φ say, where $\varphi = \theta - \varepsilon_0$. The formula for $c(\varphi)$ is thus

$$c(\varphi) = \frac{\cot\varphi \sin(\varphi + \varepsilon_0)}{\sin r \cos r} - \cot r \cos(\varphi + \varepsilon_0).$$

The values of $c(\varphi)$, —90(10)90, are given in Table I and it is clear that these values differ by at most four minutes from the corresponding values of $E(\theta)$. Nuñez states clearly that Thābit's equatio is the arc we have denoted by E, and he develops an argument to show that it differs insensibly from *c*. Carmody makes the mistake of assuming that *c* itself is Thābit's *equatio*.

4.2 The choice of constraint

In this section I offer reasons for preferring the constraint $P_1B = 90°$ rather than $B♈_0 = 90°$. In Thābit's text we read (83, p. 104)

> Et ipsius recessio a eo est secundum quantitatem medietatis diametri circuli parvi, nisi quod semper caput Capricorni et caput Cancri sunt coniuncta orbi decliui fixo non recedentia ab eo, et mouenter antecedentes et retrocedentes secundum quantitatem diametri circuli parui.

That is to say, the point B in Fig. 1, which is situated on the fixed ecliptic, is caput Cancri which situated 90° from caput Arietis, the point P_1. Neugebauer (62, p. 290) takes the same choice of constraint.

We next refer to an account[12] of this model given by Peurbach in *Novae Theoricae Planetarum*. This occurs in the last chapter of the work, following an account of Alfonsine trepidation. Peurbach writes (70, fol. 37b)

> Ubicunque etiam sectio harum eelipticarum fiat ipsam necesse est a principio Arietis & Librae mobilium per quartam circuli magni distare.

This is perfectly explicit. Peurbach goes on to refer to the variable angle of obliquity (fol. 38b):

> Unde maxima declinatio eclipticae mobilis ab equatore variabilis est, maior quandoque declinatione eclipticae fixae, quandoque minore eadem, quandoque sibi aequalis. Tunc enim aequalis est illi cum mobilis sub fixae superficie fuerit, maior uero in sitibus contactum. Unde eam Ptolomeus 23;51,20 reperit. Minor autem dum caput Arietis mobilis in sectione aequatoris & parui circuli fuerit.

[12] This particular section of Peurbach's chapter was reprinted as an appendix to the Theorica Planetarum of Gerard of Cremona in the 1480 edition (Bologna), and was mistakenly regarded as the work of Thābit or Gerard by some people including Nallino (56, vol. 1, xxxvi, n. 1), who later corrected his mistake (56, vol. 2, xvi). It is safe to say that Thābit's work was not printed during the Renaissance.

As is clear from Fig. 2, the angle ε indeed takes values both above and below ε_0 with the preferred constraint, but not with the alternative. One can see also that $\varepsilon < \varepsilon_0$ when $E \leqslant r$, in agreement with the last sentence quoted.

Amongst modern writers on this subject, only Dobrzycki has discussed the point and he is in agreement with me. Duhem (31, vol. 2, p. 242), Carmody (83, p. 95) and Neugebauer (62, p. 290) also seem to agree, whereas Goldstein (36) and North (64) take the alternative constraint.

Unexpectedly the alternative constraint would seem to be appropriate to the Alfonsine model, at least in the way in which the model is interpreted by Peurbach.

4.3 The movement of the poles and solsticial points of the eighth sphere

When the preferred constraint is used the pole of the eighth sphere describes a figure of eight curve, while the point B, *caput Cancri*, moves to and fro on the fixed ecliptic. The equations describing this figure of eight are derived in the appendix, where it is shown that

$$\begin{aligned} \xi &\doteqdot -\tfrac{1}{2} r^2 \sin 2(\theta - \varepsilon_0) \\ \zeta &\doteqdot -r \sin(\theta - \varepsilon_0) \end{aligned} \qquad (4)$$

where ξ and ζ are coordinates on the surface of the sphere measured from the pole of the fixed ecliptic in the direction of the summer solstice and spring equinox, respectively. The axis of the curve is the great circle through Υ_0 and the pole of the fixed ecliptic. Peurbach does not make any reference to this curve, although he mentions a similar curve in his discussion of the Alfonsine trepidation, which I will discuss presently.

There may be an illustration of this curve in a curious addition to the diagram of the solar eccentric in a MS of the *Theorica Planetarum* of Gerard of Cremona (Cam. Univ. Lib. Ii. 1.27 fol. 171b), a MS which ends with a note of its author Thomas de Wyndele, A.D. 1424. The figure of eight is drawn over the centres of the Zodiac and solar eccentric. This addition is essentially irrelevant to the text of the Theorica, although there is brief reference to trepidation at the end of this work. It may have been used to illustrate some contemporary discussion. The axis of the figure is at right angles to that of our curve.

When the alternative constraint is used it is not the pole of the eighth sphere, but its caput Cancri, which describes a figure of eight, identical in form to that given by eqn (4), and with its axis coinciding with the fixed ecliptic. The pole of the eighth sphere moves to and fro on a straight arc. This feature is appropriate to the interpretation of Alfonsine trepidation provided by Peurbach. In his diagram, fol. 35a, he shows the figure of eight drawn in this way, and writes (34b),

> Capita uero Cancri & Capricorni octauae sphaerae quasi figuras conoydales habentes pro basi lineas curuas utrinque a capitibus Cancri & Capricorni nouae peragere necesse est. Unde & quandoque praecedent ea, quandoque uero sequentur, quandoque autem coniunguntur, coniunguntur enim caput Cancri octauae & caput Cancri nonae dum caput Arietis octauae fuerit in maxima latitudine ab ecliptica nona & centre circulorum transeunte. Poli autem eclipticae octauae improprie dicti poli quandoque accedunt ad polos eclipticae nonae, quandoque sunt sub eis quandoque uera ab eisdem remouen-

> tur, talis tamen accessus & recessus semper est super circulo magno per polos Zodiaci nonae & centre circulorum paruorum eunte.

It is important to be clear that the locus of the pole of the eighth sphere does not completely determine the movement of the stars relative to the equinox, for one must also specify the rotation of that sphere on its axis, and the rotation is given precisely by the equation E.

Some medieval references to trepidation certainly exhibit some confusion concerning the movement of the pole and its relation to the stellar longitude. Abraham bar Ḥiyya writes for example (56, vol. 1, p. 302; 91, p. 119),

> sapientes Indiae, omnes terrae Romanorum incolas, et veteres e sapientibus Chaldaeorum, hanc tantum de stellis fixis opinionem habuisse, stellas non totum Caelum percurrere, sed octo dumtaxat gradibus zodiaci ante et retro, in orientem et in occidentem, moveri. Causam huius motus esse polum eclipticae, qui in parvo circula, (cuius diametrus) 8° zodiaci, in orientem et occidentem circumvolveretur. Illum polum circulum (totum) 1600 annis perficere; ob (hunc) poli ambitum in circulo viros doctos quosdam imaginatos esse stellas fixas totum Caelum ambire, quia arcanum illud non apparet iis manifestum et ignorarent revolutionem poli eclipticae, qua stellae fixae ab occasu ad ortum per 800 annos moverentur, et postea ad locum pristinum, in occasum versus, denuo restituerentur post 1600 annos perfectos.

Here then is an allusion to a model which we associate with Theon's account, but now bar Ḥiyya seems to imagine that the change in longitude of the star entails a displacement of the pole around a small circle, which is certainly not the case.

Bar Ḥiyya's account finds an echo in the diagram from a work of Quṭb al-Dīn al-Shīrāzī (1236–1311), K. al-tuḥfa al-shāhīya fī'l-hai'a, which Hartner reproduced (40, p. 628), for there we see a small circular locus of the pole of the eighth sphere, drawn together with several positions of the moving ecliptic, to show the range of movement of the equinox. The stars as such are not shown, and from this diagram alone we cannot discover how their longitude would change, although Quṭb al-Dīn says that the longitude change also has an amplitude of 8°. This could be the case only if a complex rotation were given to the sphere, partly cancelling the displacement produced by the oscillation of the pole.

4.4 The motus formula

Thābit's tract is followed by tables giving not only the *equatio*, which we have discussed, but also the values of the motus θ expressed in terms of Arab years. The dependence is expressed very simply by the rule

$$\theta = 1;34,2 + (2;34,58) \times (\text{Arab years}/30).$$

We take the Arab year as 354 $\frac{11}{30}$ days, counted from 14 July A.D. 622 (1948438), so that the formula may be written

$$\theta = (t - 1941987)\,\frac{360}{1481800}$$

where t is the time expressed as a Julian date. Thus the period of the motion is 1481800 days, or 4056.947 Julian years, and the angle θ vanishes when $t = 1941987 =$ 14 Nov. A.D. 604.

If we assume that the numbers 1;34,2 and 2;34,58 are obtained by truncating the results of some calculation, so that the 'true' values might be as high as 1;34,3 and 2;34,59 respectively, then we can attach to the period, and to the date when $\theta = 0$, corresponding uncertainties. Therefore the period lies in the range 1481641 to 1481800, while the date when $\theta = 0$ lies in the range $1941987.14\,{}^{-1.1}_{+0.7}$

The date when $\theta = \varepsilon_0 = 23;33$ is found to be $t = 2038944$, 29 April A.D. 870. Thābit's own dates are A.D. 826/7–901.

4.5 The situation of Thābit's model in relation to the fixed stars

In the discussion of the Toledan and Khwārizmian tables it emerged clearly that the solar longitude provided by the tables was to be measured from a point fixed in relation to the stars, a point in fact situated according to our calculation, 10′ East of ζ Piscium.

Thābit's model is situated in relation to the fixed stars in such a way that the 'moving ecliptic' AP_1B is fixed there as well as the point P_1 itself. From the modern point of view then the small circle is traced by Υ_0 moving around P_1, while the equator is carried by Υ_0 and the point which is diametrically opposed.

In all previous studies, since Delambre, of this well known model there is no mention of the question, where is P_1 situated in relation to the stars? The answer is now fairly clear, of course: since the 'Indian' tables in use in the Middle Ages with which Thābit's and other models of trepidation were used referred longitudes to ζ Piscium, or to a point very near it, *we must take P_1 as that point.*

The precise determination of this point requires some discussion, however. We do not really know how the Medieval astronomer made primary determinations of his parameters, or in what way he actually exercised his skill as an observer. It is not clear whether he simply inherited the stellar reference point along with the Khwārizmian tables, and built upon this in some way, or whether he was capable of primary stellar observations in this connection.

We therefore turn to a comparison of Thābit's *equatio* of the eighth sphere with the longitude of ζ Piscium recorded by, or attributed to various astronomers, in order to gain a further insight into the precise location of the zero of longitude. The value of the *equatio* is found by simple interpolation of the received values (Table II). Where t is the Julian Date, λ is the longitude of ζ Piscium and E is Thābit's *equatio* for this date.

To begin with we notice that Thābit's longitudes at the times of Hipparchos and Ptolemy differ by 2;38,1, and this is sufficiently close to the correct value 2;40 to enable us to conclude that his model was designed specially to provide a variable rate of precession which would accommodate these two authorities, as well as the later Arabic observations. It is not the case that Thābit introduced a 'sinuous' function for the *equatio* simply in order to make the model kinematically 'reasonable'.

The close agreement between the *equatio* and the longitudes of Hipparchos, Ptolemy and al-Ṣūfī seems at first to indicate that Thābit intended ζ itself as the

		t	λ	E
Hipparchos	28/6/—128[1]	1674485	—9;40[4]	—9;43,52
Menelaos	9/2/98[2]	1756892	—7;25[5]	—7;36,19
Ptolemy	28/8/137[3]	1771337	—7;0	—7;5,51
al-Battānī	1/3/888[2]	2045460	4;10	4;32,37
al-Ṣūfī	1/10/964[2]	2073433	5;42	5;40,31

(1) Beginning of 50th year of 3rd 76-year period of the era of Callipos, which began 28/6/—329.

(2) (56, vol. 1, p. 293).

(3) Era Antonini.

(4) Hipparchan longitudes are 2;40 less than Ptolemy's

(5) This is in accordance with al-Ṣūfī's supposition regarding Menelaos' catalogue.

reference point, while al-Battānī's figure suggests that the point is some 0;22 East of the star. Considering however that al-Battānī was a contemporary of Thābit, while al-Ṣūfī lived a century later, and that by Thābit's time doubt had certainly been cast upon the truth of the absolute longitudes of Ptolemy and Hipparchos, although not upon the difference 2;40, we conclude that Thābit meant his zero point to be some 0;22 East of ζ Piscium.

Having established the position of the point P_1 in relation to the stars, it is now natural to translate the whole model into a motion of the equator in relation to the stars. Instead of P_1 rotating about Υ_0, the reverse obtains, and the equator now passes through Υ_0, and the point diametrically opposed, intersecting the ecliptic at the point Υ. The 'fixed ecliptic' of the original model is now a moving great circle passing through Υ_0 and the solstitial point B distant 90°, and the angle $\overline{B\Upsilon_0\Upsilon}$ is maintained constant equal to ε_0. These features are shown in Fig. 3 where

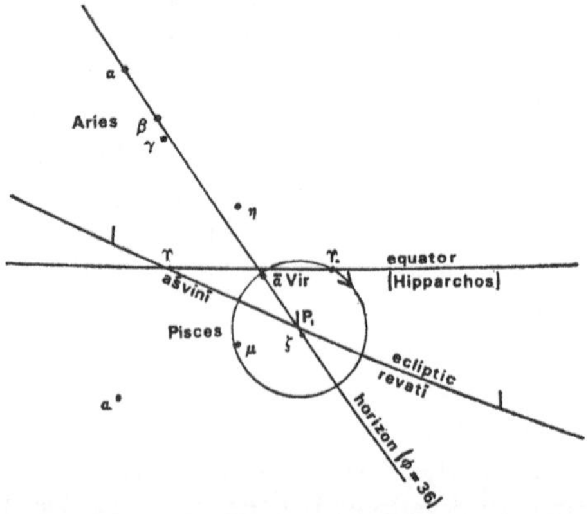

Fig. 3. Star map showing certain stars in Aries and Pisces, together with the ecliptic, the equator at the time of Hipparchos, and the position of the horizon at a latitude of 36° which serves as an alignement of α, β Arietis, α Virginis and ζ Piscium. The stars are placed according to Ptolemy. The small circle is the locus of the head of Aries in the ninth sphere, according to Thābit's model of trepidation.

the small circle is drawn in a star map, showing the stars α, β, γ Arietis, α, η, μ, ζ Piscium and 'anti-Spica', the point $\bar{\alpha}$ opposite to Spica, α Virginis, situated according to Ptolomaic coordinates. Since the orientation is that of a star map, not a celestial globe, the point Υ_0 rotates clockwise around P_1. The equator is drawn in the position it would occupy in 130 B.C.

The small circle intersects the ecliptic at the point occupied by the equinoctial point in A.D. 870, and it may be the case, as suggested earlier, that Thābit determined the radius in this way.

It is also clear from the diagram that anti-Spica lies very nearly on the circle. Indeed, according to Ptolemaic coordinates $\bar{\alpha}$ has, relative to ζ, the coordinates (3;40, —2;10), so that the distance is

$$\{(3;40)^2 + (2;10)^2\}^{\frac{1}{2}} = 4;15,32 \simeq r = 4;18,43.$$

The point lying on the ecliptic 0;9,48 West of ζ is distant exactly r from $\bar{\alpha}$.

According to the motus formula $\theta = 0$ in the year A.D. 604 and this date is quite different from the date A.D. 562–3 when the Toledan and Khwārizmian tables are free of error. Some light is shed on this discrepancy by considering the date when the equator actually passes through ζ Piscium, or alternatively through the point 0;10 east of ζ and having the same latitude. We recall that the equinoctial point in A.D. 562–3 has a longitude 0;10 greater than that of ζ. According to modern reckoning, which gives the star a latitude —0.239° and zero longitude when $t \doteqdot 1931167$ (A.D. 576), we find that the equator passes through ζ when the equinoctial point is 0.546° west of the point having the same longitude as ζ, and this occurs when $t = 1945507$ (A.D. 615), and further that the equator passes through that point which lies 0;10 east of ζ and has the same latitude as ζ, when $t = 1941134$ (A.D. 603).

It is difficult to decide whether these complex considerations have much bearing on the reconstruction of Thābit's model. Should one conclude that the centre of the circle lies south of the ecliptic, or perhaps that the observations on which the model is based require such a position, even through astronomers at the time believed it to be on the ecliptic? (According to the Sūrya Siddhānta (viii. 9) the latitude of the junction star of revatī is zero.) An exact recalculation would seem to be relevant only to observations actually carried out at the time. It is important to bear in mind that an observation of the altitude of a star at the time of meridan transit is more accurate than any other, and perhaps unknown to us, such observations were carried out around A.D. 600. Note that the equator at the time of Hipparchos passed very nearly through Spica.

The amplitude of trepidation employed by Thābit, 10;45, may have been intended to represent the interval along the ecliptic to the point culminating with the star β Arietis. In any case the distance from the point 0;20 East of ζ Pisc to the point culminating with β Arie is 10;37, if the calculations are made with Hipparchan coordinates.

The theory of trepidation seems to be appropriate to a system of star mapping framed by the horizon and meridian, in which the principle observations are of rising, setting and culmination. This will be borne out very strongly in the next section, where the historical origin of the sidereal zero point is explained. We note also that observations of the Sun on the meridian play a fundamental role in the practical determination of the movement of the eighth sphere. Indeed the use of the polar longitude depends on a conversion from the observations of the transit of the star to a point in the luni-solar motion which progresses uniformly along the ecliptic. At the moment when the star, having polar longitude L culminates, the Sun's position is known with respect to the sidereal origin, which is information of the kind required in the construction of horoscopes, in which one needs the position of the Sun, Moon and planets with respect to the ascendant.

IIb

STUDIES IN THE MEDIEVAL CONCEPTION OF PRECESSION

(Part II)

5. The Greek star catalogue and the sidereal ecliptic

We may discover through the study of Hipparchos' Commentary on the Phenomena of Aratos, how it was that a point near ζ Pisc came to be selected as the zero point of the longitude in the Indian systems and those descended from them. In this Commentary Hipparchos describes the rising of the constellations, as seen by an observer at a latitude of 36°, giving in particular the points of the ecliptic rising simultaneously with the beginning and end of the constellation, and also the stars culminating at those times. Within the format of the Commentary he does not give those stars which set simultaneously, but it may be observed that Spica sets with the rising of the beginning of Aries; Aratos only noted that Ara sets as Aries rises. Hipparchos marks this rising with the emergence of η Pisc ('the forefoot of the Ram', 41, p. 255), and in fact α, β, γ Arie and η Pisc all rise nearly at this time, while Spica sets; ζ Pisc also rises, but Hipparchos does not mention this star. He gives the point of the ecliptic which rises simultaneously as 18½ ♓ (i. e., —11;30, i. e., —8;50 with respect to Ptolemy's origin). This point is not far from ζ Pisc, and with a little study one can see that the setting of Spica is exactly simultaneous with the rising of our sidereal zero point at this latitude. Encouraged by this, I calculated the point of the ecliptic rising with each of these stars as a function of latitude, both for the positions given by Ptolemy with the longitude reduced by 2;40, and for those positions calculated for this time by Peters and Knobel for their edition of Ptolemy's catalogue. The results are shown in the graphs in figs. 4 and 5.

The longitude of this point, denoted λ_0, is given by the expression

$$\cot \lambda_0 = \frac{\tan\varphi \sin\varepsilon}{\sin\xi} - \frac{\cos\varepsilon}{\tan\xi}$$

where

$$\xi = \sin^{-1}(\tan\varphi \tan\delta) - \alpha$$

and where α, δ are the right ascension and declination of the star, φ is the geographic latitude and ε is the obliquity. The obliquity is taken as 23;51 for the Ptolomaic positions, and as 23.71 for the calculated positions. These formulae contain in fact, in modern style, the same process of calculation as developed by Ptolemy (Almagest viii. 5), who calculates first the arc ξ, which is the interval between the equinoctial point, and the point where the horizon intersects the equator. The right ascension and declination may be expressed easily in terms of the longitude λ, latitude β and polar longitude L,

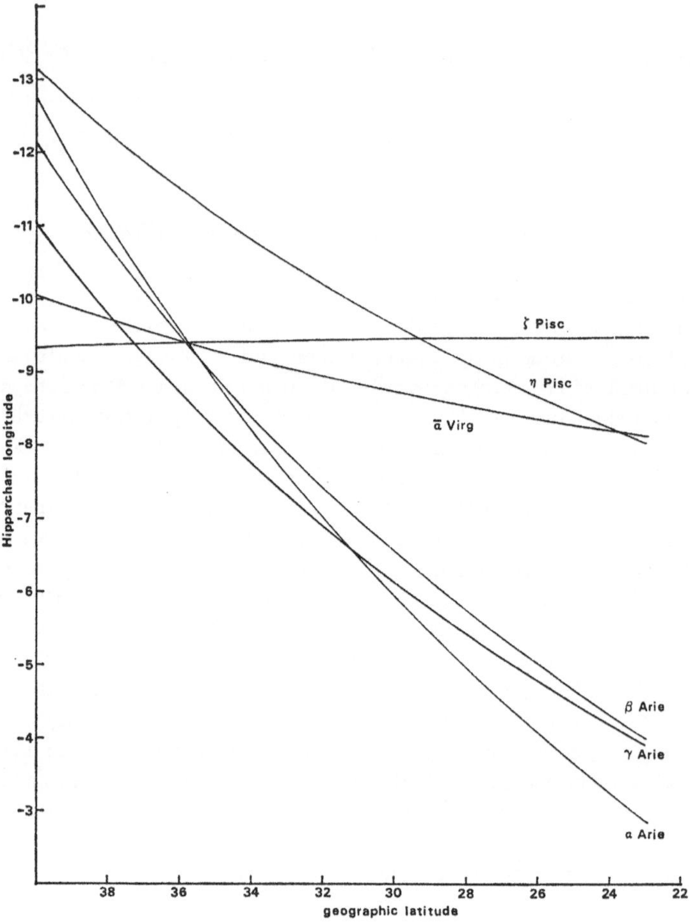

Fig. 4. Graphs of λ_0 versus φ for various stars, showing the point of the ecliptic rising with each of the stars, as a function of geographic latitude. These curves refer to the positions of the stars at the time of Hipparchos, as taken from Ptolemy's Catalogue with the longitude reduced by 2;40.

$$\tan \alpha = \tan L \cos \varepsilon$$
$$\cos \delta = \cos \beta \cos \lambda / \cos \alpha.$$

The polar longitude L is given by

$$\tan L = \tan \lambda - \tan \varepsilon \tan \beta / \cos \lambda$$

and this may be derived from a more general formula which is also needed in the discussion. If we have two stars with ecliptic coordinates λ_1, β_1 and λ_2, β_2 then the great circle through them intersects the ecliptic at a point having longitude λ_I given by

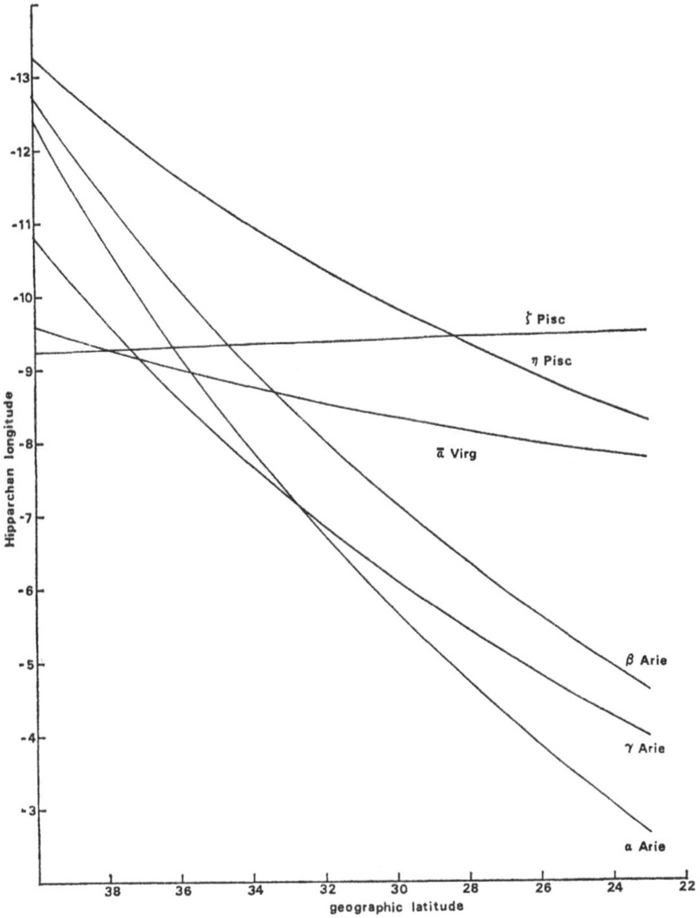

Fig. 5. These curves differ from those in fig. 4 only in referring to positions of the stars in 130 B.C. as calculated by Peters and Knobel.

$$\tan \lambda_I = \frac{\sin \beta_1 \sin \lambda_2 \cos \beta_2 - \sin \lambda_1 \cos \beta_1 \sin \beta_2}{\sin \beta_1 \cos \lambda_2 \cos \beta_2 - \cos \lambda_1 \cos \beta_2 \sin \beta_2}$$

This is most easily derived using, not spherical trigonometry, but vector methods. Let $\mathbf{n}_1$ and $\mathbf{n}_2$ be unit vectors terminating at the stars, and let $\mathbf{n}_e$ be the normal to the equator. The $\mathbf{N} = \mathbf{n}_e \times (\mathbf{n}_1 \times \mathbf{n}_2)$ is the vector in the direction of the intersection sought, and $\tan \lambda_I$ is the ratio between the appropriate components of $\mathbf{N}$.

If we let one of the 'stars' in this last argument be the pole of the equator, so that $\lambda_1 = \pi/2$ and $\beta_1 = \pi/2 - \varepsilon$, we obtain the formula for tan L, which would also be difficult to obtain otherwise.

Returning now to fig. 4 it is seen that not only does Spica set simultaneously with the rising of the origin of the sidereal ecliptic, but we also have then the rising of α,β Arie and ζ Pisc. The striking coincidence of these curves at one point shows that these four stars were assumed to be in exact alignment. We ought to say 'assumed' for it is clear from fig. 5 that the alignement does not hold for the stars in the positions calculated for that time, but is peculiar to the Greek catalogue. In the next Table the longitude λ_I is shown, calculated for the various stars taken in pairs.

pair in alignment	λ_I: Hipparchan coordinates	λ_I: calculated coordinates
α Ar, β Ar	—9;18,43	—14;36,46
α Ar, α Vi	—9;23,39	—8;56,33
β Ar, α Vi	—9;23,55	—8;40,16
γ Ar, α Vi	—9;40,45	—9;8,15
α Ar, ζ Pi	—9;23,3	—9;18,40
β Ar, ζ Pi	—9;23,2	—9;20,16
γ Ar, ζ Pi	—9;22,2	—9;17,17
α Vi, ζ Pi	—9;23,6	—9;16,28

Ptolemy's coordinates are all multiples of 0;10, indicating some rounding-off procedure. This uncertainty entails a 'thickness' for each curve of about 0;10: The intersection of the four curves may then be taken as

$$\lambda = -9;23 \pm 0;5, \qquad \varphi = 35.8 \pm 0.2$$

Ptolemy (Almagest vii.1) describes a number of alignments[1] on a much smaller scale, but gives no indication that he is aware of this one. Indeed it must be one of the best kept secrets of Greek Astronomy. In contrast to the alignments which Ptolemy describes, it would be hard actually to observe this one between diametrically opposed points except with the aid of the horizon as the circle of alignment, at the particular latitude of 36°, since Spica is so far from the close spaced pair in Aries. At an early stage in the development of stellar mapping this observation would have been especially valuable in controlling the results of triangulations over smaller patches of the sky, and their transfer to the globe. Surely such an alignment preceeds in a natural way the use of the principal great circles defined more abstractly.

This alignment would be remarkable enough simply as part of the structural history of the Greek star catalogue, concerning as it does stars of such importance near the equinoctial points. However this great circle of alignment, the Hipparchan horizon at the moment when the head of Aries rises and Spica sets, also marks the rising of the point of the ecliptic which is the zero point of our sidereal ecliptic. We have seen that this point lies between 0;12 and 0;22 East of the

[1] The Greek is σχηματισμός, which also means 'aspect' in astrology, as in the *Tetrabiblos*, i.2.

supposed position of ζ Pisc, essentially the same point as determined by the various considerations concerning trepidation and the Khwārezmian and Toledan Tables.

We have no direct documentation to show how this information was preserved in post-Hipparchan astronomy, other than the precious Indian reference to the junction star of the lunar mansion *revatī* lying 0;10 West of the zero point. This Indian statement suggests that fairly accurate work was carried out in terms of this alignment. Whether the calculation was made by geometry, or with the aid of a well made globe or planisphere, it would have been far more at home in the context of Hellenistic than in Indian astronomy. The coordinates of the junction stars in the Indian works hardly bear comparison, with regard to accuracy, with those of Greek and Arabic catalogues, and it cannot be supposed that star mapping was ever taken seriously in Indian work, in spite of their use of sidereal longitudes. Thus the sidereal longitudes which one has come to regard as 'Indian' are, like so much besides, of Hellenistic origin, and indeed find the fullest practical explanation at only one latitude, that of Hipparchos. The zero point defined in this way by the horizon reminds one of course of the construction of the ascendant in horoscopic astrology, and the zodiacal signs of the sidereal ecliptic correspond to the houses (*loci*, τόποι) measured from the ascendant.

A Babylonian text of the fifth century B.C. (Br. Mus. 86378), which was discussed in great detail by Kugler (48, pp. 1–72) gives heliacal risings, simultaneous risings and settings, and other precious information concerning stellar positions and constellations. At the end of the list of sixteen simultaneous risings and settings we have according to Kugler's reading (48, pp. 22, 32, with important corrections on p. 171),

Piscis Austrinus, especially Formalhaut } rising
Perseus

Virgo } setting
Lupus

These are references to constellations, not individual stars, but would seem to entail the recognition of Formalhaut rising while Spica sets. Note that this list gives not heliacal observations, but simply observations of the night sky, such as Hipparchos gives. The date of the heliacal rising of Formalhaut is however given elsewhere in this text (48, p. 33) as XII.15, and according to Kugler's calculation this is very nearly the time of the Spring Equinox in A.D. —500; elsewhere in this text we are told that the heliacal rising of α, β Arie takes place on I.1 (48, pp. 3, 5, 48). This text preceeds Hipparchos by some 300 years, and does not present observations of the degree of accuracy needed in a star catalogue, but it does testify to a practise of observation of risings and settings, etc., and of α Arie and Spica in particular, well established by the time of Hipparchos.

This Babylonian list of heliacal risings stands at the head of an ancient tradiation of calendars in which risings, settings and culminations are listed against the days of the year, sometimes with associated weather records. These include the Greek parapegmata, the calendar in Geminos' Eisagoge, Ptolemy's Phaseis and the Arabic *kitabu'l-'anwa'* (*'anwa'* = pl. of *nau'*, dawn rising or setting). For example in 'Harib's Calendar', or the 'Calendar of Cordova', we have the setting of Spica and the rising of *Baṭn al-Ḥūt* on 6th April (69, p. 66) (*Baṭn al-Ḥūt*, the Belly of the Fish, a conspicuous ring of stars in Andromeda and Pisces consisting of β, μ, ν, π, δ, ε, ζ, η Andr and ψ^1, ψ^2, ψ^3, χ, φ, ν, τ, 82 Pisc, much larger than the northern Fish of Piscium. See al-Ṣūfī, (78, pp. 118, 196)[2].

The star Spica stands in a special relationship to the Indian calendar. The Indian solar year begins when the Sun has zero longitude, measured in the sidereal ecliptic, while their luni-solar year begins with the last new Moon in the solar year, with the month *Caitra* after the lunar mansion *Citra* (= Spica) with which the full Moon is in conjunction (on the average) in the middle of the month. The last month of the solar year is also called *Caitra.*

Finally, note the remarks made in 9.1 on the possible roles of η Pisc and Spica in determining the zero point of the sidereal longitude in the later Babylonian ephemerides.

5.1 The identification of the stellar coordinates in the Indian sources

We have seen that the star ζ Pisc occurs in the alignement discovered above, and there can therefore be no doubt that it was deliberately made part of the convention which determined the sidereal longitude. However the first reference to this star in this connection was made by Colebrooke as part of his masterful study of the Indian asterisms or nakṣatras (22, vol. 2, p. 281 seq). The star is listed by Ptolemy as the thirteenth in Pisces, with magnitude four. It is also mentioned, by implication, in Geminos' Eisagoge (35, p. 39), who gives the number of stars in the two 'threads' leading down from the Fishes to their junction point α Pisc, viz., five in the Northern thread and nine in the Southern. These numbers include α Pisc in both cases, and in the Southern the 'detour' through 80, 89 Pisc (Ptolemy's 14, 15) is omitted, so that the Southern one consists of α, ξ, ν, μ, ζ, ε, δ, 51, 41 Pisc. It appears of course in all the Medieval Catalogues derived from Ptolemy, those of al-Battānī, al-Ṣūfī and al-Bīrūnī, as well as Ulūgh Beg's. The latitude is given accurately by Ptolemy as —0;10, but in the Arab catalogues this appears as —6;0, through a simple omission of the prime on the letter ϛ' representing one sixth. In his preface (78, p. 195) to this part of his catalogue, al-Ṣūfī notes that the latitude can hardly be —6;0, and explains that in fact it has the same declination as ε Pisc, which would make its latitude +0;22. Nevertheless he gives —6;0 in his table, and in this, as in the rest, he is followed by al-Bīrūnī.

It is therefore unlikely that Arab or Latin astronomers, who were so dependent on these versions of Ptolemy, would actually have understood that this star marked the zero point used in the systems derived from India. Indeed al-Bīrūnī

[2] al-Ṣūfī (78, p. 118) remarks that this last mansion is sometimes called *al-risha,* the 'thread' which properly includes the thread of five stars descending through η Pisc, referred to as *rikis* in the Arsacid Babylonian texts (46, pp. 29, 231).

Table I.

θ or φ	E_I	ε_I	E_{II}	ε_{II}	c
—90	—10;39,43	23;58,40	—10;42,42	23;51,39	—10;41,1
—80	—10;30,02	23;58,19	—10;31,51	23;53,58	—10;32,5
—70	—10;01,41	23;56,16	—10;02,09	23;54,53	—10;05,2
—60	—9;15,23	23;52,44	—9;14,35	23;54,53	—9;19,1
—50	—8;12,17	23;48,10	—8;10,35	23;53,24	—8;15,3
—40	—6;54,03	23;43,05	—6;51,56	23;50,48	—6;56,1
—30	—5;22,50	23;38,07	—5;20,51	23;47,24	—5;23,5
—20	—3;41,17	23;33,52	—3;39,51	23;43,35	—3;41,2
—10	—1;53,07	23;30,51	—1;51,51	23;39,49	—1;52,1
0	0	23;29,26	0	23;36,34	0
10	1;52,36	23;29,48	1;52,15	23;34,14	1;51,4
20	3;41,34	23;31,53	3;41,24	23;33,05	3;39,3
30	5;23,24	23;35,27	5;23,52	23;33,17	5;20,1
40	6;54,53	23;40,03	6;56,21	23;34,48	6;51,0
50	8;13,17	23;44,28	8;15,49	23;37,26	8;09,1
60	9;16,22	23;50,04	9;19,47	23;40,53	9;13,0
70	10;02,28	23;54,17	10;06,20	23;44,42	10; 0,0
80	10;30,28	23;57,16	10;34,11	23;48,27	10;30,1
90	10;39,43	23;58,40	10;42,42	23;51,39	10;41,1

Table II.

θ	E	δ	ε
5	0;55,52	0;22,40	23;56,17
10	1;50,36	0;44,31	23;44,17
15	2;45,16	1;06,45	23;49,48
20	3;39,23	1;27,20	23;28,23
25	4;31,12	1;48,04	23;30,18
30	5;22,30	2;09,21	23;40,37
35	6;09,06	2;28,06	23;41,49
40	6;53,12	2;45,55	23;43,30
45	7;36,35	3;02,38	23;38,24
50	8;14,00	3;17,45	23;40,12
55	8;47,48	3;31,40	23;43,34
60	9;17,44	3;44,46	23;51,32
65	9;43,53	3;54,19	23;45,43
70	10;05,30	4;02,48	23;44,56
75	10;22,47	4;09,08	23;41,46
80	10;35,01	4;14,28	23;44,38
85	10;42,13	4;17,30	23;45,40
90	10;45,00	4;18,43	23;46,19

expresses his perplexity on this point when attempting to identify the Indian asterisms (13, vol. 2, pp. 83–5). He remarks also that Indians themselves whom he consulted were of no help.

In the one Arab catalogue, that of Abū'l-Ḥasan, in which this sidereal longitude is employed, there is no mention of ζ Pisc (see § 8).

In the Sūrya Siddhānta the twenty-eight 'junction stars' (yogatārā) of the asterisms are given in terms of the polar longitude and polar latitude. We are told also the number of stars in each asterism, together with certain vague information about the relative position of the junction star in the asterism. From this meagre data Colebrooke made the first attempt to identify precisely each of these stars. His work was reviewed, and modified slightly, by Burgess in his translation of the Sūrya Siddhānta (19, p. 181 seq). Burgess converted the Indian coordinates L, ℓ to ecliptic coordinates λ, β and made his comparison with the ecliptic coordinates of stars according to Flamsteed's *Catalogus Brittanicus*, having subtracted 15;42 from the longitudes of the latter so as to reduce the epoch to A.D. 560, which was supposed to be the approximate date of the Indian 'observations'.

In Table III we give the Indian coordinates, the corresponding values of λ, β,[3] together with the Hipparchan coordinates (Ptolemy's with the longitude reduced by 2;40) of the star supposed to be the correct identification of that in the Indian list. The Indian coordinates marked (1), (2) are quoted by Colebrooke from the Brāhmasphuṭasiddhānta and the Siddhānta Śiromaṇi, respectively. The number following the name of the star is serial number in Ptolemy's list.

The identification given follows Burgess except in the case of no. 1, *aśvinī*. Both Colebrooke and Burgess expressed great uncertainty as to whether α or β Arie were intended, neither star corresponding at all nearly to the Indian position. It is suggested now that while the list generally gives correct polar coordinates derived approximately from Hipparchos, in the case of ásvinī the Hipparchan ecliptic coordinates of α Arie are preserved.

The choice of α Arie is interesting also because the true polar longitude is essentially equal to the sidereal motion of the Moon. In the uniform divisions of the ecliptic corresponding to the nakṣatra then ζ Pisc begins the first, and α the second; see fig. 3.

Believing that the Indian coordinates were genuine observations made in the sixth century, Burgess and Colebrooke did not make any comparison with the Greek catalogue. We have however seen that the sidereal longitude is determined by a convention peculiar to Hellenistic methods, and this is the reason for the inclusion of the Hipparchan coordinates in Table III. This argument is reinforced by two points: 1) the clarification of the coordinates of α Arie; 2) the precession curve determined by the Sūrya Siddhānta passes exactly through the Hipparchan origin (see § 9.4.1).

The coordinates of Spica, no. 14 *citra*, given by the Sūrya Siddhānta are such as to place it in longitude diametrically opposite to the zero point, although the later Indian works place it much nearer to its correct position.

Considering only the entries for the Sūrya Siddhānta, and omitting nos. 1, 14,

[3] The calculation is made most conveniently by making use of Burgess' formulae

$$\lambda = L \mp \sin^{-1}(\tan\beta \cot\psi),\ \beta = \sin^{-1}(\sin\ell \sin\psi)$$

where $\psi = \tan^{-1}(\cot\varepsilon \sec L)$. Burgess' values were in error by a few minutes occasionally, although this has no effect on the process of identification.

Table III.

	Indian				Hipparchos		Modern	
No. naksatra	*L*	*ℓ*	λ	β	λ	β		Δλ —9;20
1 aśvinī	8	10	12;3	9;9	8;0	10;0	α Ari ext. 1	—5;17
2 bharaṇī	20	12	24;40	11;4	17;0	11;10	35 Ari ext. 4	—1;40
3 kṛttikā	37;50	5	39;10	4;43	39;30	4;30	19 Tau 30	0;20
1,2	37;28	4;31	38;58	4;16				0;8
4 rohiṇī	49;30	—5	48;7	—4;48	40;0	—5;10	α Tau 14	—1;13
1	49;28	—4;33	48;12	—4;22				—1;8
2	49;28	—4;30	48;13	—4;19				—1;7
5 mṛgaśīrṣa	63	—10	61;1	—9;48	54;20	—13;50	λ Ori 1	—2;39
6 ārdrā	67;20	—9	65;49	—8;52	59;20	—17;0	α Ori 2	—2;51
1,2	67	—11	65;6	—10;50				—3;34
7 punarvasu	93	6	92;52	6;0	84;0	6;15	β Gem 2	—0;28
8 puṣya	106	0	106;0	0;0	98;40	—0;10	δ Cnc 5	—2;0
9 āśleṣā	109	—7	110;0	—6;56	100;40	—13;10	δ Hya 2	0;0
1,2	108	—7	108;57	—6;56				—1;3
10 maghā	129	0	129;0	0;0	119;50	0;10	α Leo 8	—0;10
11 pūrva phalgunī	144	12	139;54	11;17	131;30	13;40	δ Leo 20	—0;56
1,2	147	12	142;46	11;14				1;56
12 uttara phalgunī	155	13	150;6	12;3	141;50	11;50	β Leo 27	—1;4
13 hasta	170	11	174;26	—10;5	164;0	—12;30	δ Crv 5	1;6
14 citrā	180	—2	180;49	—1;50	174;0	—2;0	α Vir 14	—2;31
1	183	2	183;48	—1;50				0;28
2	183	—1;45	183;42	—1;36				0;22
15 svātī	199	37	182;48	33;44	174;20	31;30	α Boo ext. 1	—9;52
16 viśākhā	213	—1;30	213;31	—1;24	201;20	—1;40	ι Lib 5	2;51
1	212;5	—1;23	212;34	—1;18				1;54
2	212;5	—1;20	212;33	—1;15				1;53
17 anurādhā	224	—3	224;55	—2;52	213;0	—1;40	δ Sco 2	2;35
1,2	224;5	—1;44	224;36	—1;39				2;16
18 jyeṣṭhā	229	—4	230;7	—3;50	220;0	—4;0	α Sco 8	0;47
1,2	229;5	—3;30	230;3	—3;22				0;43
19 mūla	241	—9	242;54	—8;48	234;50	—13;20	λ Sco 20	—1;16
1,2	241	—8;30	242;48	—8;19				—1;22
20 purva-aṣāḍhā	254	—5;30	254;40	—5;28	245;0	—6;30	δ Sgr 2	0;20
1,2	254	—5;20	254;39	—5;18				0;19
21 uttara-aṣāḍhā	260	—5	260;23	—4;59	252;40	—3;10	σ Sgr 6	—1;37
22 abhijit	266;40	60	264;7	59;58	254;40	62;0	α Lyr 1	0;7
1,2	265	62	260;51	61;55				—3;9
23 śravaṇa	280	30	282;32	29;54	271;10	29;10	α Aql 3	2;2
1,2	278	30	280;2	29;56				—0;28
24 śraviṣṭha	290	36	296;12	35;32	285;50	32;0	β Del 4	1;2
25 śatabhiṣaj	320	—0;30	319;50	—0;28	312;10	0;10	λ Aqr 24	—1;40
1	320	—0;18	319;54	—0;17				—1;36
2	320	—0;20	319;54	—0;19				—1;36
26 purva-bhadrapada	326	24	334;43	22;27	324;2	19;40	α Peg 4	1;23
27 uttara-bhādrapadā	337	26	347;25	23;57	345;10	26;0	δ Peg 1	—7;5
28 revatī	359;50	0	359;50	0	350;20	—0;10	ζ Psc 13	0;10
1,2	360	0	360	0				0;20

27 (which is subject to some gross error), the average value of $\Delta\lambda - 9;20$ is $-0;6$, with a variance of 1;33. This result is entirely consistent with the view that the zero point coincides with that determined by the alignement found in § 5, the great circle passing through α, β Arie, ζ Pisc and Spica.

Burgess (19, p. 211) found that his longitudes for A.D. 560 were less than those of the Sūrya Siddhānta by a mean interval 0;56, and concluded that the latter were observed around A.D. 490. One must always be wary of any attempt to infer a date of observation from sidereal longitudes, and although we can now be sure, explained in § 3, that important Sassanian observations were made in A.D. 556 which led to the adoption of 564 as a 'canonical' year for the Khwārezmian Tables, in the case of the junction stars we take the view that the coordinates were calculated at the time of Hipparchos.

There is not in fact much coherence between the Indian star longitudes, and the sidereal longitudes found in the various Indian systems, in view of the dates of zero deviation calculated so thoroughly by Billard. Only in the case of the Brāhmasphuṭasiddhānta, and then only as a gross average, are the mean longitudes actually measured from a point near to ζ Pisc[4].

This is not the place at which to enter into a discussion of the extremely rich and difficult subject of the origin and diffusion of the 'lunar mansions'. Historians originally debated the problem of the origin in terms of Indian and Chinese sources, and later introduced the question of an ultimate Babylonian source. In fact the fifth century cuneiform tablet we have mentioned already provided Kugler with evidence for a Babylonian origin of the mansions, for we find there a list of seventeen constellations, headed by the Pleiades and entitled 'stars in which the path of the Moon lies' (48, p. 70). This part of the debate concerns the earliest stage, and naturally Greek astronomy has no bearing on it. However the eventual development leading to the choice of 28 well defined asterisms and their junction stars comes later, perhaps as late in Indian and Chinese literature as the time of Hellenistic astronomy. The realisation therefore that Hellenistic methods are very clearly involved with the Indian junction stars adds a new dimension to the whole argument, and one is no longer free to argue for a simple generation of the mansions in either Indian or Chinese astronomy, or even for a direct transmission from Babylon. Either the junction stars were selected in Hellenistic astronomy and then transmitted to India, or the selection was made in India and assigned coordinates by people schooled in the detailed trigonometrical techniques of the Hipparchan school. Although Weinstock (99) has argued for a Greek knowledge of the mansions[5], the latter alternative seems to fit better. It has long been clear that some knowledge of Hellenistic astronomy reached India, and again

[4] Note also that the star ζ Pisc is not part of the large oval of stars constituting *Revatī*, an oval which must be identical with the Arabic *Baṭn al-Ḥūt*, and deriving, one would suppose, from the Babylonian *Anunitum* (as described by van der Waerden in his study of the tablet Br. Mus. 86378, 94, p. 14, fig. 3).

[5] He argues from a linguistic analysis of a twelfth century Greek text, and from a Greek magical papyrus of the third or fourth century A.D., in which 28 animal symbols are given in association with the Moon.

we point out that the model of precession in the Sūrya Siddhānta exactly agrees with the Hipparchan origin.

6. *The role of Jupiter in trepidation and the Khwarezmian Tables*

The importance of the years A.D. 562/3 in the Khwarezmian Tables cannot be discovered from the internal structure of the solar tables. However the outstanding astronomical event of that time is the conjunction of Jupiter with the Sun at the time of the Spring Equinox of 564, and I had already argued at some length in an earlier version of this paper that this event was crucial in fixing the parameters of the Tables. This conjecture is now fully justified by the passage in the Qānūn'l-Mas'ūdī given in § 3.1, where we saw that observations of Jupiter were carried out under Khusrau Anūshirwān in 556 by way of fixing a cycle of Jupiter for the interval 552–564.

The mean position of the planet is easily found from the Khwārezmian Tables.

JUPITER	Khwārezmian	Toledan
mean rate	0.0830 9791 0722	0.0830 9082 9754
radix at 1948438	330;16,49	331;39,37
period in days	4332.2388	4332.6080
time of zero long. in 564	1927134.45	1927116.03
long. at Equinox in 564 (1927135.80)	—0;6,42	—1;38,33

The times of zero longitude are respectively 1.3 and 19.8 days before the Equinox of 564.

According to Tuckerman's Tables the true longitude vanished at 1927130.75, while the conjunction was at 1927137.45, respectively 5.05 days before and 1.65 days after the Equinox.

Observations of Jupiter pay a major role in ancient and medieval calendars, and seem to have served to link observations of the day and night sky, so important obviously in the phenomenon of precession.

Kugler (47, pp. 166–9) in his studies of Babylonian tablets of Jupiter's motion found that the planet was used as a 'calibration marker' in a sense, regulating the lunar calendar in Babylonian astronomy, and later Pannekoek (68) and van der Waerden (93) discovered in these tablets the Babylonian use of 'tithis', the uniform lunar day so familiar from Indian astronomy.

Heliacal risings of Jupiter were the basis of a twelve year cycle used in India in the fifth and sixth centuries, and inscriptions of that time have been found in which the date was expressed in terms of this cycle, as Diksit (28) and Fleet (34) have shown. Might there not be some historical connection between this calendar, and the contemporary methods used by Khusrau's astronomers, since both involve a twelve year cycle, and conjunctions of Jupiter with the Sun?

Considerations relating to this planet seem also to be relevant to Thābit's model of trepidation. Essentially five parameters are needed to fix the model in relation to empirical data, the sidereal location of the point P_1, the period and epoch of the *motus*, the radius (or alternatively the maximum excursion of the *equatio*), and the obliquity ε_0. We have seen in § 4.5 that the model fits remarkably well at three points at least which were known to him, and if he demanded agreement with Hipparchos, Ptolemy and also some Arabic measurement, his model would be fixed, for we ought to assume that the sidereal reference point was given, and not open to adjustment, while the obliquity similarly is a received quantity. He may also, as we have seen, have wished to make the maximum excursion coincide with the point of the ecliptic culminating with β Arietis.

This ought to be a sufficient explanation therefore of the parameters, and yet it is possible to argue that there is a relation between some aspects of the model and the motion of Jupiter. The twelve year cycle mentioned in § 3 was geared to the synodic period of the planet, but there was another Jupiter calendar also used in India, a sixty year cycle, based on the sidereal period of the planet. In fact successive years of the cycle are marked by the passage of the mean planet through successive signs at sidereal longitudes 0°, 30°, 60°, . . . (24;81, p. 32).

In order to understand better the bearing this might have on the development of models of trepidation we compare the actual 60 year calendar with one which would be obtained if precession were taken into account. The aim is to show what might have been done by people who attempted to take into account precession in terms of its effect on this calendar.

If the planet is referred to the moving equinox, then the dates of passage through the signs are progressively advanced and after some 1000 years, when the precession amounts to 15°, the disparity between the new-year dates of the sidereal and tropical Jupiter calendars is greatest, and afterwards decreases linearly to zero, as the precession increases to 30°. At that point the year numbers of the two Jupiter calendars will differ by one, but the 'local error' due to precession will be zero. This displacement of the new year date, with respect to the new year date of the sidereal Jupiter calendar might have been conceived as a zig-zag function ranging from 0° to 15° and back to 0° on a base of some 2000 years. Various models of trepidation using such functions are well known, and will be mentioned later, and it is possible that all such models arose in this way, or related ways. In other cases one might have studied the progression through lunar mansions instead of signs, and one might have interpreted precession in terms of its effect on conjunctions of Jupiter and Saturn (Great Conjunctions), instead of Jupiter and the Sun. In any case zig-zag functions arise naturally when the accumulated precession is projected out onto signs or lunar mansions.

If we take as a prototype of Thabit's model, a zig-zag function of range 15° on a base of 2028.5 years, the slope would be 53.24″ p.a. This is quite close to the value 54″ per sidereal year quoted in the Sūrya Siddhānta, where one has a zig-zag function of range 54° on a base of 7200 years.

One would therefore look for a relation between the sidereal period of Jupiter and the period of trepidation, and indeed a simple calculation confirms that the period 1481800 days is very nearly 342 sidereal periods and 4057 tropical years. Corresponding to the range of values allowed for the period, as found in § 4.4, we would have the ranges,

4332.284–4332.748: sidereal period of Jupiter
365.206– 365.245: tropical year.

The values for the sidereal period of Jupiter known from Indian and Arabic tables fall within this range, although the hypothesis would fit best a value around 4332.7 somewhat larger than those actually encountered. (The tropical period is around 4330.365 days, which is al-Battānī's value.)

Periods of 71 and 83 years[6] were associated with the motion of Jupiter in Babylonian astronomy (46, pp. 43–4), but there one finds also a large period, mentioned in at least one tablet in the British Museum, Sp. II. 985, described by Kugler (46, pp. 48–50). This period is 344 years, a period related in a simple way to the small periods, for we have $344 = 4 \times 71 + 60$, but related also toThābit's period, for we have,

$$4057 = 12 \times 344 + 71 = 47 \times 71 + 60 \times 12.$$

The period of 344 years corresponds to 29 sidereal revolutions of the planet, and if one takes as a Babylonian year 365.26 days (which is approximately the value entailed by the tables, 45, pp. 72, 91), there is implied a sidereal period

$$\frac{344 \times 365.26}{29} = 4332.74,$$

which is the sort of value entailed by our own hypothesis.

Thus the theory of trepidation was not simply an exercise in empirical curve fitting, but an attempt to *explain* the phenomenon of precession in terms of the movement of Jupiter, the calendar marker of the night sky, in its relation to the Sun, the time keeper of the daily sky.

7. Trepidation and azimuthal oscillation

In the course of precession the declination of a star will oscillate between the limits $\beta \pm \varepsilon$, and therefore the azimuth of that point of the horizon at which the star rises will also oscillate. Thus Eisler was led to suggest that the notion of trepidation arose from ancient observations of the changing azimuth of stars on the horizon. As he relates (32), he persuaded Böker to produce, with the aid of a precession globe, some curves of azimuthal oscillation for a number of important stars. These results were published in a study which Eisler reviewed for these *Archives*, a review which included many of Böker's interesting curves.

The notion was taken up by Michel (53) in an interesting paper where he attempted to fix a specific connection with Thābit's trepidation. An observer at geographic latitude φ finds that the star rises at a point of the horizon having azimuthal coordinate A, measured northward from the point due East, where,

$$A = \sin^{-1} (\sin \delta \, / \cos \varphi).$$

[6] The period of 71 years is the basis of the Hipparchan and other formulations of Jupiter's mean motion. Ptolemy, for example (Almagest ix. 3), states that the angle between Jupiter and the Sun increases by 65×360 in 71 years less 4.9 days.

If the star is situated on the ecliptic, then $\sin \delta = \sin \varepsilon \sin \lambda$. According to his calculation, if $\varphi = 33;19$ as for Baghdad, and $\varepsilon = 23;33$, then A will be 10;45 (the extreme value of Thabit's trepidation) when $\lambda = 23;33$. He remarks that this is 'plus qu'une coincidence'. But it is not even that, for one must take $\lambda = 22;58$ in order to make $A = 10;45$. This is of course just inconsequential number fiction, and it would prove nothing even if the arithmetic were correct.

Michel is on more interesting ground however when he draws attention to the 'horizon' mentioned in Thābit's treatise and to the circle thus labelled in his diagram. Certainly the horizon plays no role in the geometric analysis leading to the dependence of E upon the *motus*. On the other hand we have seen why the horizon is important in understanding trepidation, for the horizon at a latitude of 36° determines the alignement of the stars which defines the fixed points in the stellar sphere which are carried around the small circles in the model. This may be the reason for Thābit's reference to it, but one must remember that so far we have had no evidence at all an awareness of this alignement in post-Hipparchan astronomy.

On the other hand, there was another, less common, usage of the term 'horizon', when it meant simply the equinoctial colure. This is mentioned by Isaac Israeli in the Yĕsōd hā 'Ōlām (43, p. iv of the first part of the German translation), who carefully distinguishes it from the more usual sense. This usage also occurs in Pedro Nuñez' treatise on the sphere (65, p. 112), where he gives a diagram which includes a circle at right angles to the equator, labelled 'horizonte'. This sense seems to fit rather better in Thābit's work, which opens with a definition of the coodinate frame,

> Imaginabor speram equatoris diei et tres circulos in ea signatos qui sunt circulus orizontis et circulus linee medii celi transiens per duos polos et secans circulum orizontis in ipsis polis, deinde circulum orientis et occidentis, et est ille qui est in medietate eius quod est inter duos polos et dividit speram in duo media. (83, p. 102)

It is clear that no firm argument has been made out for Eisler's suggestion.

8. *The star catalogue of Abū'l-Ḥasan 'Alī ibn al-Marrākushī*

It might be concluded that only Indian works actually give stellar longitudes measured from ζ Piscium, but there is one Arabic work at least to show that that would be wrong. This is the star list included in a work by the thirteenth century astronomer of Western Islam, Abū'l-Ḥasan 'Alī ibn al-Marrākushī, a work translated by J. J. Sédillot (79; the star list is on pp. 146–9). Sédillot unfortunately provided no identification of the stars, complaining that too much work would be involved in making the comparison with Bayer's catalogue. Nevertheless it is easy enough to identify all the 240 stars by comparison with Ptolemy's list. It is then clear that we have a list in which the longitudes are 6;38 greater than Ptolemy's, in other words they are measured from a point 0;22 East of ζ Piscium. They are also listed strictly in order of increasing longitude, without regard to latitude. The list begins:

ζ Ceti, δ Andr, θ Erid, α Pisc, β Andr, γ Arie,
β Arie, β Cass, ν Ceti, δ Ceti, . . .

The star ζ Pisc is not listed. The list is dated A.D. 622 (Hijra)[7].

Abū'l-Ḥasan in the same work gives tables of trepidation whose parameters are very similar to Thābit's and which are meant to be used with the solar and planetary tables he provides (the solar tables use the sidereal year 365.26 days). The *motus* formula is

$$\theta = 3;51 + (5409') \times (\text{Arab years}/1000)$$

$$\theta = \frac{360}{1415108}\,(t - 1933304)$$

while the table of the *equatio* is given very nearly by $E = 10° \sin \theta$. The date of zero *equatio* is 5th Feb A.D. 581. The curve of $E(t)$ is shown in fig. 6.

From the examples which he gives it is clear that he intends that this *equatio* should be added to the longitudes given in his list of stars. However the *equatio* is 0;40 in A.D. 622, and so there is an inconsistency with the date carried by the list of stars. The rectification would be simplest if the date on the list were ignored, or changed to A.D. 581.

Why should Abū'l-Ḥasan have chosen such an unlikely zero point for the sidereal longitude? We recall that the year A.D. 564 is of special importance in the calibration of those Arabic tables following in the Indian/Sassanian tradition, and we know also that al-Ṣūfī's star catalogue had by Abū'l-Ḥasan's time superseded all others. Now according to al-Ṣūfī's model of precession (9.8) we find that the longitude of ζ Pisc was —0;22 in A.D. 564, and conclude therefore that Abū'l Ḥasan calculated his reference in this way, taking as his zero point the equinoctial point of A.D. 564.

The extreme position of the equinoctial point has the Ptolomaic longitude 3;22, which is very nearly equal to the polar longitude of γ Arie. We have seen that in Thābit's model, the extreme point culminated very nearly with β Arie. In § 9.7 there is a reference to late Medieval lists of lunar mansions in which the coordinate of the first mansion is the Alfonsine correction to Ptolemy's longitude plus 3;24.

9. *Notes on models of trepidation and precession*

1. Babylonian zero point. 2. Ptolemy. 3. Theon of Alexandria. 4. Sūrya Siddhānta. 5. Ārya Siddhānta. 6. Siddhānta Śiromaṇi. 7. al-Battānī. 8. al-Ṣūfī. 9. Alfonsine. 10. Copernicus. 11. Modern formula.

The movements of the eighth sphere according to a number of historically important sources are briefly discussed, and all the results are collected in graphical form in fig. 6. The zero point of the graph is the point lying 0;20 East of the Ptolomaic position of ζ Pisc, which we take as the zero point of Thābit's

[7] There may be a reflection of this list in a list of astrolabe stars credited to Messahalla in his Tractatus Astrolabii, Camb. Univ. MS Ii III.3, fol. 71r (38, p. 162), in so far as this list also begins with ζ Ceti. The longitudes are about 15;24 greater than Ptolemy's and refer the stars to the equinox of about A.D. 1220.

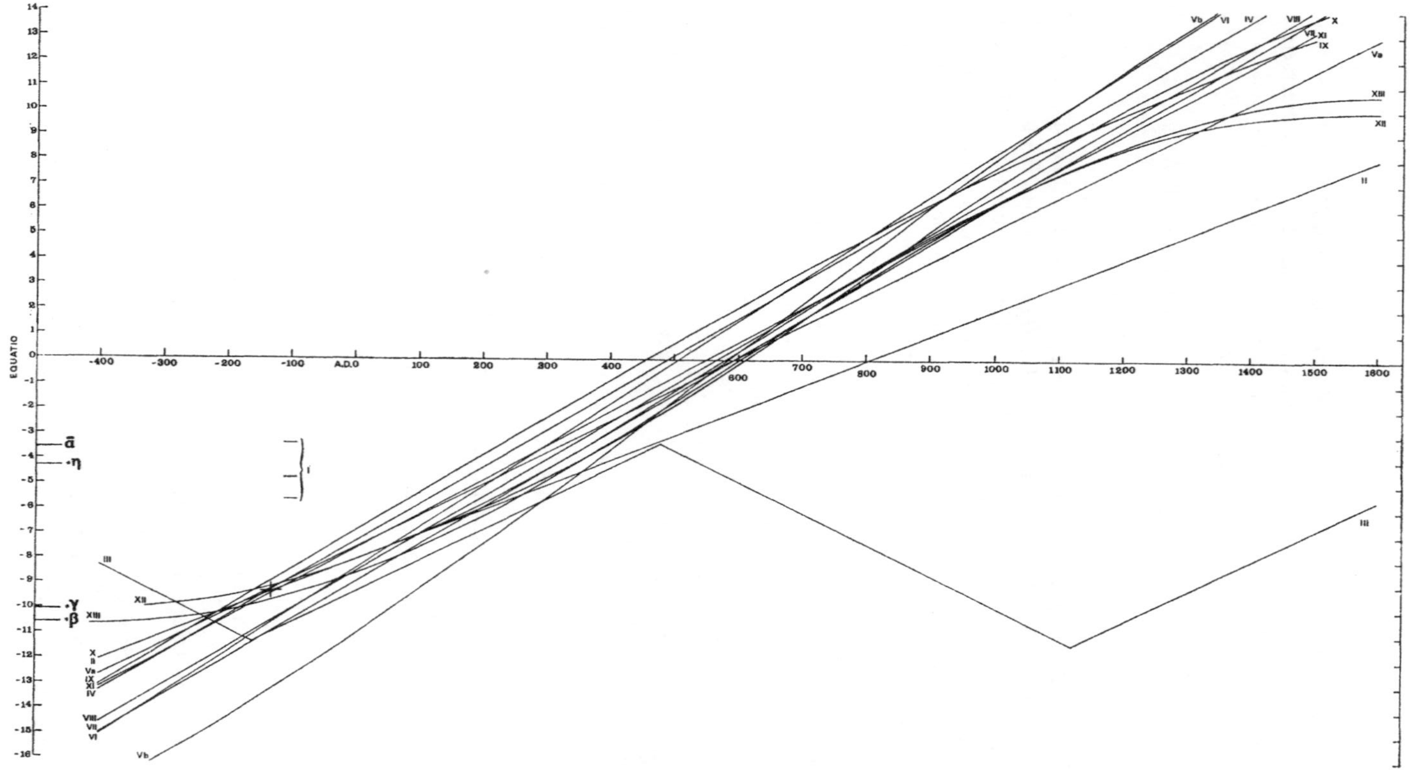

Fig. 6. Curves showing the *equatio E* of precession for the various models discussed in the paper. The reference level is the point of the ecliptic lying 0;20 East of Ptolemy's position of ζ Pisc. Legend: I Babylonian zero of longitude, II Ptolemy, III Theon, IV Sūrya Siddhānta, Va Ārya Siddhānta (linear), Vb Ārya Siddhānta (sine), VI Siddhānta Śiromaṇi, VII al-Bāttanī, VIII al-Ṣūfī, IX Alfonsine, X Copernicus, XI Modern, XII Abū'l-Ḥasan, XIII Thābit ibn Qurra. The positions of anti-Spica (ᾱ), and the points culminating with β, γ Arietis and η Piscium (*β, *γ, *η) are marked, as well as the Hipparchan origin (+).

model, and in the following this point will be denoted E_0. The equinoctial point according to modern calculations coincided with E_0 in A.D. 545, when its longitude was 18;52 (A.D. 1900), since the calculated position of ζ Pisc is 0;6 West of the Ptolomaic position. The positions of anti-Spica (—3;36), and the positions culminating with β, γ Arie (—10;37, —10;4, Ptolomaic) and η Pisc (—4;20) are also shown.

The graphs representing Thābit's and Abū'l-Ḥasan's models are taken from the formulae already quoted.

9.1 Babylonian zero point

Lunar and planetary tablets of Babylonian origin all record longitudes as measured from some sidereally fixed reference point, and Kugler made many calculations in an attempt to fix this point (45, p. 87; 46, pp. 28, 172–4; 47, pp. 513–21). I have made no recalculation of his results, which were based on P. V. Neugebauer's tables, but it is most unlikely that a repetition, making use of Tuckerman's tables would yield anything essentially new. Kugler calculated the position of the equinoctial point of date in terms of the Babylonian longitude, and reduced his result to the standard date A.D.—100. A calculation, using the modern formula, of the equinoctial point of A.D. —100 gives the position 9;2 East of E_0. Thus we may translate Kugler's positions of the equinoctial point, λ_0 measured on the Babylonian scale, into a position E relative to E_0 where

$$E = 360 - 9;2 - \lambda_0.$$

Kugler's positions fall into three separate groups around $E = -3;22$, —4;46, —5;37 (47, pp. 513–21). More recently Huber (42) has reviewed the topic in the light of texts more recently brought to notice, but without really fixing the point more precisely. His result may be expressed as $E = -4;34 \pm 0;20$, with a standard deviation of 1;8.

Kugler compared (in effect) the figure —3;22 with the longitude of anti-Spica, with which it nearly agrees, and made other comparisons also with the longitudes of μ, ν Pisc. One possibility hitherto overlooked, it seems, is that the culmination of η Pisc marked the meridian passage of the zero point. If one uses the coordinates of this star calculated by Peters and Knobel, we find that its polar longitude is —5;0 (A.D. —129), which implies that $E = 5;0 - 9;20 = -4;20$, which is well within the range of Huber's result. This star is the first in the list of thirty-three Babylonian 'normal stars'. This list was first established by Epping, and is given by Kugler in various parts of his work, 46, pp. 29, 321; 47, p. 550 seq (where he discusses the identification of the description of η Pisc, given in the list as 'the star reached by the band of the fish', a reference found only in late material of the Arsacid period). This list is also examined by van der Waerden (94) and Huber.

We have already seen that Hipparchos marks the beginning of the rise of Aries by the emergence of this star on the horizon. Finally note that the zero point of the Babylonian ecliptic lies very nearly on the circle of Thābit's model.

9.2. *Hipparchos und Ptolemy*

The phenomenon of precession was discovered by Hipparchos who, as we understand it, determined stellar longitudes in his own time, and discovered the change in the longitudes when compared with those fixed earlier by Timocharis. This discovery raises important questions about Hipparchos' ability to determine the position of the ecliptic and the equinoctial point in the night sky. The exact practical location of the ecliptic can be determined only with the aid of observations of lunar eclipses, as the name ecliptic correctly suggests. It is true that a less direct method of locating it could have been through a consideration of the stars exactly at the time of solstice when the ecliptic passed through the E–W points of the horizon, but this method depends on the determination of midnight and the date of the solstice, and is fraught with error. The equinoctial points could be determined as the intersections of the ecliptic with the equator, as drawn for example on a globe. This much would serve for an initial orientation for the reduction of observations, but we have Ptolemy's example (Almagest vii. 1–3) to show that in making precise determinations of the positions of the stars, they were observed in relation to the Moon, and then lunar theory was invoked in the final determination of the coordinates. Dicks has given an interesting critical survey of the problem of fixing the equinoctial points in earlier Greek science (27).

It is essential to bear in mind therefore the following fundamental assumption, and indeed fundamental weakness, of Greek Astronomy, that the 19 year cycle of lunar motion was perfectly coherent with the tropical tear. For Hipparchos is quoted as saying (Almagest iii.1; 72, p. 145), 'We find in 19 years as many months (235) as those people maintained, but on the other hand the year has an additional term, which is at most[8] 1/300 day less than the quarter, . . .'. Thus Hipparchos and Ptolemy assume that a full Moon, for example, in congruent points of the 19 year cycle, occurs always at the same position in relation to the equinoctial point. This is not in fact true, for the equinoctial point precesses from these positions at the rate of about 0.5° per century. It is for this reason that Hipparchos' year is slightly too long, for it is designed to fit the 19 year cycle, this length being

$$\frac{126007\frac{1}{24}}{4267} \times \frac{235}{19} = 365\tfrac{1}{4} - \frac{1}{313.68} \text{ days,}$$

since Hipparchos determined the synodic month from the 345 year eclipse cycle. We can see why Hipparchos says 'at most 1/300'; Censorinus (*de die natali*, ch. 18; 20, p. 38) gave the additional term as 1/304, when describing the Hipparchan year.

It was in this way then that Hipparchos and Ptolemy located the equinoctial point in the night sky. The relation between the lunar theory and the equinoctial point had to be completed by a precise fixing of the date of at least one equinox, presumably done by Hipparchos, and with considerable accuracy. When the stellar longitudes were determined with the aid of the lunar theory, a precession rate of 1°

[8] Manitius, who has here 'at least' (mindestens), should be corrected; the text is . . . ἔλασσον τριακοσιοστῷ ἐπιλαμβάνοντα μάλιστα μέρει . . .

per century would be obtained, for the lunar positions at congruent points of the 19 year cycle do indeed precess at that rate. Delambre (25, vol. 2, p. 254) tried to show from an analysis of Hipparchos' observations that a proper comparison between his and Ptolemy's would have yielded a precession rate of 0;49 p.a. This view was firmly refuted by Vogt (92) who determined from the Commentary on Aratos the coordinates of 122 stars mentioned by Hipparchos, and concluded that the differences in longitude from those of Ptolemy well justified the rate of 0;36 p.a.[9].

In Babylonian astronomy the date of the equinox was fixed according to the rule first clarified by Neugebauer (57), by which the tropical year is 12 months 11;3,10 tithis (1 tithi = 0;02 month). Since we have 235/19 = 12 months 11;3, 9,28, . . . tithis, the Babylonian value is a rounding off of the same proportion used in Greek astronomy. More sophisticated developments in both Babylonian and Greek astronomy concern the inequalities related to the eccentricity of solar and lunar orbits, but do not give any improvement over this fundamental assumption about the mean motion.

These features are preserved in the Jewish calendar, for it also is based on the assumption of an exact 19 year cycle, in which the mean synodic month is 29 days $12^{793/1080}$ hours (43, second part, p. 29), and this is essentially a rounding off of the Hipparchan value, which may be expressed as 29 days $12^{792.979/1080}$ hours. There are actually two versions of the calendar using respectively the years $365\frac{1}{4}$ and 365 days $5\frac{3791}{4104}$ hours, and the latter is exactly 235/19 times the synodic month (43, 2nd part, p.i.). With this latter value, which is virtually the same as that of Hipparchos, and obtained by the same method, the average date of 1 Nisan drifts slowly with respect to the Spring Equinox. During the present cycle, 1959–1977, it occurs some 6.5 days later. Coincidence would have obtained around A.D. 300, about the time when this calendar was consistuted in its present form.

The epoch of Ptolemy's catalogue is the era of Antoninus which we take as 28 August 137 = 1771337. The equation of the eighth sphere is, at that time, —6;40.

9.3 Theon of Alexandria

A model of precession, or rather trepidation, attributed to Hellenistic astrologers, is reported by Theon of Alexandria (86, p. 53), and precise accounts are also given by Abū Maʿshar (2, Tract II Diff 8; 31, vol. 2, p. 503) and al-Battāni (56, vol. 1, p. 126–7). Many historians have had occasion to discuss this report (31, vol. 2, p. 625; 12, p. 86–7; 52, pp. 446, 485; 58, p. 7). In this model the equation of the eighth sphere measured from some as yet undetermined sidereal reference point, oscillated between 0° and —8°, reaching —8° 128 years before Augustus,

[9] Ptolemy has been accused of outright plagiarism, and of a dishonest fabrication of evidence of precession in support of his stellar coordinates, for example by Böker (15, p. 46) who made a rather superficial attempt to show that the Greeks used a fixed sidereal ecliptic which was taken over by Ptolemy. One might of course criticize Ptolemy for failing to make a determination of the equinoctial point independent of the lunar theory.

and returning to 0° 640 years later in AD 483, so that in Theon's time it is about —1;31. Abū Ma'shar tells us that the amplitude is 5;12,45 at the beginning of the year 265 Yez. This era began on 16/6/632 = 1952063, so that the beginning of 265 is 1952063 + 264×365 = 2048423. This fixes the date of zero amplitude as

$$2048423 - 5;12,45 \times 80 \times 365 = 1896218 = 25/7/479.$$

One may make an estimate of the stellar situation of this model, but I know of no clear argument leading to a unique position. There is no reason at all to suppose that the Indian-Arabic origin E_0 was intended to be used, and one would be more likely to find some association with the Babylonian ecliptic. This model of precession might have been used to compare predictions of the longitude of the Sun and the planets made according to astrological methods with those of Ptolemy or Theon, these being the only tables known at the time which gave tropical longitudes. We know in fact of a Greek ephemeris of A.D. 349/50, analysed by Neugebauer (59) and Burckhardt (18) in which the longitudes listed for both inner and outer planets are all nearly 1;40 greater than those calculated from Theon's tables[10]. This difference is equal to the amount given by this model of trepidation for A.D. 350, which shows, in case it were doubted, that there were astrological tables with this model was used. However we cannot find the sidereal reference point exactly unless we know the basis of calculation of the emphemeris.
We can argue indirectly however by finding the error in the tropical tables. For the solar tables of Ptolemy one can determine easily an approximate error formula. The true sun, according to Ptolemy or Theon, at the time of the Spring equinox (according to modern calculation) has the longitude

$$-0.00430°n - 0.923°$$

where n = A.D. year number. Taking A.D. 350 as a representative date, the error is then —2;26, and the error in the astrologer's table would be —2;26 + 1;40 = —0;46, which is the distance from the true equinoctial point of A.D. 350 to the sidereal reference point. The former is located —2;47 with respect to E_0, so the desired point is at —3;33. If we had taken an earlier date, say A.D. 1, we would obtain in the same way —2;30. Both μ Piscium (—3;10) and anti-Spica (—3;20) as well as one of the Babylonian zero points fall in this range. In Fig. 6 the zero level is taken as anti-Spica.

Biot claimed to find μ Piscium as the centre of an oscillation of ± 8°, writing,

> Ce mouvement avait atteint sa limit occidentale, *in consequentia*, 128 ans avant l'ère d'Auguste; de sorte que, 28 ans plus tard, au temps d'Hipparque, le point équinoxial était entré dans sa phase de rétrogradation, comme ce grand astronome l'a observé. En partant de ces données, je trouve, par un calcul fort simple, que le mouvement oscillatoire ainsi défini devait s'opérer autour d'une petite étoile de la constellation des Poissons que nous désignons par la lettre μ . . . (12. p. 86)

[10] The Greek horoscopes analysed by Neugebauer and van Hoesen always measure longitude from some point placed at a distance of the order of 4° West of the equinoctial point of date. Although this material is very interesting, it is difficult to derive clear numerical results; Theon's trepidation is brought into the commentary on horoscope no. 81 (60, p. 27).

Perhaps he calculated the longitude corresponding to a Hipparchan precession rate for the interval 640–28 = 612 years, which would put the reference point at a Ptolomaic longitude 2;40 — 6;7 = —3;27, about the same as Ptolemy's longitude (—3;30) of μ Piscium. This is not very logical, but I do not see any other way in which he might have obtained this result. With his usual boldness, Biot may have come nearer the truth than he deserved. Martin (52, p. 485) accepted Biot's claim.

One obvious defect of this version of trepidation is that it is in flagrant disagreement with Ptolemy's report that precession had progressed 2° from the time of Timocharis (c. 295 B.C.) to Hipparchos.

The practice of setting the equinoctial point at 8° of Aries, which is found in the Babylonian system B, and apparently in much of the horoscopic material gathered by Neugebauer and van Hoesen, is presumably related to this model of trepidation. This position of the equinoctial point is situated therefore near $E = -11;30$; the only stellar alignement one might associate with it is the line joining β Arie with α Pisc.

9.4 Sūrya Siddhānta

A model of trepidation is mentioned in rather cryptic terms in the Sūrya Siddhānta (iii 9.12;10;19 pp. 99–105), a model first brought to the notice of European historians by Davis (23, p. 266) and Colebrooke (22, vol. 2, p. 329). The amplitude oscillates between the limits ±27°, passing through zero in the year 3600 complete of the Kali yuga. The rate is 54″ per year (the year being the sidereal year as used in the Sūrya Siddhānta), and therefore the period of a complete cycle is 7200 years. The Julian date of the time when the amplitude is zero is[11]

$$588463.12 + 3600 \times 365.2587564 = 1903394.6 = 19/3/499$$

The sidereal reference point is 0;10 East of ζ Piscium, according to the text (viii.1 seq.; 10, p. 175) where the positions of the nakṣatras are specified, and this point is +0;10 on the graph. The formula is therefore,

$$E = (t - 1903395)\frac{0;0,54}{365.2587564} + 0;10$$

It is interesting that there is very close agreement between this curve and the Hipparchan position of the equinox. For we can take Hipparchos's date as 1674484 = 27/6/—128[12], and his position of the equinoctial point on the graph is

[11] To find the Julian date corresponding to a Kali yuga date we need to know the epoch of the Kali yuga and the length of the year. The epoch is midnight 17/18 Feb. A.D. —3101 at Ujjain, and allowing a longitude correction *(kṣepa)* of 2;10,14,30 days by which the epoch is set back when working out the solar calendar (81, § 26; the kṣepa allows for the equation of the orbit of the Sun at the time of the equinox), and this reduces the epoch to 588463.12. According to the Sūrya Siddhānta the length of the year is 1577917828/4320000 = 365.2587564 days.

[12] According to Ptolemy (Almagest vii.2), Hipparchos' longitudes are fixed for the beginning of the 50th year of the 3rd cycle of Callipos, that is 201 years after 28/6/—329 = 1601069, which comes to 1674484.

—6;40 —2;40 = —9;20. On the other hand the value from the above formula is —9;14,2.

The extreme position of the equinoctial point, 27° East of E_0 may be related to the heliacal rising of α Arie. The *arcus visionis* may be taken as 14°, and when the calculation is made using Hipparchan coordinates and the Hipparchan latitude of 36°, the Sun is situated 28;25 East of E_0. This result would be only slightly different if one calculated for another epoch, or indeed for another near-by latitude. One may suppose that the intention of this model of trepidation was to relate the heavens to the earlier time when the heliacal rising of this star marked the commencement of the year. Kugler for example argued very strongly for this location of the New Year, in his discussion of the fifth century tablet Br. Mus. 86378 (48, p. 5), a tablet frequently mentioned in this paper already. We recall that the list of *nakṣatras* in this work begins with α Arie.

9.5 Ārya Siddhānta

A model of trepidation is mentioned by Bentley (9, pp. 138–141) and Colebrooke (22, vol. 2, p. 332) who refer to a certain work called the Ārya Siddhānta. This was at first confused with the Āryabhaṭīya, the fifth century work by Āryabhaṭa I. The Ārya Siddhānta in question however is later, perhaps of the tenth century (80, p. 8), and is attributed to an otherwise unknown author named for convenience Āryabhaṭa II. It is available in a number of Manuscripts, and from two of these it was printed under an alternative title Mahāsiddhānta, by M. S. Dvivedi (6), together with a commentary in Sanskrit and an introduction in English. Colebrooke's information is drawn from Munīśvara's Marīci (A.D. 1603), while Bentley, who as usual does not explain the nature of his sources, must have had access to one of the MSS.

In the following, references for the various parameters are made to the accounts of Bentley and Colebrooke, as well as to the text itself.

There are 578159 cycles of trepidation per kalpa (i. 11; 22, vol. 2, p. 332; 9, p. 139), with an excursion equal to the maximum declination of 24° (iii. 13; 22, vol. 2, p. 332; 9, p. 141). The epoch is defined in relation to the Sindhind system (71, p. 28) which places the origin of the kalpa, whose length is 4320000000 years, 1972944000 (= 4567×432000) years before the Kali yuga (i. 7; 9, p. 139). The creation, at the end of which the universal conjunction takes place, lasts 3024000 years[13] (i. 7; 9, p. 139), so that the conjunction, and in particular zero trepidation, occurs 1969920000 years before the Kali yuga. By the time the Kali yuga is reached the motion is

$$\frac{1969920000 \times 578159}{4320000000} = 263640.504^{\text{rev}}, \text{exactly,}$$

[13] In the Sūrya Siddhānta the universal conjunction takes place 17064000 years after the beginning of the kalpa, leaving 195588000 years to the Kali yuga. See Burgess's remarks on this figure (19, p. 19).

and the remaining 0.496^{rev} is accomplished in

$$\frac{0.496 \times 4320000000}{578159} = 3706.1085 \text{ years,}$$

which is the Kali yuga date of zero trepidation. (Note that since we have *accessio* at this date, the cycles must have begun with *recessio*.) According to the calendar of the Sūrya Siddhānta the Julian date is

$$588463.12 + (365.2587564 \times 3706.1085) = 1942152 = 28/4/605$$

which is very near to the corresponding date of Thābit's model. The motus θ which increases by 360° during the course of one cycle, is

$$\theta = \frac{360 \times 578159}{1577917542000} (t - 1942152)$$

since there are 1577917542000 days per kalpa (i. 13; 6, p. 139) according to the Ārya Siddhānta.

Now it is difficult to interpret the passage of the work (iii.13) which explains the relation between the equation of trepidation and the motus. Colebrooke and Bentley take the passage to mean

$$\text{equation} = \frac{24}{90}\theta, \ -90 < \theta < 90, \qquad \text{(a)}$$

whereas according to Dvivedi,

$$\text{equation} = \sin^{-1}(\sin 24° \sin \theta). \qquad \text{(b)}$$

The passage reads,

> *ayanagrahadoḥ krāntijyācāpā kendravaddhanarṇa syāt.*
> The *bhuja* of the equinoctial point is the arc of the sine of the declination, and is added or subtracted as in the case of the anomaly *(kendra)*.

The *bhuja* of an angle A is that part which determines the sine, the zig-zag function which takes the values A, $\pi - A$, $-\pi + A$. $2\pi - A$ respectively in the four quadranst. The 'arc of the sine of the declination' is a quantity which obviously ranges from 0° to 24° since the maximum declination according to this work is 24°. It is not clear however in what way this is understood to depend on the motus. The usual rule for expressing the declination δ in terms of the longitude λ is

$$\sin \delta = \sin 24° \sin \lambda,$$

and if this formula is to be understood to apply to the motus, then of course Dvivedi's interpretation is justified. Indeed, it is hard to see, if the linear motion is understood, why the passage should follow immediately after a section describing the calculation of sines.

The equation of the equinoctial point is to be added or subtracted like the equation of the anomaly of a planet, that is positive in the range $0 < \theta < 180$, negative if $180 < \theta < 360$.

The sidereal origin of the longitude is the junction star of *revatī*, identified as usual with ζ Piscium (xi.2; 9, p. 153). From formula (b) we have therefore

$$E = \sin^{-1}(\sin 24° \sin \theta) + 0;20.$$

which is shown in the graph, and which is calculated with the value $\sin 24° = \frac{1397}{3438}$ given in the text (iii.4).

Now the version (a), while not satisfactorily based on the text may well be of historic interest, particularly if it is used together with longitudes measured from the point 0;10 East of ζ Piscium. In this case one has

$$E_0 = \frac{96 \times 578159}{1577917542000}(t - 1942152) + 0;10$$
$$= 0.00003517501\,(t - 1937414).$$

This is interesting because like the model in the Sūrya Siddhānta, it very nearly agrees with the Hipparchan value —9;20 when t = 1674484, giving in fact $E = -9;14,55$ (for the Sūrya Siddhānta at this date $E = -9;14,1$). It is therefore worth considering the possibility that both these linear models were constructed to agree with the Hipparchan star catalogue, while differing only in the rate used. The rate of change of E in the linear interpretation is 46.25″ p.a., and is closer to the rate entailed by the Babylonian and Hipparchan luni-solar data (46.807″ p.a.) than is any other medieval model.

Version (b), which appears to be the correct reading of the text, agrees well with the other Indian curves, at least from the tenth century, which is supposed to be the date of origin of the work, when it intersects the Sūrya Siddhānta curve. The rate is exceptionally high at earlier times, reaching a maximum of 70.5″ p.a.

The extreme position 24° East of E_0 may refer to the heliacal setting of the Pleiades which occurs when the Sun is situated some 22°–24° East of E_0 (depending on the choice of the *arcus visionis*), and this nearly coincides with the (daily) rising of α Pisc at $E = 23;48$[14].

The heliacal setting of the Pleiades may be the proper role of this star in those older lists of the nakṣatras which begin with it, rather than to suppose that they refer to some remote epoch when the equinoctial point was near this star. On the possible role of this heliacal setting in the Harranian Moon-cult, see Lewy (50), and for a very original discussion of its importance in the 'Lion and Bull' motif, see Hartner (39).

9.6 Siddhānta Śiromaṇi

A model of precession, in the proper sense of continuous rotation, is described in Bhāskara II's treatise Siddhānta Śiromaṇi, in the second book entitled Golādhyāya (on the sphere), translated by Wilkinson and Śāstrī (10). Bhāskara, when describing the model (vi.17;10, p. 157), is actually quoting Muñjāla (A.D. 662:

[14] This happens at the 'Spring Equinox' according to Callipos, as quoted in Geminos' Calendar (35, p. 228). That a daily, not a heliacal rising is here intended by Geminos is shown by his use of the verb ἀνατέλλει, instead of ἐπιτέλλει (35, p. 147). The position 24° East of E_0 is actually the midpoint of the Hipparchan sign of Aries, from which point, according to Hipparchos (41, p. 132), Eudoxos measured his longitudes.

10, p. 141). The rate of movement of precession is 199669 revolutions per kalpa (4320000000 years), which amounts to a rate 0.016639083 = 0;0,59,54,2,31,12 p.a.

The epoch of the motion, when the equinoctial point coincided with the stellar reference point, is the beginning of the kalpa, which is taken according to the Sindhind system (71, p. 28) to be 1972944000 (= 4567 × 432000) years before the Kali yuga (10, p. 108). At the beginning of the Kali yuga therefore, the motion amounted to $91188^{\text{rev}}299.628°$, leaving 60.372° of the current revolution, which is completed at the date

$$\frac{60.372}{0.016639083} = 3628.325 \text{ Kali yuga.}$$

To convert this to the Julian date, one has to choose one of the Indian calendars and according to the Sūrya Siddhānta we have

$$588463.12 + (365.2587564 \times 3628.325) = 1913741 = 16/7/527.$$

The length of the year according to the Siddhānta Śiromaṇi itself is 365.2584375 (1577916450000 days per kalpa), which is slightly smaller than in the Sūrya Siddhānta, but the difference would make only 2 days difference in the above date.

Colebrooke (22, vol. 2, p. 332) quotes a commentary on this work, the Marīci by Munīśvara (A.D. 1603), who says that in Bhāskara's own time, at the date 1105 Śaka complete = 4284 Kali yuga, the motion amounted to $91189^{\text{rev}}10;54,35,23,55,40,48$. This is quite exactly confirmed as the result of the calculation

$$\frac{199669 \times (1972944000 + 4284)}{4320000000} = 91189^{\text{rev}}10;54,35,23,55,40,48,$$

which is important in confirming Bhāskara's use of the Sindhind system of placing the epoch of the motion at the beginning of the kalpa.

Concerning the location of the sidereal reference point we have Colebrooke's information taken from the Marīci that it was assumed to be ζ Piscium itself (22, vol. 2, plate facing p. 284). Bhāskara apparently followed Brahmagupta; certainly the latter in his Khaṇḍakhādyaka takes Revatī to begin at ζ (16, p. 149).

Finally therefore,

$$E = \frac{199669 \times 360}{1577916450000}(t - 1913741) + 0;20$$
$$= 0.000045554275\,(t - 1906424),$$

the date when E vanishes being 4/7/507[15].

[15] The date on which the correction vanishes is of course still a problem for the modern Hindu astrologer. In a popular modern handbook (75, p. 51) we are told that the dates A.D. 361, 394, 397, 498, 559 and others have been proposed. See 'Notes added in proof' at end.

9.7 *al-Battānī*

This model of precession uses the rate of 1° in 66 years, where one year = 365.24056 days (56, vol. 2, p. 124), while the epoch of the star catalogue is 1/3/888 = 2045460 (56, vol. 2, p. 369), and ζ Pisc has the longitude 4;10. Therefore

$$\begin{aligned} E \ & (t - 2045460) \times 0.00004148366 + 4;30 \\ & = (t - 1936984) \times 0.00004148366, \end{aligned}$$

the date when $E = 0$ being 5/3/591.

In Ptolemy's time this puts the equinoctial point 0;11 East of Ptolemy's position.

Luni-solar intercalation, and the 19-year cycle in particular played no role in Islamic culture, and were indeed proscribed in the Qur'ān (ix.36), so that perhaps Islamic astronomers were predisposed to view in a fresh light the problem of determining the tropical year In any case it was shown first of all in the Baghdad school that this year was shorter than the Greek value.

9.8 *al-Ṣūfī*

This uses the same rate as al-Battānī, although it is not clear exactly what value one should assume for the length of the year for his usage, and one might assume al-Battānī's value. At the date of his catalogue 1/10/964 = 2073432 (56, vol. 1, p. 194), the longitude of ζ Piscium was given as 5;42, so that

$$\begin{aligned} E & = (t - 2073432)\ 0.00004148366 + 6;2 \\ & = (t - 1927993)\ 0.00004148366 \end{aligned}$$

the date when $E = 0$ being 23/7/566.

At Ptolemy's time this puts the equinoctial point 0;10 West of Ptolemy's position.

9.9 *Alfonsine*

The Alfonsine tables include a more complicated model than those described so far in these notes, for it is the sum of linear and sinusoidal terms, although the latter is approximated closely by the sine function and so is simpler in this respect than Thābit's formula. Strictly speaking the tables given are for the longitude of the solar apogee, which is assumed to be fixed in relation to the stars.

The linear term *(motus medius augis solis)* increases at the rate of one revolution in 49000 Julian years, and is given by

$$a_1 = \frac{360}{49000 \times 365.25} (t + 1829349)$$

where the constant has been adjusted so as to give agreement with the values listed in the table for the various eras (4, fol a7^r).

The sinusoidal term *(equationum motus accessus et recessus sphaere stellate)* is

$$a_2 = 9° \sin \theta$$

where θ *(medius motus accessus et recessus octave sphaere)* is given by

$$\theta = \frac{360}{7000 \times 365.25}(t - 1727038)$$

the zero point at 16/5/16 (not A.D. 15 as in Dreyer, 30, p. 247) being adjusted so as to reproduce the values given for the various eras (4, fol a7r). It is to be understood that the operative date for any era is noon of the day preceeding, so that the Alfonsine era itself runs from noon 31/5/1252. These formulae for a_1 and θ will reproduce exactly all the numbers given for the eras, but the table of the *equationum* differs slightly from the sine function.

The quantity $a = a_1 + a_2$ is the longitude of the solar apogee. To find the equation of precession for our graph note that the Alfonsine star table is constructed from Ptolemy's with a difference of longitude of 17;8, and so ζ Piscium has a longitude of 10;8. Thus our formula is

$$E = a(t) - a(t_0) + 10;28$$

where $t_0 = 2178502$ is the Alfonsine era. The value of $a(t_0)$ is given in the tables as 88;41,1, although a direct evaluation using an exact sine function gives 88;40,35 and an interpolation of the equationum table gives 88;40,8. It makes no perceptible difference to the graph which one uses.

We see from the graph that there is general disagreement with the Arabic results, but that the curve lies nearer to the Indian curves up to c. A.D. 1000. The agreement with Hipparchos is reasonable, since here $E = -8;54$ compared with the Hipparchan value —9;20.

The intersection with al-Ṣūfī's curve is near A.D. 1260. At the Alfonsine era, al-Ṣūfī's curve reaches 10;23,31, only some four minutes greater than the Alfonsine value 10;28. One would obtain the Alfonsine value at 2180301 = 4/5/1257. We know that al-Ṣūfī's tables were translated in 1256 for King Alfonso; see Wegener's discussion of the correct date of Alfonso's star tables (98, pp. 143–7).

The Alfonsine model of precession was of course widely used in the late Middle Ages, and one finds many star tables in which the original Alfonsine coordinates, which differ from Ptolemy's by 17;8, are adapted by the use of this model to a later date. Just to give an example, in Saxl's survey of astrological manuscripts in Rome and Vienna (77) there are incipits in three manuscripts which supply a number of precession constants for various years:

source	date	correction to Ptolemy	corresponding value of E	value of E from formula
(1)	1438	19;4	12;24	12;22,32
(2)	1424	18;56	12;16	12;14,21
(3)	1424	18;56	12;16	12;14,21
(3)	1434	18;59,10	12;19,10	12;20,12
(3)	1464	19;19	12;39	12;38,7
(3)	1500	19;36,4	12;56,4	12;57,48

The sources are (1), Rome, Bibl. Vat., Cod Vat. Latinus 8174, fol 449, (77, I, p. 99); (2), Vienna, Nationalbibliothek, 5318, fol 2, (77, I, p. 133); (3), Vienna, Nationalbibliothek, 5415, fol 217, (77, I, p. 153).

All these figures are in clear agreement with the Alfonsine formula, and many other examples might be found of this application. Apart from star lists proper however, one encounters in some manuscripts lists of lunar mansions, and in these the longitude of the first mansion al-naṭh (cornu Arietis) may be equal to the correction to the Ptolomaic longitude, as in the lists above, but there are also lists of mansions in which the longitude has a somewhat different meaning.

In a fifteenth century manuscript (Br Mus Royal 12. D. 6, fol 63v) for example, there is a list of coordinates of initial points of the 28 mansions, beginning 19;26♈ 2;17♉, 15;9♉, ... The first coordinate is the Alfonsine correction to Ptolemy for the year A.D. 1480, approximately.

On the other hand in the same manuscript (fol 45v) there is a carefully composed list of 28 uniformly spaced mansions, with the names given 'secundum hebreos, syros, latinos' and beginning with the longitude 22;5♈. At the foot of the table is written 'Anno chtisti 1400 principium prima mansio lune in 22,5 Arietis fixi secundum tables Alfonsi ...' Now in A.D. 1400 the Alfonsine correction to Ptolemy is 12;0,8 + 6;40 = 18;40,8 which leaves an excess of 3;24,52. This puts the first mansion 22;5 — 12;0,8 = 10;4,52 East of E_0.

The same thing is found in a late fifteenth century manuscript (Trinity College Cam., 0.II.40, fol 118v) where the coordinate of the first mansion is given as 22;59 in A.D. 1490. The Alfonsine correction to Ptolemy for that year is 12;52,13 + 6;40 = 19;32,13 which leaves an excess of 3;26,47 and puts the mansion 22;59 — 12;52,13 = 10;6,47 East of E_0. These last two examples are in agreement as to the sidereal location of the first mansion, and the significance would appear to be that the point in question on the ecliptic has the same longitude as the circle of declination passing through γ Arietis in the time of Ptolemy, for the polar longitude of that star is 3;24 as calculated from Ptolomaic coordinates.

9.10 Copernicus

The Copernican model of precession is described by him in the de Revolutionibus (iii 5–12), and in ch. 12 an example is given which displays the complete formula. In common with the Alfonsine model the Copernican motion is a sum of linear and oscillatory terms, but all the parameters differ from the Alfonsine model. The linear and oscillatory terms have periods of 25816 and 1717 Egyptian years (365 days) respectively. Copernicus measures his longitudes from γ Arietis (which is the first lunar mansions) and his formula for its longitude λ is

$$\lambda = 5;32 + \frac{360\,n}{25816} - 1;10 \sin\left(\frac{360\,n}{1717} + 13;30\right)$$

where n is the time measured in units of 365 days from 31/12/0 = 1721423. Our formula for the equation is simple $E = \lambda - 13;20$. If we now replace n by $(t - 1721423)/365$, we obtain

$$E = -73.5670 + \frac{360\mathrm{t}}{9422840} - \frac{7}{6}\sin\left[\frac{360\,(t - 1697921.56)}{626705}\right]$$

The model was plainly intended to fit Ptolemy's and Hipparchos's longitudes, for at these times we have $\lambda = 4;0,42$ and $6;39,25$ respectively, very close to their values 4;0 and 6;40. See Moesgaard's discussion (54) of the observational basis of this formula.

In Thābit's time (29/4/870 = 2038944), $\lambda = 17;59,1$ and $E = 4;39,1$, whereas Thābit's equatio of that date is equal to the radius of the small circle $4;18,43 = 4;39,1 - 0;20,18$. This indicates that Copernicus might have understood the sidereal reference point of Thābit's model to lie 0;20 West of our adopted reference point, in other words just the star ζ Piscium itself.

Copernicus of course differed from his predecessors in so far as he explained the phenomenon of precession in terms of a movement of the earth rather than the stellar sphere. Duhem has drawn attention however to the fact that this was not really a point of originality with Copernicus, for already in 1386 Albert of Saxony (Lib. II, Qu. VI, fol E3va; 31, vol. 9, p. 359) had reported that such a view was current, and since his work was printed in 1492, was it not likely that Copernicus came upon it? There seems to be little purpose therefore in Ravetz's attempt to explain Copernicus's originality on this point (76). His contribution lay rather in his clear grasp of the distinctive character of the observational and empirical tradition of astronomy that developed continuously from Babylonian origins, which had always emphasized sidereal rather than tropical or 'terrestrial' periods. This tradition had not been adequately comprehended in the Ptolemaic theoretical framework, and he realised that if the sidereal periods of the planets were really of fundamental significance, then inevitably they must be seen as rotating about the Sun in relation to the stars.

9.11 Modern precession formula

The modern formula for the rate of precession (7, p. 530) is

$$p = 50.2564'' + 0.0222''\, T \text{ p.a.}$$

where T[16] is a measure of time expressed in units of 36525 days, measured from A.D. 1900.0 = 2415020. The quantity E plotted in the graph must equal the longitude of the point 20′ East of ζ Piscium, which is $18;26 + 0;20$ when $T = 0$. The formula for the rate is integrated to obtain

$$E = 18.77 + \frac{5025.64}{3600}\, T + \frac{1.11}{3600}\, T^2$$

where $T = (t - 2415020)/36525$.

[16] While our reference (7) specifies the use of the tropical year here, there is no point in maintaining this distinction. Although the tropical is, properly speaking, the interval between passages of the (true) Sun through the equinoctial point of date, in this reference 'tropical year' means the interval in which the 'geometric mean longitude of the Sun,' as defined by Newcomb's formula, increases by 360°. Whatever the exact meaning, it is hardly a suitable unit in which to measure the passage of time, since it is not constant. In the interval of 2000 years since Hipparchos the accumulated difference would be some 14 days, entailing a precession of merely 0;0,2, which is entirely unobservable.

This formula agrees well with the Hipparchan figure —9;20, for it gives $E = -9;24$ when $t = 1674484$.

Appendix. Analysis of Thābit's model

The use of vector algebra rather than spherical trigonometry allows a more flexible analysis. Let the unit vectors **i k j** be chosen so that the terminus of **k** is $\mathbf{n}_0$, the terminus of **j** is the pole of the equator, and the terminus of **i** is the point on the equator having right ascension 90°. Let the pole of the fixed ecliptic be $\mathbf{n}_0$,

$$\mathbf{n}_0 = -\mathbf{i} \sin \varepsilon_0 + \mathbf{j} \cos \varepsilon_0 \qquad 1$$

where $\varepsilon_0 = 23;33$ is the obliquity of the fixed ecliptic, and let **n** be the pole of the moving ecliptic having direction cosines cos α, cos β, cos γ,

$$\mathbf{n} = \mathbf{i} \cos \alpha + \mathbf{j} \cos \beta + \mathbf{k} \cos \gamma \qquad 2$$

The small circle of radius $\mathbf{r} = 4;18,43$ is the terminus P_1 of the unit vector ρ

$$\rho = \mathbf{i} \sin r \cos \theta + \mathbf{j} \sin r \sin \theta + \mathbf{k} \cos r \qquad 3$$

where θ, the *motus*, is shown in Fig. 1.

Let B be the intersection of the fixed and moving ecliptics, at the terminus of ρ_1. With either constraint on the point B we have

$$\mathbf{n}_0 \cdot \rho_1 = \mathbf{n} \cdot \rho = \mathbf{n} \cdot \rho_1 = 0 \qquad 4$$

while the preferred constraint (I) is expressed by

$$\rho \cdot \rho_1 = 0 \qquad 5.\mathrm{I}$$

and the alternative (II) by

$$\mathbf{k} \cdot \rho_1 = 0. \qquad 5.\mathrm{II}$$

Now since $\mathbf{n}_0 \cdot \rho_1 = 0$ we can assume the following form for ρ_1,

$$\rho_1 = \mathbf{i} \cos \varepsilon_0 \cos \varphi + \mathbf{j} \sin \varepsilon_0 \cos \varphi + \mathbf{k} \sin \varphi \qquad 6$$

where φ is some angle related to the *motus* θ.

The conditions $\mathbf{n} \cdot \rho = \mathbf{n} \cdot \rho_1 = 0$ are equations which may be solved for cos α, cos β,

$$\cos \alpha = \frac{\cos r \sin \varepsilon_0 - \tan \varphi \sin r \sin \theta}{\sin r \sin (\theta - \varepsilon_0)} \cos \gamma \qquad 7$$

$$\cos \beta = -\frac{\cos r \cos \varepsilon_0 - \tan \varphi \sin r \cos \theta}{\sin r \sin (\theta - \varepsilon_0)} \cos \gamma \qquad 8$$

This completely determines the three direction cosines because we have the additional relation $\cos^2\alpha + \cos^2\beta + \cos^2\gamma = 1$. In this way cos γ is found,

$$\cos \gamma = -\frac{\sin r \sin (\theta - \varepsilon_0)}{\sqrt{[1 - \sin^2 r \cos^2 (\theta - \varepsilon_0) + \tan^2\varphi \sin^2 r - 2 \tan \varphi \sin r \cos r \cos^2 (\theta - \varepsilon_0)]}} \qquad 9$$

the sign being chosen so that **n** is directed upward.

When the first constraint is used we have from (5.I)

$$\sin r \cos\theta \cos\varepsilon_0 \cos\varphi + \sin r \sin\theta \sin\varepsilon_0 \cos\varphi + \sin\varphi \cos r = 0$$

$$\tan\varphi = -\tan r \cos(\theta - \varepsilon_0), \qquad 10$$

and when substituted in (7) — (9) we obtain

$$\cos\alpha = -\frac{\cos r \sin\varepsilon_0 + \tan r \sin r \cos(\theta-\varepsilon_0)\sin\theta}{\sqrt{[1+\tan^2 r \cos^2(\theta-\varepsilon_0)]}}$$

$$\cos\beta = \frac{\cos r \cos\varepsilon_0 + \tan r \sin r \cos(\theta-\varepsilon_0)\cos\theta}{\sqrt{[1+\tan^2 r \cos^2(\theta-\varepsilon_0)]}} \qquad 11.I$$

$$\cos\gamma = -\frac{\sin r \sin(\theta-\varepsilon_0)}{\sqrt{[1+\tan^2 r \cos^2(\theta-\varepsilon_0)]}}$$

When the second constraint is used, (5.II), we have

$$0 = \mathbf{k} \cdot \rho_1 = \sin\varphi,$$

in other words $\varphi = 0$. Therefore,

$$\cos\alpha = -\frac{\cos r \sin\varepsilon_0}{\sqrt{[(1-\sin^2 r \cos^2(\theta-\varepsilon_0)]}}$$

$$\cos\beta = \frac{\cos r \cos\varepsilon_0}{\sqrt{[(1-\sin^2 r \cos^2(\theta-\varepsilon_0)]}} \qquad 11.II$$

$$\cos\gamma = -\frac{\sin r \sin(\theta-\varepsilon_0)}{\sqrt{[(1-\sin^2 r \cos^2(\theta-\varepsilon_0)]}}$$

The equinoctial point ♈ is at the terminus of the vector

$$(\mathbf{j}\times\mathbf{n}) \,/\, |\mathbf{j}\times\mathbf{n}| = \frac{\mathbf{i}\cos\gamma - \mathbf{k}\cos\alpha}{\sqrt{[\cos^2\alpha + \cos^2\gamma]}} \qquad 12$$

so that the angle E between ♈ and P_1, the *equatio*, is determined by

$$\sin E = \frac{|\rho\times(\mathbf{j}\times\mathbf{n})|}{|\rho|\;|\mathbf{j}\times\mathbf{n}|} = \frac{|\mathbf{j}\,\rho\cdot\mathbf{n} - \mathbf{n}\,\rho\cdot\mathbf{j}|}{|\mathbf{j}\times\mathbf{n}|}$$

$$= \frac{|\rho\cdot\mathbf{j}|}{|\mathbf{j}\times\mathbf{n}|} = \frac{\sqrt{[\cos^2\alpha + \cos^2\gamma]}}{\sin r \sin\theta} \qquad 13$$

If we now substitute from (11.I) and (11.II) respectively we obtain equations (1a) and (2a) given previously (part I, p. 212).

The obliquity of the moving ecliptic, denoted by ε, is determined by

$$\cos\varepsilon = \mathbf{j}\cdot\mathbf{n} = \cos\beta$$

$$\varepsilon = \beta \qquad 14$$

This yields directly equations (1b) and (2b) given previously.

The angle between ♈ and ♈$_0$, which Carmody (83, p. 95) denoted by c, is given by his formula if one uses the first constraint rather than the second. Therefore in spite of numerous obvious divergences between his interpretation and ours, he has at least read the text correctly on this point. We have

$$\cos c = \frac{\mathbf{k}\cdot(\mathbf{j}\times\mathbf{n})}{|\mathbf{j}\times\mathbf{n}|} = \frac{\cos\alpha}{|\mathbf{j}\times\mathbf{n}|}$$

$$\sin c = \frac{|\mathbf{k}\times(\mathbf{j}\times\mathbf{n})|}{|\mathbf{j}\times\mathbf{n}|} = \frac{|\mathbf{k}\cdot\mathbf{n}|}{|\mathbf{j}\times\mathbf{n}|} = \frac{\cos\gamma}{|\mathbf{j}\times\mathbf{n}|}\,.$$

Substituting from (11.I) will give

$$\cot c = \frac{\cos\alpha}{\cos\gamma} = \frac{\cos r\sin\varepsilon_0 + \tan r\sin r\cos(\theta-\varepsilon_0)\sin\theta}{\sin r\sin(\theta-\varepsilon_0)} \qquad 15.\mathrm{I}$$

In our notation Carmody's formula is

$$\cot c = \frac{\cot(\theta-\varepsilon_0)\sin\theta}{\sin r\cos r} - \cot r\cos\theta \qquad 16$$

and one may show the identity of (15.I) and (16).

With the second constraint one obtains,

$$\cot c = \frac{\cos\alpha}{\cos\gamma} = \frac{\cot r\sin\varepsilon_0}{\sin(\theta-\varepsilon_0)}. \qquad 15.\mathrm{II}$$

In order to discuss the locus of the pole of the moving ecliptic, the terminus of **n**, we find the components of **n** in a reference frame obtained by rotating from **i j k** through an angle ε_0 about **k**. The components of **n** are then

$$\begin{aligned}\xi &= \cos\varepsilon_0\cos\alpha + \sin\varepsilon_0\cos\beta\\ \eta &= -\sin\varepsilon_0\cos\alpha + \cos\varepsilon_0\cos\beta\\ \zeta &= \cos\gamma\end{aligned} \qquad 17$$

The locus of **n** may be followed in terms of ξ and ζ, which are measured from the pole of the fixed ecliptic towards the Summer Solstice and Spring Equinox respectively, while η is essentially constant.

For the first constraint,

$$\xi = \frac{\tan r\sin r\cos(\theta-\varepsilon_0)\sin(\theta-\varepsilon_0)}{\sqrt{[1+\tan^2 r\cos^2(\theta-\varepsilon_0)]}} \simeq -\frac{1}{2}r^2\sin 2(\theta-\varepsilon_0)$$

$$\eta = \frac{\cos r + \tan r\sin r\cos^2(\theta-\varepsilon_0)}{\sqrt{[1+\tan^2 r\cos^2(\theta-\varepsilon_0)]}} \simeq 1 \qquad 18.\mathrm{I}$$

$$\zeta = -\frac{\sin r\sin(\theta-\varepsilon_0)}{\sqrt{[1+\tan^2 r\cos^2(\theta-\varepsilon_0)]}} \simeq -r\sin(\theta-\varepsilon_0)$$

while for the second,

$$\xi = 0$$

$$\eta = \frac{\cos r}{\sqrt{[1 - \sin^2 r \cos^2 (\theta - \varepsilon_0)]}} \simeq 1 \qquad 18.II$$

$$\zeta = -\frac{\sin r \sin (\theta - \varepsilon_0)}{\sqrt{[1 - \sin^2 r \cos^2 (\theta - \varepsilon_0)]}} \simeq -r \sin (\theta - \varepsilon_0)$$

The locus according to (18.I) is a figure of eight, and according to (18.II) it is a straight arc.

The locus of the moving solsticial point, located 90° from P_1, is at the terminus of

$$\begin{aligned} \mathbf{n} \times \rho = \mathbf{i}\,[\cos \beta \cos r - \cos \gamma \sin r \sin \theta] + \mathbf{j}\,[\cos \gamma \sin r \cos \theta - \cos \alpha \cos r] \\ + \mathbf{k}\,[\cos \alpha \sin r \sin \theta - \cos \beta \sin r \cos \theta] \\ \equiv \mathbf{i} \cos A + \mathbf{j} \cos B + \mathbf{k} \cos C, \end{aligned}$$

and the components of this in the same rotated frame as before are

$$\begin{aligned} \lambda &= \cos \varepsilon_0 \cos A + \sin \varepsilon_0 \cos B \\ \mu &= -\sin \varepsilon_0 \cos A + \cos \varepsilon_0 \cos B \\ \nu &= \cos C \end{aligned}$$

The locus is expressed in terms of ν and μ which are measured from *caput cancri* of the fixed ecliptic towards $♈_0$ and the pole of the fixed ecliptic, respectively, while λ is nearly constant. Under constraint I we find,

$$\lambda = \frac{1}{\sqrt{[1 + \tan^2 r \cos^2 (\theta - \varepsilon_0)]}} \simeq 1$$

$$\mu = 0 \qquad 19.I$$

$$\nu = -\frac{\tan r \cos (\theta - \varepsilon_0)}{\sqrt{[1 + \tan^2 r \cos^2 (\theta - \varepsilon_0)]}} \simeq -r \cos (\theta - \varepsilon_0)$$

so that the locus is a straight arc.

Under constraint II,

$$\lambda = \frac{1 - \sin^2 r \cos^2 (\theta - \varepsilon_0)}{\sqrt{[1 - \sin^2 r \cos^2 (\theta - \varepsilon_0)]}} \simeq 1$$

$$\mu = -\frac{\sin^2 r \cos (\theta - \varepsilon_0) \sin (\theta - \varepsilon_0)}{\sqrt{[1 - \sin^2 r \cos^2 (\theta - \varepsilon_0)]}} \simeq -\frac{1}{2} r^2 \sin 2 (\theta - \varepsilon_0) \qquad 19.II$$

$$\nu = -\frac{\sin r \cos r \cos (\theta - \varepsilon_0)}{\sqrt{[1 - \sin^2 r \cos^2 (\theta - \varepsilon_0)}} \simeq -r \cos (\theta - \varepsilon_0)$$

so that the locus is a figure of eight, extended East-West. The figure of eight locus described by (18.I) and (19.II) is very narrow, the ratio of length to width being $2 / r \simeq 26.6$.

LIST OF STAR COORDINATES

Hipparchan coordinates

	λ	β	L	l	α	δ
a Arie	8;0	10;0	3;32,15	10;56,28	3;14,10	12;22,15
β Arie	5;0	8;20	1;17,17	9;7,2	1;10,41	9;38,17
γ Arie	4;0	7;20	0;44,19	8;1,20	0;40,32	8;19,15
η Pisc	—2;20	5;20	—4;4127	5;49 47	—4;17,30	3;56,5
ζ Pisc	—9;40	—0;10	—9;35,38	—0;10,55	—8;47,17	—4;2,45
α Virg	174;0	—2;0	173;7,18	—2;11,3	173;42,15	0;35,29
ᾱ Virg	—6;0	2;0	—6;52,42	2;11,3	—6;17,45	—0;35,29

Coordinates calculated for 130 B.C.

	λ	β	L	l	α	δ
α Arie	8;2	9;54	3;38,47	10;49,11	3;20,21	12;17,6
β Arie	4;24	8;24	0;40,55	9;10,50	0;37,28	9;27,18
γ Arie	3;36	7;5	0;28,18	7;44,24	0;25,55	7;55,47
η Pisc	—2;42	5;15	—5;0,14	5;43,55	—4;35,1	3;43,19
ζ Pisc	—9;46	—0;16	—9;39,4	—0;17,26	—8;51,0	—4;9,22
α Virg	174;16	—1;55	173;25,48	—2;5,28	173;58,49	0;32,45
ᾱ Virg	—5;44	1;55	—6;34,12	2;5,28	—6;1,11	—0;32,45

SUMMARY

These studies are concerned largely with that tradition of empirical astronomy which makes use of the 'fixed' zodiac (fixed in relation to the stars), in contrast to the Hipparchan ecliptic measured from the equinox of date. This tradition, with its associated techniques, extends with a remarkable degree of continuity from early Babylonian observations of risings and culminations, through Hipparchan techniques of star mapping, Indian astronomical systems, Thābit ibn Qurra's model of trepidation, the Khwārezmian and Toledan Tables in Arabic and Latin, down to Copernicus' *de Revolutionibus.*

When a fixed zodiac is used two points need to be settled: (i) the determination of the point of zero longitude, (ii) the nature of the particular model of precession or trepidation used to calculate the position of the equinox of date. It is found to begin with that in both the Toledan and Khwārezmian Tables the true Sun is predicted correctly for the year A.D. 564, at a time when the equinoctial point is found a little to the East of ζ Pisc. Since the model of trepidation elaborated by Thābit ibn Qurra was used with the Toledan Tables, it is concluded that the oscillation of the equinoctial point in that model was centred on the same point, near ζ Pisc. It is well known that this point marked the beginning of the fixed zodiac used in the Indian systems, such as the Sūrya Siddhānta and others.

By making an analysis of the positions given in Ptolemy's star catalogue it has been found that there is a perfect alignement of the stars α Arie, β Arie, ζ Pisc and α Virg; this circle of alignement is the horizon for the latitude of Hipparchos' observatory on Rhodes (36°), and thus it was assumed that at that latitude one would have simultaneously the rising of α Arie, β Arie, ζ Pisc and the setting of α Virg. Moreover the point of the ecliptic rising then is some 0;16 East of ζ Pisc, and so identical with the zero point of the ecliptic as used in the Indian and Arabic Tables mentioned above.

Some additional notes are provided to cover various models of precession and trepidation in detail, including various Indian and Arabic models, and a graph is given in which is shown the movement of the equinoctial point according to each model.

TABLE OF CONTENTS

REFERENCES

(1) A. Aaboe, 'On the Babylonian Origin of some Hipparchan Parameters,' *Centaurus*, 4 (1955), 122–5.
(2) Albert of Saxony, *Quaestiones subtillissimae in libros de Caelo et Mundo*, Venice, 1492.
(3) Albumasar, *De magnis coniunctionibus: annorum revolutionibus: ac eorum profectionibus: octo continens tractatus*, Venice, 1515.
(4) *Tabule Astronomice Alfonsi Regis*, edited by Joh. Santritter, together with *Canones sive Propositiones in Tabulas Alfonsinas*, Venice, 1492.
(5) A. Schott and R. Böker, *Aratos, Sternbilder und Wetterzeichen*, München, 1958.
(6) *Mahāsiddhānta, a Treatise on Astronomy by Āryabhaṭ*, edited with his own Commentary by Mahāmahopādhyāya Sudhākara Dvivedi, Benares, 1910.
(7) *The Astronomical Ephemeris for the Year 1975*, H. M. Nautical Office, London, 1973.
(8) R. Steele, *Compotus Fratris Rogeri, Opera hactenus inedita*, fasc. VI, Oxford, 1920.
(9) J. Bentley, *A Historical View of the Hindu Astronomy*, London, 1825.
(10) *Siddhānta Śiromani of Bhāskara*, translated by L. Wilkinson, revised by B. Śāstrī, Benares, 1866.
(11) R. Billard, *L'Astronomie indienne, investigations des textes sanskrits et des données numériques*, Paris, 1971.
(12) J. B. Biot, *Études sur l'astronomie indienne et sur l'astronomie chinoise*, Paris, 1862.
(13) E. C. Sachau, *Alberuni's India*, 2 vols., London, 1910.
(14) Abū Raiḥān Muḥ. b. Aḥmad al-Bīrūnī, *al-Qānūn al-Mas'ūdī* (Canon Masudicus), 3 vols., Hyderabad, 1954–6.

(15) R. BÖKER, *Die Entstehung der Sternsphäre Arats, Berichte über die Verhandlungen der Sächsischen Akademie der Wissenschaften zu Leipzig*, math.-naturw. Kl., 99 (1952), Heft 5.
(16) P. C. SENGUPTA, *The Khaṇḍakhādyaka, an Astronomical Treatise of Brahmagupta*, Calcutta, 1934.
(17) *Brāhmasphuṭasiddhānta and Dhyānagrahopadeṣādhyāya by Brahmagupta*, edited with his own Commentary by Mahāmahopādhyāya Sudhākara Dvivedi, Benares, 1902.
(18) J. J. BURCKHARDT, 'Zwei griechische Ephemeriden, '*Osiris* 13 (1958), 79–92.
(19) E. BURGESS and G. WHITNEY, Translation of the Surya Siddhanta, with Notes and an Appendix, *J. America Orien. Soc.* 6 (1860), 141–500.
(20) *Censorinus, de die natali*, edited by F. Hultsch, Leipzig, 1867.
(21) D. A. CHWOLSON (Khvol'son), *Die Ssabier und der Ssabismus*, St. Petersbourg, 1856.
(22) H. T. COLEBROOK, *Miscellaneous Essays*, 2 vols., London, 1837.
(23) S. DAVIES, 'On the astronomical computations of the Hindus,' *Asiatic Researches*, 2 (1790), 225–287.
(24) S. DAVIES, 'On the Indian cycle of sixty years,' *Asiatic Researches*, 3 (1791), 209–227.
(25) M. DELAMBRE, *Histoire de l'Astronomie Ancienne*, 2 vols., Paris, 1817.
(26) M. DELAMBRE, *Histoire de l'Astronomie au Moyen-Age*, Paris, 1819.
(27) D. R. DICKS, 'Solstices, Equinoxes, and the Presocratics,' *J. Hellenic Studies*, 86 (1966), 26–40.
(28) SH. BALAKRISHNA DIKSHIT, 'The Twelve-year cycle of Jupiter,' *The Indian Antiquary*, 17 (1888), 1–7, 312–317.
(29) J. DOBRZYCKI, 'Teoria precesji w astronomii średniowiecznej', *Studia i Materialy z dziiejow nauki polskiej*, seria C, Z. 11, 1956, 3–47.
(30) J. L. E. DREYER, 'On the original form of the Alfonsine Tables,' *Monthly Notices of the RAS*, 80 (1920), 243–262.
(31) P. DUHEM, *Le Système du monde*, tomes I–X, Paris, 1914, 1958–9.
(32) R. EISLER, Review of 'Robert Böker, *Azimuthpendelungen der Fixsterne und die Überlieferung. Berechnungen zur vorgriechischen Astronomie* Nr. IV, Leipzig (1949),' *Archives Int. d'Histoire des Sciences*, 1949, 1184–1189.
(33) BAYARD DODGE, *The Fibrist of al-Nadīm*, Columbia University Press, New York, 1970.
(34) J. F. FLEET, 'The use of the twelve year cycle of Jupiter in records of the early Gupta period,' *The Indian Antiquary*, 17 (1888), 331–339.
(35) *Gemini Elementa Astronomiae ad codicum fidem recensuit Germanica interpretatione et commentarius instruxit Carolus Manitius*, Teubner, Leipzig, 1898.
(36) B. R. GOLDSTEIN, 'On the theory of trepidation, according to Thābit b. Qurra and al-Zarqāllu and its implications for homocentric planetary theory,' *Centaurus* 10 (1964), 232–247.
(37) H. H. GOLDSTINE, *New and Full Moons, 1001 B. C. to A. D. 1651 (Mem. Am. Phil. Soc.* no. 94], Philadelphia, 1973.
(38) R. T. GUNTHER, *Early Science in Oxford*, vol. 5, *Chaucer and Messahalla on the Astrolabe*, Oxford, 1929.
(39) W. HARTNER, 'The Earliest History of the Constellations in the Near East and the Motif of the Lion-Bull Combat,' *J. Near Eastern Studies*, 24 (1965), 1–16.
(40) W. HARTNER, 'Trepidation and planetary theories, common features in late Islamic and early Renaissance astronomy,' *Proc. of International Conference "Oriente e Occidente nel Medioevo: Filosofia e Scienze," Accad. Naz. dei Lincei*, Rome, 1971.
(41) *Hipparchi in Arati et Eudoxi Phaenomena Commentariorum libri tres, ad codicum fidem recensuit Germanica interpretatione et commentarius instruxit Carolus Manitius*, Leipzig, 1894.
(42) P. HUBER, 'Über den Nullpunkt der babylonischen Ekliptik', *Centaurus* 5 (1958), 192–208.
(43) R. ISAAC BEN JOSEPH ISRAELI, *Liber Jesod Olam seu Fundamentum Mundi opus astronomicum celeberrimum, ex codice manuscripto denuo ediderunt, textum emendarunt notas adjecerunt, nec non versionem epitomarium vernaculam addendam curaverunt B. Goldberg & L. Rosenkranz* Berlin (1848).
(44) E. S. KENNEDY, 'The Sassanian Astronomical Handbook *Zīj-i-Shāh* and the Astrological Doctrine of "transit" *(Mamarr)*,' *J. Amer. Or. Soc.* 78 (1958), 246–262.

(45) F. X. KUGLER, *Die babylonische Mondrechnung. Zwei Systeme der Chaldäer über den Lauf des Mondes und der Sonne. Auf Grund mehrerer von J. N. Straßmaier S. J. copierten Keilinschriften des Britischen Museums, mit einem Anhang über chaldäische Planetentafeln*, Freiburg im B. (1900).
(46) F. X. KUGLER, *Sternkunde und Sterndienst in Babel. Assyriologische, astronomische und astral-mythologische Untersuchungen.* I. Buch: *Entwicklung der babylonischen Planetenkunde von ihren Anfängen bis auf Christus, nach zumeist ungedruckten Quellen des Britischen Museums, mit 24 Keilinschriftlichen Beilagen*, Münster in W. (1907).
(47) IBID: II. Buch: *Natur, Mythus und Geschichte als Grundlegen babylonischen Zeitordnung nebst eingehenden Untersuchungen der älteren Sternkunde und Meteorologie, mit drei Figurtafeln und zahlreichen Keilinschriftlichen Beilagen.* Teil I, Münster in W., 1909/10; Teil II, 1912.
(48) IBID: *Ergänzungen zum ersten und zweiten Buch.* I. Teil, I–VIII Abhandlungen: *Astronomie und Chronologie der älteren Zeit*, Münster in W., 1913; II. Teil, IX–XIV Abhandlungen: *Sternkunde und Chronologie der älteren Zeit*, Münster in W., 1914.
(49) *3. Ergänzungsheft zum ersten und zweiten Buch, von Johann Schaumberger*, Münster, 1935.
(50) H. LEWY, 'Points of Comparison between Zoroastrianism and the Moon-cult of Harran,' in *A Locust's Leg, Studies in Honour of S. H. Taqizadeh*, ed., W. B. Henning and E. Yarshater, London, 1962.
(51) E. LITTRÉ, 'Guillaume de Saint-Cloud,' *Histoire Littéraire de la France*, 25 (1869), 63–74.
(52) T. H. MARTIN, 'Mémoire sur cette question: La Precession des équinoxes, a-t-elle été connu des Egyptiens ou de quelque autre peuple avant Hipparque?', *Mém. Présentée par divers savants à l'Acad. des Insc.*, 8 (1869), 303–522.
(53) H. MICHEL, 'Sur l'origine de la théorie de la trepidation', *Ciel et Terre* (Bruxelles) for 1950, 227–234.
(54) K. P. MOESGAARD, 'The 1717 Egyptian years and the Copernican theory of precession,' *Centaurus* 13 (1968), 120–138.
(55) K. P. MOESGAARD, 'Tychonian observations, perfect numbers, and the date of creation: Longomontanus's solar and precessional theories,' *J. History of Astronomy*, 6 (1975), 84–99.
(56) C. A. NALLINO, *al-Battani sive albatenii Opus Astronomicum*, pars I Milan, 1903; pars II, Milan, 1907.
(57) O. NEUGEBAUER, 'Solstices and Equinoxes in Babylonian Astronomy during the Seleucid Period,' *J. Cuneiform Studies*, 2 (1948), 209–222.
(58) O. NEUGEBAUER, 'The alleged Babylonian discovery of the precession of the Equinoxes', *J. American Orien. Soc.* 70 (1950), 1–8.
(59) O. NEUGEBAUER, 'An astronomical almanac for the year 348/9 (P. Heid. Inv. No. 34),' *Historisk-filologiske Meddelelser, Det Kong. Danske Vid. Selskab* 36 (1956) nr. 4.
(60) O. NEUGEBAUER and H. B. VAN HOESEN, *Greek Horoscopes (Mem. Am. Phil. Soc. no. 48)*, Philadelphia, 1959.
(61) O. NEUGEBAUER, 'The Astronomical Tables of al-Khwārizmī,' *Historisk-filosofiske Skrifter, Det Kong. Danske Vid. Selskab*, 4 (1962).
(62) O. NEUGEBAUER, 'Thābit ben Qurra "On the solar year," and "On the motion of the eighth sphere," translation and commentary,' *Proc. Am. Phil. Soc.* 106 (1962), 264–299.
(63) P. V. NEUGEBAUER, *Sterntafeln von 4000 vor Chr. bis zur Gegenwart nebst Hilfsmittels zur Berechnung von Sternpositionen zwischen 4000 vor Chr. und 3000 nach Chr.*, Leipzig, 1912.
(64) J. D. NORTH, 'Medieval star catalogues and the movement of the eighth sphere,' *Arch. Int. d'Hist. des Sciences* for 1967, 71–83.
(65) P. NUÑEZ, *Tratado da Sphere, com a theorica do sol e da luna e ho primeiro livro da geographia de Caludio Ptolomeo*, Lisbon (1537); (facsimile) München, 1915.
(66) P. NUÑEZ, *In Theoricas planetarum Georgii Purbachii annotationes aliquot, per Petrum Nonium Salaciensem (Opera)*, Basle, 1592.
(67) TH. VON OPPOLZER, *Canon der Finsternisse*, *Öster. Akad. Wiss. Denkschriften, math-naturw. Kl.*, 52 (1887).
(68) A. PANNEKOEK, 'Calculation of dates in the Babylonian Tables of the Planets,' *Proceedings, Koninklijke Nederlands Akademie van Wetenschapen*, 19 (1916), 684–703.
(69) C. PELLAT, *Le Calendrier de Cordoue publié par R. Dozy*, nouvelle edition accompagnée d'une traduction francaise annotée, Leiden, Brill, 1961.

(70) Georg Peurbach, *Novae Theoricae Planetarum*, Venice, 1537.
(71) D. Pingree, *The Thousands of Abū Ma'shar*, Warburg Institute, London, 1968.
(72) Ptolomaeus, *Handbuch der Astronomie* [Deutsche Übersetzung und erlautende Anmerkungen von K. Manitius, Vorwort und Berichtigungen von O. Neugebauer], Leipzig, 1963.
(73) C. H. F. Peters and E. B. Knobel, *Ptolemy's Catalogue of Stars, a Revision of the Almagest*, The Carnegie Institution of Washington, 1915.
(74) Ptolemy, Φάσεις ἀπλανῶν ἀστέρων καὶ συναγωγὴ ἐπισημασιῶν, *Opera Astronomica Minora*, ed., J. L. Heiberg, Leipzig, 1907.
(75) B. V. Raman, *A Manual of Hindu Astrology (Correct casting of Horoscopes)*, 9th ed., Bangalore, 1972.
(76) J. R. Ravetz, *Astronomy and Cosmology in the Achievements of Nicolaus Copernicus*, Warsaw, 1965.
(77) F. Saxl, *Verzeichnis astrologischer und mythologischer illustrierter Handschriften des lateinischen Mittelalters;* I, *In Römischen Bibliotheken, Sitz. Heidelberger Akad. Wiss. (Phil.-hist. Klasse)*, 1915 (no. 5) – II, *Die Handschriften der National-Bibliothek in Wien*, ibid., 1925/6 (no. 2).
(78) H. C. F. C. Schjellerup, *Description des étoiles Fixes (Kitāb al-kawākib al-thābita al-musawwar) of Abū'l-Ḥusain ᶜAbdu'l-Rahmān ibn ᶜUmar al-Ṣūfī*, St. Petersbourg, 1874.
(79) L.-A. M. Sédillot, *Traité des Instruments Astronomiques des Arabes composé au 13me siècle par Aboul Hhassan Ali, de Maroc (jāmiᶜu'l-mabādī wa'l-ghāyāti)*, trad. par J. J. Sédillot, t. 1, Paris, 1834.
(80) S. N. Sen, *A Bibliography of Sanskrit Works on Astronomy and Mathematics*, Part I: *Manuscripts, Texts, Translations & Studies*, National Institute of Sciences of India, New Delhi, 1966.
(81) R. Sewell and S. B. Dīksit, *The Indian Calendar, with tables for the conversion of Hindu and Muhammedan dates, and vice versa*, London, 1896.
(82) H. Suter, *Die astronomischen Tafeln des Muhammed ibn Mūsā al-Khwārizmī in der Bearbeitung des Maslama ibn Aḥmad al-Madjrīṭī* etc., København, 1914.
(83) F. J. Carmody, *The Astronomical Works of Thābit b. Qurra*, Berkeley, 1960.
(84) Hasan Taqīzādeh, *Gāh Shumārī dar Īrān-i Qadīm*, Tehran, 1938.
(85) Hasan Taqīzādeh, *Old Iranian Calendars*, Royal Asiatic Society, London, 1938.
(86) Theon of Alexandria, *Commentaires de Théon d'alexandrie sur les Tables Manuelles astronomiques de Prolemée.* ed. N. Halma. 1re partie contenant les prolegomènes de Ptolomée (pp. 1–26), les commentaires de Theon (pp. 27–83), et les tables preliminaires (pp. 84–147), terminées par les ascensions des signes du zodiaque dans la sphère droite (pp. 149–155), Paris, 1822.
(87) G. J. Toomer, 'A survey of the Toledan Tables,' *Osiris 15* (1968), 1–174.
(88) B. Tuckerman, *Planetary, Lunar and Solar positions, A.D. 2 to A.D. 1649, at five-day and ten-day intervals*, [*Mem. of Am. Phil. Soc*, No. 59], Philadelphia, 1964.
(89) J. M. Millás Vallicrosa, El Libro de los Fundamentos de las Tablas astronómicas de R. Abraham ibn ᶜEzra, edición critica, con introducción y notas, Madrid–Barcelona (1947).
(90) J. M. Millás Vallicrosa, 'La Traducción latina del "Liber de motu octave sphere" de Tābit ibn Qurra,' *Al-Andalus 10* (1945) 89–108. This is reprinted in *Nuevos estudios sobre historia de la ciencia española*, J. M. Millás Vallicrosa, Barcelona, 1960.
(91) J. M. Millás Vallicrosa, *La obra Forma de la Tierra de R. Abraham bar Ḥiyya ha-Bargel ni, traducción del Hebreo, con prólogo y notas*, Madrid–Barcelona, 1956.
(92) H. Vogt, 'Versuch einer Wiederherstellung von Hipparchs Fixsternverzeichnis', *Astronomischen Nachrichten*, 224 (1925), cols. 17–54.
(93) B. L. van der Waerden, 'Zur babylonischen Planetenrechnung', *Eudemus* 1 (1941), 23–48.
(94) B. L. van der Waerden, 'Babylonian Astronomy. II. The Thirty-six Stars,' *J. Near Eastern Studies*, 8 (1949), 6–26.
(95) B. L. van der Waerden, 'Diophantinische Gleichungen und Planetenperioden in der indischen Astronomie', *Vierteljahrschrift der Naturforschenden Gesellschaft in Zürich* for 1955, 153–170.

(96) B. L. VAN DER WAERDEN, 'Ausgleichspunkt, "Methode der Perser" und indische Planetenrechnung', *Archive for History of Exact Sciences.* 1 (1961), 107–121.
(97) B. L. VAN DER WAERDEN und J. J. BURCKHARDT, 'Das astronomische System der persischen Tafeln I', *Centaurus* 13 (1968), 1–28.
(98) A. WEGENER, 'Die astronomische Werke Alfons X', *Bibliotheca Mathematica* 3rd Ser 6 (1906), 129–185.
(99) S. WEINSTOCK, 'Lunar Mansions and early Calendars,' *J. Hellenic Studies*, 69 (1949), 48–69.
(100) E. ZINNER, 'Die Tafeln von Toledo (Tabulae Toletanae)', *Osiris* 1 (1936), 747–774.

Notes added in proof

1. When the present work was completed in the Autumn of 1975, I learned from Prof. O. Gingerich that he also had attempted to determine the sidereal situation of Thābit's model. He made use of the assumption that the star lists accompanying some MSS of the Toledan Tables were intended for the year A.D. 1070. The result agreed broadly with mine. ('The accuracy of the Toledan Tables', Owen Gingerich and Barbara Welther, preprint series no. 140, October 1975, Centre for Astrophysics, Cambridge, Mass., and ΠΡΙΣΜΑΤΑ, Festschrift für Willy Hartner, Ed. Y. Maeyama & W. G. Saltzer, Steiner Verlag Wiesbaden, 1977, 151–164.)

2. The authors of the modern Indian popular calendars, the *pañcāṅga*, almost all now use anti-Spica as the origin of the sidereal longitudes. The year of zero precession is in most cases near A.D. 290, but those *pañcāṅga*-makers who follow 'older' methods, use dates near A.D. 499 or A.D. 522. There is no sign that the dates quoted in § 9.6 n. 15 have actually been used by *pañcāṅga*-makers in recent years. A most useful survey is provided in the *Report of the Calendar Reform Committee*, published by the Government of India, Council of Scientific and Industrial Research, New Delhi (1955). The national calendar determined by the Reform Committee adopted, in effect, anti-Spica, and defined the precession of this origin to be 23;15 on 21/3/1956.

III

ACCESSION AND RECESSION: RECONSTRUCTION OF THE PARAMETERS

Introduction

The theory of accession and recession was a model of precession which came to be known in the Latin world at an early stage in the process of the transmission of Arabic materials into Latin. The theory is represented principally by the tract *De motu octave spere*. That tract, with its tables, stands on its own in numerous manuscripts, but the same tables, along with brief instructions for their use are found in many of the copies of the *Toledan Tables*, generally regarded as the work of the Spanish Islamic astronomer al-Zarqāla, (*fl. ca* A.D. 1070). It might at least be surmised that the model was intended to accompany the *Toledan Tables*, but nevertheless from the time of the translation of the tract by Gerard of Cremona (or a member of that school of translators), it was attributed to Thābit ibn Qurra, who lived some 200 years earlier. In an earlier summary discussion I gave several distinct reasons why one should reject that attribution, and in this paper the details will be given in full[1]. It will be seen that the accession and recession, as defined in the tables just mentioned, is designed to work only with the *Toledan Tables*, so that an attribution to any author before the late eleventh century is altogether out of the question.

The Toledan Tables

The *Toledan Tables* are known from a great many Latin manuscript sources[2]. Their origin in eleventh century Islamic Spain is proclaimed in the text itself, although no Arabic source is extant. The scientific content is a complex mixture of earlier parameters together with observational and theoretical arguments made in the eleventh century. First of all we draw attention to the relation to the mean parameters of the

---

[1] Mercier (1987a).

[2] Toomer (1968); Mercier (1987a).

earlier *zīj* of al-Battānī[3]. The mean longitudes according to al-Battānī are tropical, measured in other words from the equinox of date, while in the *Toledan Tables* those longitudes are sidereal, that is measured from a point fixed in relation to the stars. From these mean parameters we can derive the extremely important fact that the synodic longitudes are the same in each of these *zījes*. That is, the differences between the tropical longitudes of the Sun and planets in the *zīj* of al-Battānī are respectively equal to the difference between the sidereal longitudes of the Sun and planets in the *Toledan Tables*. In the following table, the parameters of the synodic differences are given, reduced to the meridian and initial date of the *Toledan Tables*, and to the long interval, 212620 days (600 × 354;22) used there. The only serious discrepancy is in the case of Mars, where the difference between the synodic motions amounts to some 8 seconds per year.

Comparison of Synodic motions

	radix at 1948438		motion in 600 x 354;22 days	
	Toledan	al-Battānī	Toledan	al-Battānī
Moon	7;17, 7	7;17, 6.5	7199^r 356;36,51	7199^r 356;36,51.7
apogee	280;51,32	280;51,33.3	−514^r 120; 4,35	−514^r 120; 2,28.0
node	12; 8,54	12; 8,51.9	−613^r 147;42,22	−613^r 147;42,15.9
Mercury	73;46,18	73;47,33.9	1834^r 307; 0,20	1834^r 307; 5,56.1
Venus	45;28,37	45;10,16.4	364^r 45;36,10	364^r 45;35,25.3
Mars	97;43,48	97;47,24.2	−272^r 219; 6,41	−272^r 220;24,24.3
Jupiter	217;58,26	217;58, 4.3	−533^r 13;20,31	−533^r 13;21,34.8
Saturn	2;10, 4	2;10,23.7	−562^r 126;36,42	−562^r 126;40,43.4

The synodic deviation curves show the same result with an immediate graphic force, as one can see from Figs. 1, 2[4].

[3] All the parameters in question are listed in the appendix.

[4] The 'deviation' is the medieval longitude reduced by the same longitude as computed from modern parameters. On the use of deviation curves see Billard (1972), Mercier (1987b).

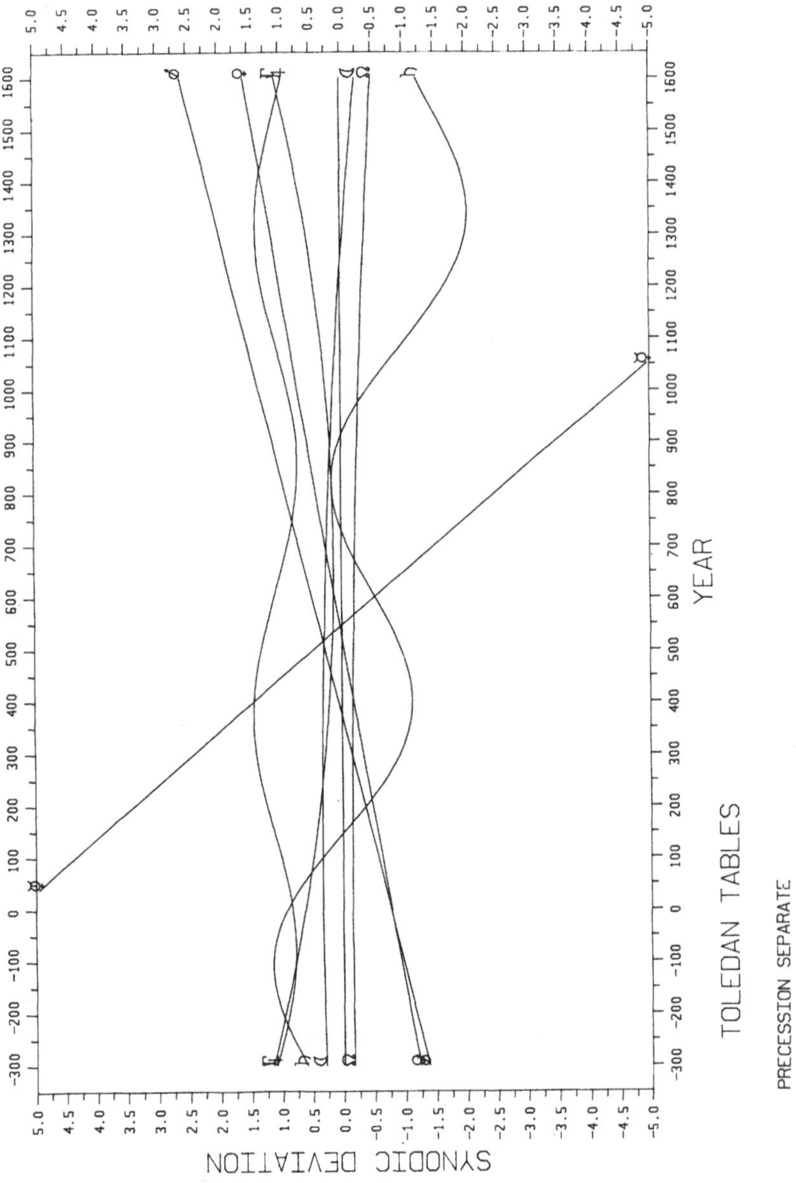

Fig. 1 Synodic deviation curves, Toledan Tables.

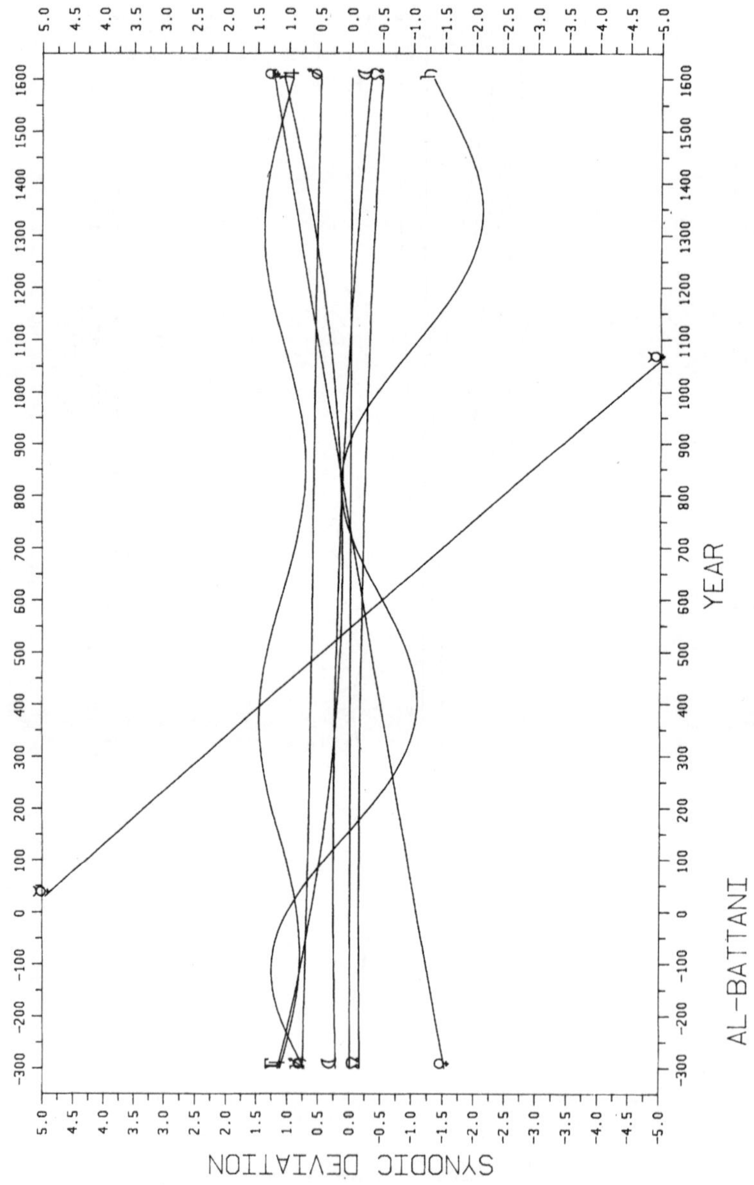

Fig. 2 Synodic deviation curves, al-Battānī.

Parameters of Accession and Recession

The table of the motus is given in two parts, for individual Arabic years, and Arabic years grouped in multiples of 30. The reference year is the Hijra Era, 622 July 14, 1948438. The value of accession is tabulated for values of the argument in steps of 5 degrees. The argument for any year is calculated from the first two tables, and then the corresponding value of accession is taken from the third part.

expanded years		collected years		accession and recession	
1	0;5,9	radix	1;34,2	5	0;55,52
2	0;10,20	30	4;9,0	10	1;50,36
3	0;15,29	60	6;43,58	15	2;45,16
4	0;20,39	90	9;18,56	20	3;39,23
5	0;25,49	120	11;53,54	25	4;31,12
6	0;30,59	150	14;28,52	30	5;21,30[a]
7	0;36,9	180	17;3,50	35	6;9,6
8	0;41,19	210	19;38,48	40	6;53,12
9	0;46,29	240	22;13,46	45	7;34,35[b]
10	0;51,39	270	24;48,44	50	8;13[c]
11	0;56,49	300	27;23,42	55	8;47,48
12	1;1,59	330	29;58,41	60	9;17,44
13	1;7,9	360	1,2;33,39	65	9;43,53
14	1;12,19	390	15;8,37	70	10;5,30
15	1;17,29	420	1,7;43,35	75	10;22,47
16	1;22,39	450	1,10;18,33	80	10;35,1
17	1;27,49	480	1,12;53,31	85	10;42,13
18	1;32,59	510	1,15;28,29	90	10;45
19	1;38,9	540	1,18;3,27		
20	1;43,18	570	1,20;38,25		
21	1;48,29	600	1,23;13,23		
22	1;53,39	630	1,25;48,22		
23	1;58,48	660	1,28;23,20		
24	2;2,59	690	2,0;58,18		
25	2;9,8	720	2,3;33,16		
26	2;14,18	750	2,6;8,14		
27	2;19,28	780	2,8;43,12		
28	2;24,38	810	2,11;18,10		
29	2;29,48	840	2,13;43,8		
30	2;34,58	870	2,16;28,6		

a MS: 5;22,30 b MS: 7;36,35 c MS: 8;14

The underlying parameters are fixed with the aid of least mean squares analysis. The argument as a function of the date is expressed as

$$\theta = \beta(t - t_T)$$

where
$$t_T = 1941987.6$$
$$\beta = 360/B$$
$$B = 1481785.4 \text{ days.}$$

The date corresponding to t_T is 604 Nov 14. The period equals 4056.9 Julian years, 4181.5 Arabic years.

The amplitude as a function of the argument θ is assumed to be derived from the following expression. Complications arising from the geometrical model will be discussed below.

$$A = \arcsin(\alpha \sin \theta).$$

The least squares determination yields a value of $\alpha = \sin(10;45 \pm 0;0,30)$, and there is no harm apparently in simply choosing 10;45. Three suggested emendations to the table as found in the numerous manuscripts are included in the listing above, but whether these are correct or not the value of α is beyond question.

Various formulae for this movement of the eighth sphere have been considered, including those derived by Goldstein, Dobrzycki, North, and myself. All these are fairly complicated, perhaps prohibitively so from the point of view of the medieval calculator, and perhaps we should regard the above approximation for the movement of the eighth sphere as representing the procedure actually used.

Variable Obliquity

In an earlier study I derived formulae for the equation of the eighth sphere, that is 'accession and recession'[5], and the obliquity of the

---

[5] I have indicated elsewhere that this expression, rather than 'trepidation' is to be preferred. 'Accession and recession' translates the Arabic *al-iqbāl wa'l-idbār*, in turn derived from the Sanskrit *prāk-paścāt*.

ecliptic[6], formulae which followed to the letter the procedures prescribed in the latin text *De motu octave spere*. The same interpretation was assumed by Dobrzycki[7]. In this model of accession and recession there are two ecliptics, the moving and the fixed. According to the Latin text the interval from the intersection of these to the controlling point on the circumference of the small circle is 90°, and that was the assumption which Dobrzycki and I made. The simpler constraint fixes the interval from the intersection to the centre of the small circle as 90°. That was the assumption made by North (1967)[8]. It is clear to me now that the latter approach leads to the correct formula for the obliquity. This can best be demonstrated by examining the variation of the obliquity as a function of time. It is interesting however that Dobrycki and I were not alone in this reading of the text, for the procedure was interpreted by Regiomontanus in just the same way. He gave the step by step working in the case where the angle in the small circle is 45°, with quite enough detail to make it clear that the constraint on the motion was taken as Dobrzycki and I had earlier assumed[9].

According to the simpler constraint, the formula for the obliquity ϵ is

$$\cos\,\epsilon = \frac{\cos\,\epsilon_0}{\sqrt{1+\tan^2 r\,\sin^2(\theta-\epsilon_0)}}$$

[6] Mercier (1976, 1977); the formulae are in Mercier (1976), p. 212.

[7] Dobrzycki (1965), p. 12.

[8] In a brief earlier study Goldstein (1964) described the model without fixing such details. His account however is erroneous with regard to his derivation of the amplitude 10;45 from the parameters ϵ and r; in fact his calculation gives not 10;45 as he claims, but 10;42,37.

[9] Regiomontanus (1514), ch. 63; Regiomontanus (1557), ch. 55. The earlier discussion gives the calculations of accession and obliquity when the argument is 45°. In the later account he gives only the steps without any example. Neither of these works of Regiomontanus in included in the collection Schmeidler (1972).

where $r = 4;18,43$
$\epsilon_0 = 23;33.$
$\theta = \alpha \sin \beta(t - t_T)$

As the argument θ varies the obliquity ϵ decreases from the maximum arccos(cos r cos ϵ_0) = 23;55,9.7 to the minimum 23;33, and back again. The important step in understanding that this is the correct formula is the consideration of ϵ as a function of time. From the graph of ϵ, Fig. 3, we see that the obliquity takes the Ptolemaic value 23;51,20 *ca* A.D. 130 and the minimum value 23;33 in the year 870[10]. However when computed according to the interpretation which follows the Latin text literally, the minimum value of the obliquity falls considerably below 23;33, which contradicts the text. We have to conclude from this that the correct constraint is the one assumed by North.

Precession and the Toledan Tables

It is clear that the *Toledan Tables* were constructed in a context of astronomical research whose orientation was determined by the *Zīj* of al-Khwārizmī. That was a work which derived ultimately, as we know well, from an Indian source, the *Brāhmasphuṭasiddhānta*, and which used sidereal longitudes. When the *Zīj* of al-Battānī later came to be known in Spain, it might be said that people reacted to it by converting all the longitudes to sidereal form. As a first step towards this, the synodic differences were constructed. The next step would be to add to these a new, sidereal, mean Sun, so producing sidereal longitudes for the Moon and the planets. The question of what precisely was taken for the new sidereal Sun is a difficult one, and is the subject of this paper, but light is thrown on the matter by a consideration of the deviation curve of the Sun in the *Toledan Tables*. If the precessional correction which was intended to be used with the *Toledan Tables* is that of the *De motu octave spere*, then we should add that amount of precession to the Toledan sidereal mean Sun when finding the deviation. The curve is shown in Fig. 4. Now with this result we can make the vital observation that the

[10] In this figure I show also the variation of the obliquity which follows from the model given by Zarqāla in his tract on the motion of the fixed stars, Millàs Vallicrosa (1952), p. 325 seq.

deviation of the tropical mean Sun not only vanishes ca 1076 and -145, but also agrees accurately with the deviation of the Ptolemaic Sun at the time of the pseudo-observation of the Autumn Equinox in the year 132. We are therefore compelled to see these as the intended goals of the coordination of the Toledan mean Sun with this particular model of accession and recession. Our aim now is to reconstruct all the controlling parameters as a consequence of this assumed strategy. One feature of the oscillatory model is at once clear: *the year 604, when the accession vanishes, is just mid-way between 132 and 1076.* Evidently then, the oscillation was postulated to vanish at this convenient mid-point, leaving one to fix a certain period and amplitude so that when this oscillation is added to the sidereal sun, there is not only agreement with the sound observation *ca* 1076 and that of 132, but also agreement with the early Hipparchian observation in -145.

Further, a vital intermediate step is suggested by the following consideration. If we consider only the deviation represented by the straight line passing through the points for the years 132, 1075, we get a slope of about -4 seconds per annum. Since the true rate of precession is about 50.25 seconds per annum, this straight line represents a precession of about 54 seconds per annum, or one degree in about 66 years, added to the Toledan mean Sun. Thus we may hypothesize an initial proposal for a sidereal mean Sun and precession, based on sound observations ca 1075 together with Ptolemy's pseudo-observation of the Autumn equinox in 132. These two points define a mean tropical longitude for the Sun. A mean sidereal Sun was then derived by subtracting from this precession at the rate of 1 degree per 66 years, and assuming that this precessional correction vanished at the mid-point between the two observations, A.D. 604. The deviation curve for this model is shown marked '2' in Fig. 4. This curve shows clearly the two triple intersections in the years 132 and 1075.

There would next be a second stage in the development, in which the sidereal mean Sun just obtained was retained, but with the linear precession replaced by the oscillatory model, designed to maintain agreement with the data for the years 132, 1075, but agreeing in addition with the Hipparchian data for the year -145. Of course agreement in either 132 or 1075 implies ageement with the other, since 604 lies midway between 132 and 1075. In Fig. 4 there is also shown the deviation curve

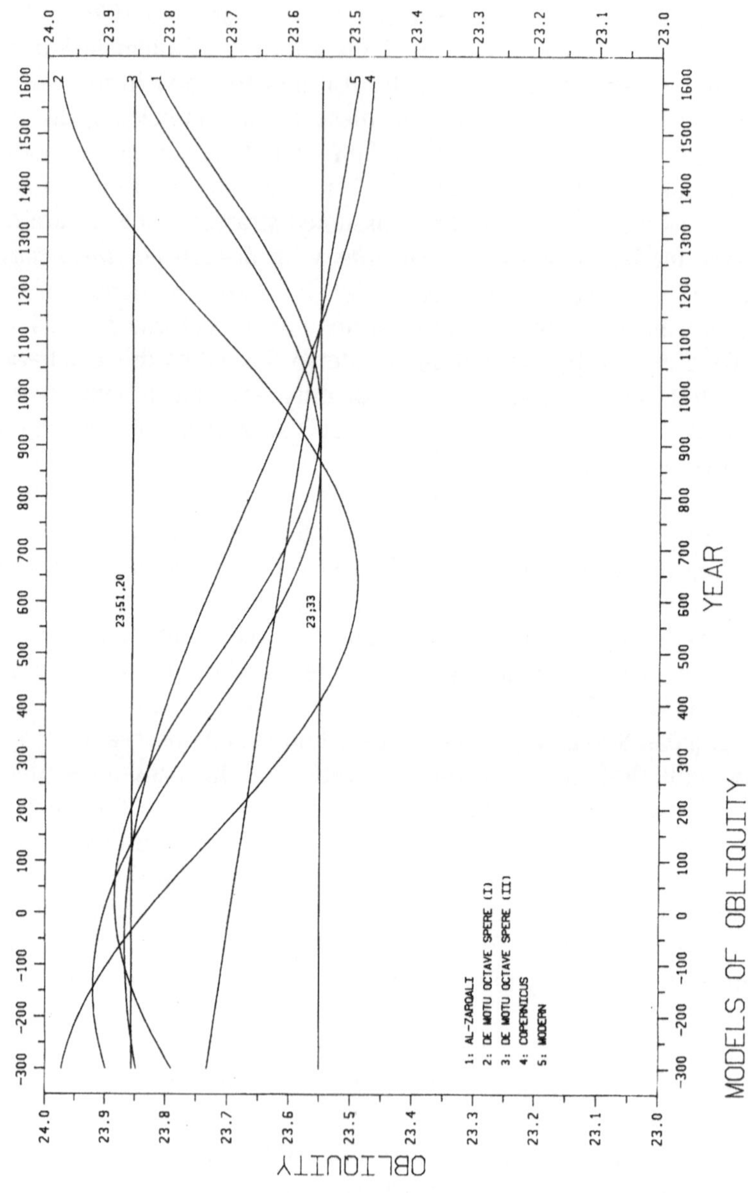

Fig. 3
Models of obliquity: al-Zarqāla (Tract on the Fixed stars);
De motu octave spere *I (Mercier) II (North); Copernicus; Modern*

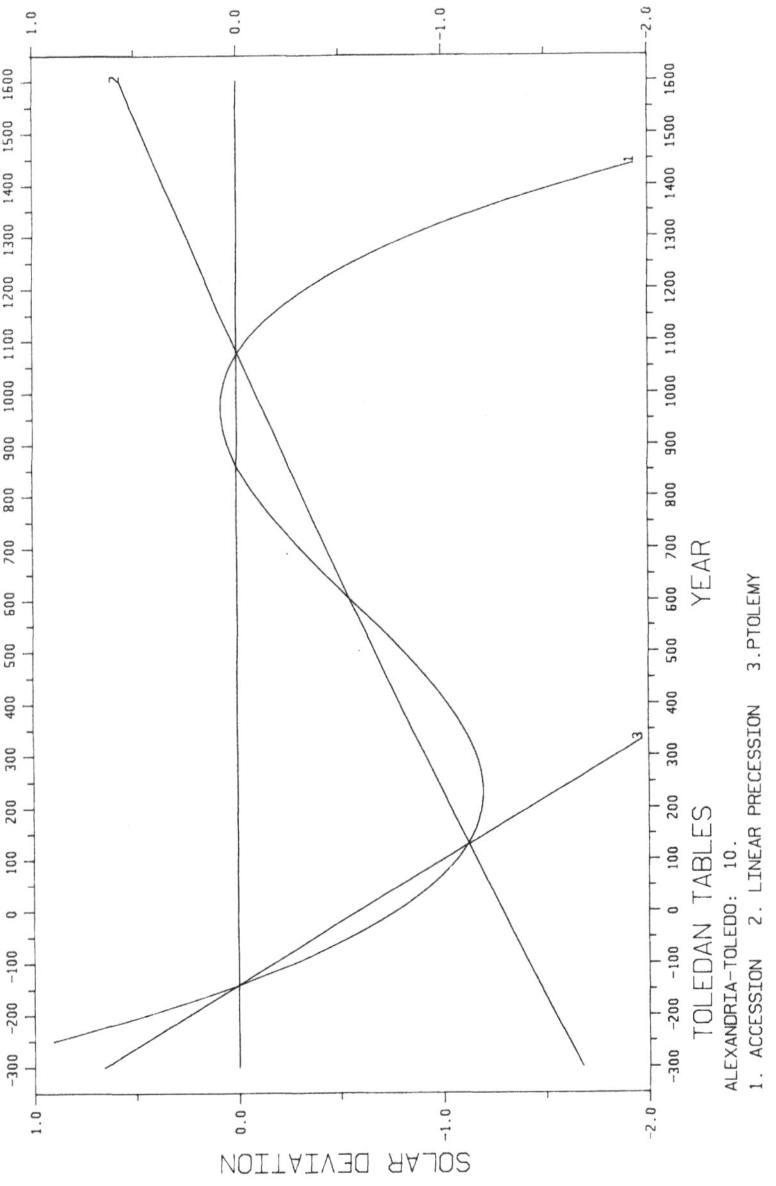

Fig. 4
Solar deviation, Toledan Tables *with models of precession;*
Solar deviation, Ptolemy.

for the Sun from the *Almagest* itself, with its triple intersection at the year 132. One appreciates easily how the introduction of the sinusoidal curve balanced between the years 132 and 1075 enabled the earliest data to be fitted without destroying the agreement with the later dates.

Exact Analysis

We now demonstrate the exactness of these remarks. Let the Toledan sidereal Sun be

$$L_\sigma = (t - t_H)m + c \qquad (1)$$

while the linear precession, and accession and recession, are given respectively by

$$P = (t - t_T)p \qquad (2)$$
$$A = \arcsin(\alpha \sin \beta(t - t_T)) \qquad (3)$$

The numerical values of the controlling parameters are as follows

$$t_H = 1948438 \quad \text{(J.D. of Hijra Era)}$$
$$m = (582^r\ 40;7,1)/(600 \times 354;22)$$
$$c = 113;41,11$$
$$p = 1/(66 \times 365.25)$$

With the linear precession at the rate one degree per 66 years, the tropical Sun is

$$L = L_\sigma + P,$$

while with accession it is

$$L' = L_\sigma + A.$$

The intersections of the two functions P and A, represented by the curves 1 and 2 in Fig. 4, determine the positions near the years 132 and 1075. An exact solution of the equation $P = A$ has the roots

1769322.3, 2114652.9,

apart of course from t_T. The calendar dates for 1769322, 2114653 are A.D. 132 Feb 20 and 1077 Aug 9. We must reflect that the parameters on which P and A depend must be approximations resulting from various calculations, and that our reverse reconstruction is most unlikely to land on the exact day in the years 132 and 1075 which had been chosen originally. Thus a sensible interpretation obliges us to select the Autumn Equinox in 132, for which time Ptolemy reports a position of the Sun. The condition that the mean values L, L' are each equal to 180 in that year fixes the times:

$$L\text{: } t = 1769537.113$$
$$L'\text{: } t = 1769537.068.$$

If instead we look for the times when the true, rather than the mean, longitudes are each 180, we get 1769539.083, 1769538.963, respectively. In any case, these two times differ insignificantly by about one hour, even though they are some 200 days later than the exact solution of $A = P$. On the other hand, according to Ptolemy the longitude of the true Sun equals 180 when $t =$ 1769539.167, and the mean Sun equals 180 when $t =$ 1769536.8725.

The date corresponding to the other root, 1077 Aug 9, suggests that the Autumn Equinox in 1077 was used. In line however with the interpretation that the date of zero accession is midway between the two equinoxes, we should look to that of 1075. According to modern calculations, then true equinox in that year is 1075 Sep 17 17^h;46,40 (UT), 2113961.24074, and the mean at 1075 Sep 15 17^h;49,38 (UT), 2113959.2428. The point midway would then however be in March 604, not November.

In order to establish the points of intersection between the Ptolemaic deviation curve and L, that is the curves marked 1 and 3 in Fig. 4, we must find the intersections for a range of differences between the meridians of Toledo and Alexandria. In the following table the precise points of intersection are listed, selected from calculations made for a range of meridians in steps of 0.01, which are such that the dates of intersection are those of the Autumn equinoxes used by Hipparchus or Ptolemy. It is odd that the difference between the meridians is not 40°. In fact the meridian of Alexandria seems to have been set 10 East of

Toledo. Elsewhere in the analysis of this problem we will see that the difference 40;20, as defined in the geographical tables of the *Toledan Tables*, was used, so that the difference 10 can only be an error, perhaps an error arising in some odd way from the difference between the two familiar longitude norms.

In this table, the dates nearest to the historical observations are marked *; these are not of course the exact dates of the observations, only the dates given by the roots of the equation for that value of the meridian difference. For example the observation in -146 was at the midnight of Sep 26/7.

Ideally we want a pair of roots which are simultaneously the dates of observations of Hipparchus and Ptolemy. This indicates that the best choice would be the Hipparchian dates in -145 or -146. At this point one should be guided further by the fact that Ibn Yūnis[11], and others whom he reports, had used the A.E. of -146 in their determinations of the length of the year. Along with this we must have the Ptolemaic pseudo-observation of 132. The precise times reported for these are respectively

-146 Sep 26/27	midnight	1668000.5
132 Sep 25	2 pm	1769539.083.

Meridian	Hipparchus	Ptolemy
9.93	1668364.99 -145 Sep 25 *	1769435.95 132 Jun 13
10.20	1668275.41 -145 Jun 28	1769539.71 132 Sep 25 *
11.03	1668001.11 -146 Sep 27 *	1769846.21 133 Jul 29
17.27	1665988.38 -151 Mar 24	1772101.34 139 Oct 1 *
23.75	1663983.03 -157 Sep 27 *	1774359.02 145 Dec 6
24.96	1663617.35 -158 Sep 26 *	1774771.89 147 Jan 22

[11] Caussin (1804), p. 144.

In earlier Islamic determinations of the length of the year, Ptolemy's Autumn Equinox of 132 was used by al-Battānī and al-Bīrūnī. Ibn Yūnis notes that of 139, but did not use it to fix the length of the year; instead he chose the Hipparchian A.E. of -146, which he combined with his own observation.

The statements made above about the relation between *L* and *L'*, and their relations with the various observations, are thus confirmed with very considerable precision, so showing in what way both the sidereal mean Sun, and the model of accession and recession, were determined from a shrewd combination of contemporary and ancient observations.

Observations by al-Zarqāla

In his famous 'Études sur Zarkali' Steinschneider[12] included various details of the work and views of Al-Zarqāla as recorded by Isaac Israeli in his *Yesod 'Olam*, written in Hebrew in Toledo in 1310. The theory of accession and recession is attributed to him, and without reference to Thābit, the two ecliptics being said to have coincided 750 years before the time of composition, *i.e.* A.D. 560[13]. In particular we have the details of al-Zarqāla's observation of the time of the Autumn Equinox of 1075. According to the printed text we have[14].

> I find that Abraham ha-Zarqāl wrote in the *Maᵃmār ha-Šemeš*, that he was an assistant and testified that he observed with great precision, with perfected instruments set up for the purpose in Toledo, and found that the moment of the true Tᵉkūfāh Tišrī in the year 468 of the

---

[12] Steinschneider (1887) §122.

[13] Book II, ch. 6. See Goldberg (1848), Hebrew text I, 22[a]; German summary I, ix[b].

[14] Book IV, ch. 15. See Goldberg (1848), Hebrew text II, 29[d]; German summary II, viii. However I have translated this passage on the basis of the MSS Cambridge University Library Oo.6.65, and Oxford Bodley Hunt 299, which are certainly more exact in some details. The work is preserved also in Oxford Bodley, Oppenheim 408 & 182, Pocock 368, Reggio 6, and London BL Add 15972. In Hunt 299 name Zarqāl is written as זרקא׳ , where in this case the abbreviation א׳ (אל) resembled יל, so that it may have been read as Zarqīl, a form often found elsewhere.

Ismaelites, that is the year 4836 of the World, 10th year in the cycle 255, occurred 4 hours 324 *ch^{e}laqīm* after Noon on day 5 of the week [Thursday] in Toledo, Tišri 4, and the 4th of Ṣafar of the Ismaelites; therefore the observation at the navel of the earth (for which they determined the calculations of the *ʿIbbūr*, as we explained), is Friday 2 hours 486 *ch^{e}laqīm*, because Toledo, according to this man, is West of the navel by the amount 4 hours, 162 *ch^{e}laqīm*.

Note that the T^{e}qufa Tišrī is the Autumn equinox, the unit *ch^{e}laq* is 1/1080 of an hour, the year 4836 is measured in 19-year cycles, so that $254 \times 19+10=4836$, and the 'navel' is of course Arin/Ujjain. The day in question is, without difficulty, A.D. 1075 Sept 17 Thursday (J.D. 2113961). The details of the times are consistent[15]. Since the moment in Toledo is measured from Noon, while the moment at the 'navel' is from the night of the following 24-hour period - the 6th day, Friday - that is 6pm on the Thursday, we have

$$4^h324^{ch} \quad + 4^h162^{ch} \quad = 6^h \quad + 2^h486^{ch.}$$

The conversion to the Hebrew units was presumably made from fairly simple numbers in terms of hours and minutes, or degrees and minutes. Indeed we have

$$\begin{aligned} 4^h\,324^{ch} &= 4;18 \\ 4^h\,162^{ch} &= 2;9 \\ 2^h\,484^{ch} &= 2;27 \end{aligned}$$

We may therefore take the time of the original observation as

---

[15] At least the times are consistent if an error in the printed text is corrected, for it has incorrectly תפ"ד ('484'). However, all the manuscripts which I have been able to consult have תפ"ו ('486').

1075 Sept 17 4;18 after Noon, 2113961.17917[16]. Steinschneider remarks that the interval Arin-Toledo is given earlier in the work as 62, *i.e.* $4^{h}144^{ch}$, and elsewhere as $4^{h}108^{ch}$ (61;30), so there may be some occasion for emendation. The correct time of this A.E. is 1075 Sep 17 5;30,40[17], reckoning the time according to MST at Toledo. The error is therefore 1;12,40, an error very much in line with errors in equinox times measured by earlier Islamic astronomers.

At this time the position of the true Sun according to the *Toledan Tables* (with accession and recession), is 179;55.7.

In his *Tract on the Motion of the Fixed Stars*, known to us only in a Hebrew version[18], and written after he had composed his work on the Sun, which is frequently cited here as *Tract on the Sun* (Hebrew: *Ma^{a}mār ha-Šemeš*), al-Zarqāla refers to further observations which he has made. He relates that he has made observations for a period of 25 years before writing on the Sun. He moved at some point from Toledo to Córdoba where he continued observations, relating it seems more to the stars than to the Sun, and for which he developed various models of accession and recession designed to fit determinations of the positions of fixed stars. Here he reports observations for the years 455/1063, 467/1074 (α Leonis), 473/1080.

He refers to Thābit's *Tract on the Solar Year*, giving the observation that α Leonis is at 13;14 in a year which is not stated, but which may be determined to be 829-830. The year agrees with both the

---

[16] Steinschneider added to the confusion by giving the cycle number as 258 (misreading ה=5 for ח=8), and the date as September 27. He erred also, along with the German summary of the Yesod 'Olam, in assuming חצות to mean midnight instead of noon. The present context clearly requires 'noon', even though 'midnight' is the more common sense; 'noon' is a Talmudic usage, according to the *Megiddo Modern Dictionary Hebrew-English*, Tel Aviv, 1975. The errors continue with Millàs (1950), p. 245 n. 2, (accepted by Toomer (1969), 333 n. 37) who while reading 'noon', persists in accepting 27 September, and introduces $2^{h}\,284^{ch}$ in place of $2^{h}\,484^{ch}$.

[17] The modern calculations are based on the new VSOP87 parameters, employed, for example, in the algorithms used by Meeus (1991).

[18] Millàs-Vallicrosa (1950), pp. 250-273 (Hebrew), 274-342 (translation).

Arabic text of Thābit on the Solar Year[19] and with Ibn Yūnis[20], but the two latter agree in that the position is 13;0, not 13;14.

Observations by al-Zarqāla are also indicated by al-Marrakūshī[21], who mentions one for the year 453/1061, as well as the one for 473/1080.

The Toledan Mean Sun

The A.E. of A.D. 132 reported by Ptolemy is A.D. 132 Sep 25 2 p.m., 1769539.083, Alexandrian mean time. Correspondingly, the time when the mean Sun is 180° is 1769536.873. From the previous section the time fixed by al-Zarqāla for the A.E. in 1075 is 1075 Sept 17 4;18 p.m. The difference of longitude between Alexandria and Toledo is 51;20 - 11;0 = 40;20 according to the coordinates given in the *Toledan Tables*. Thus the interval between these two Equinoxes is

$$2113961.17917 - 1769539.083 + 40;20/360 = 344422.2082$$

In these 943 years, the mean year is 365.2409419 days. If this tropical year is converted to the sidereal mean motion, allowing 1 degree per 66×365.25 days we have

$$943 \times 360/344422.2082 - 1/(66 \times 365.25) = 0.985609244$$

In order to compare this with the mean motion in the *Toledan Tables*, consider the motion in 600 lunar years, 212620 days, 582r 40;14,15, which is rather greater than the correct amount 582^r 40;7,1. Clearly the Toledan mean motion was not derived from the time of the true Equinox. In order to consider instead the time of the mean equinox in 1075, we have to estimate the equation of the Sun. Before the *Toledan Tables* were established, we might suppose that the *Zīj* of al-Battānī would have been used for this purpose. The equation of the Sun would be determined by the maximum 1;59,10 and, for this date, the apogee 85;13,

---

[19] Morelon (1987), p. 55.

[20] Caussin (1804), pp. 160,170.

[21] Sédillot (1835), pp. 127, 139

making the equation -1;58,20. To convert this to time we may use either the exact mean solar motion according to al-Battānī, or perhaps the approximation 0;59,8, 2113959.1782. In any case we derive a sidereal motion as in the previous paragraph, but this time agreeing extremely well with that of the *Toledan Tables*, for in 600 lunar years the excess over 582 is now 40;6,38. Some very slight differences from this estimate of the equation could very well have led to the figure 40;7,1. We may conclude then that the rate of motion of the Sun according to the *Toledan Tables* could have been derived from the A.E. observations of Ptolemy for 132 and that of al-Zarqāla for 1075, assuming the longitude difference between Alexandria and Toledo to be 40;20, and reduced to sidereal motion by assuming linear precession at the rate one degree in 66 Roman years. Al-Zarqāla's observation errs by 1;12,40, and the calculated time of mean equinox errs similarly. Hence any calculation along these lines from truly correct equinox times would not give the same result. Naturally we can assume that many other observations would have been recorded, but if the one for 1075 was not that used in the development of the Toledan mean Sun, it must have been another observation erring to much the same extent.

The mid-point between the equinoxes of 132 and 1075 is (1769536.7605 + 2113959.1782)/2 = 1941747.969, that is 604 Mar 20. This differs from that implied by the table of the argument of accession, 1941987, which is Nov 14 in that year. Another problem arises when we seek to fix the constant term in the rule for the mean Sun. This is clear from calculations from the *Toledan Tables* for the true or mean A.E. in 1075 at the dates recorded by al-Zarqāla. For we have seen above that at that date the true Sun from the *Toledan Tables* is 179;55.7, while at 2113959.1782 the mean Sun is 179;56.1. This means that any determination of the constant will differ from 113;41,11 by some four minutes. Thus between the midpoint and 2113959.1782 the sidereal motion is 172;51,23, and the linear precession is 7;8,37, which together make 180;0, as one would expect from the preceeding derivation. The interval from 1948438 to the midpoint contributes 59;6,16, so that the constant term is 172;51,23 - 59;6,16 = 113;45,7. The correct result is 113;41,11, about 4 minutes smaller. This interval of 4 minutes is certainly somehow related to the fact that the correct interval between Alexandria and Toledo was used in processing the date of the A.E., while, as we saw above, an interval of only about 10;0 is appropriate to the inclusion of the

A.E. dates of Ptolemy and Hipparchus. The discrepancy of 30 degrees produces an error of (30/360) × mean Sun = 0;4,56.

To see the problem from another point of view, consider the position of the Sun at the times of true and mean A.E. calculated from modern parameters, for a range of years near 1075, compared with the results from the *Toledan Tables*. The modern calculations are referred to the Greenwich meridian, but the shift of 4;0 to the meridian of Toledo is included in the calculation from the *Toledan Tables*.

Modern True A.E.	Toledan True Sun	Modern Mean A.E.	Toledan Mean Sun
1070 Sep 17 12;45,11	179;59.3	1070 Sep 15 0;45, 4	179;59.7
1071 Sep 17 18;41,19	179;59.5	1071 Sep 15 6;33,59	179;59.6
1072 Sep 17 0;21,56	179;59.1	1072 Sep 14 12;22,54	179;59.5
1073 Sep 17 6;10,53	179;59.0	1073 Sep 14 18;11,48	179;59.4
1074 Sep 17 12; 0,19	179;58.9	1074 Sep 15 0; 0,43	179;59.3
1075 Sep 17 17;46,40	179;58.7	1075 Sep 15 5;49,38	179;59.3
1076 Sep 17 23;39,43	179;58.8	1076 Sep 14 11;38,32	179;59.2
1077 Sep 17 5;21,24	179;58.4	1077 Sep 14 17;27,26	179;59.1
1078 Sep 17 11; 7,49	179;58.2	1078 Sep 14 23;16,13	179;58.7

Now it is clear from this list that both mean and true positions found from the tables are really very accurate. This shows first of all that the *Toledan Tables* (with accession and recession) were fitted very accurately indeed to the observations made at that time, and not merely to the one observation known to us, in 1075, even though that would have been sufficient to fix the sidereal motion used in the *Toledan Tables*. It is not inconceivable that the accuracy just noted is an accidental result of the confusion about the difference of longitudes between Alexandria and Toledo. Otherwise, the calculation which has just been given for the constant must have been applied not to the A.E. for 1075 but to another observation, or more probably to several.

Parameters of Accession

Let t_1 and t_2 be the times of the Autumn equinoxes which were used to fix the sidereal year of the *Toledan Tables*, *i.e.* 132 and 1075 or solar observations were inferior to those of Hipparchus, Ibrāhīm clearly hopes to integrate these, as well as the other variations just mentioned, into a theoretical framework. This framework depends on the concept of moving and fixed ecliptics. The pole of the moving, or visible, ecliptic is carried on a small circle whose centre is the pole of the fixed ecliptic. No parameters are given, but rather the discussion is developed as a preparation for the reduction of observations extended over a suitable period. He has in mind observations made by the astronomers of al-Ma'mūn by 'his father' Sinān ibn Thābit, and himself. He attributes neither observations nor theoretical arguments to Thābit.

Al-Zarqāla on the Motion of the Fixed Stars

In al-Zarqāla's *Tract on the Motion of the Fixed Stars* which as remarked above is known in a Hebrew version, he explains his own determination of a model of accession and recession. The strategy of this approach differs essentially from that followed in the reconstruction of the parameters of the *De motu octave spere*, for that depended on the times of equinox and solstice, while here al-Zarqāla proceeds by considering the longitude of α Leo at various times, that is as determined by Hipparchus, Ptolemy, al-Battānī and himself.

He explains three models which vary according to the orientation of the epicycle. The third model, the one ultimately accepted by him, is similar to that known from the *De motu octave spere*, in that the epicycle is in the tangent plane to the sphere. The numerical details are somewhat different, however. At one point in his analysis he gives numerical details of the model as they relate to observations of α Leo made by Hipparchus, Ptolemy, al-Battānī and himself. The variables used in the model include the angle of rotation around the epicycle (θ), the declination of the point on the epicycle from the equator (δ), the accession or recession (E), and the radius of the epicycle (r). These are related as follows

$$r = \sin\delta / \sin\theta$$
$$\sin E = \sin\delta / \sin\epsilon$$

where ϵ is the obliquity. These relations may be inferred from the account of the geometry, but are also given explicitly later in the tract[43]. Although I have had to make some slight emendations to the Hebrew text we may confidently list the the following details; the first three columns represent al-Zarqāla's information, while the remaining columns give my calculations.

observer	θ	E	δ	δ (calc)	ϵ (calc)	JD (calc)	date (calc)
Hipparchus	-67;27,48	-9;28,30	(absent)	-3;48,59.5	23;50,59	1668143	-145 Feb 17
Ptolemy	-40;58,30	-6;42,45	-2;42	-2;42,30.7	23;50,45	1772256	140 Mar 4
al-Battānī	28;10	4;52,56	1;56,58a	1;56,57.9	23;33,33	2044017	884 Mar 19
al-Zarqāla	46;0,40b	7;27,15	2;58,20	2;58,19.4	23;33,24	2114154	1076 Mar 28

In the manuscript the readings are as follows:
a) fol. 16v, l. 23: '1;36,18'.
b) fol. 16v, l. 25-6: '46;40,30';
fol. 17v, l. 9: '46 deg 40 sec 46;0,40'.

For the time of Hipparchus no value of δ is provided. Emendations of two of these values are plainly required. In case a) there has evidently been the usual confusion in the reading of the Arabic letters ي(10), ل(30), ن(50), where so often the dots were omitted. In case b), the confusion in the value of θ for the time of al-Zarqāla can be confidently put right, for the value is repeated inconsistently, and provides us with the emendation needed: 46;0,40[44].

The calculations are performed with the use of al-Zarqāla's value of r = 0;4,19,26[45]. With this value of r a value of δ can be found for each value of θ, and from that value of δ,and al-Zarqāla's value of E, the value of ϵ is found. He gives also his rule for θ as a function of time, as follows,

---

[43] Millàs Vallicrosa (1950), p. 335 seq.

[44] In his Spanish translation Millàs has wrongly put 46°40 in the second passage; Millàs-Vallicrosa (1950), p. 322.

[45] This is the value as quoted in text, and is clearly meant to used in the formula as shown. The related angle *arcsin(r)* = 4;7,58 is employed later by Zarqāla.

solar observations were inferior to those of Hipparchus, Ibrāhīm clearly hopes to integrate these, as well as the other variations just mentioned, into a theoretical framework. This framework depends on the concept of moving and fixed ecliptics. The pole of the moving, or visible, ecliptic is carried on a small circle whose centre is the pole of the fixed ecliptic. No parameters are given, but rather the discussion is developed as a preparation for the reduction of observations extended over a suitable period. He has in mind observations made by the astronomers of al-Ma'mūn by 'his father' Sinān ibn Thābit, and himself. He attributes neither observations nor theoretical arguments to Thābit.

Al-Zarqāla on the Motion of the Fixed Stars

In al-Zarqāla's *Tract on the Motion of the Fixed Stars* which as remarked above is known in a Hebrew version, he explains his own determination of a model of accession and recession. The strategy of this approach differs essentially from that followed in the reconstruction of the parameters of the *De motu octave spere*, for that depended on the times of equinox and solstice, while here al-Zarqāla proceeds by considering the longitude of α Leo at various times, that is as determined by Hipparchus, Ptolemy, al-Battānī and himself.

He explains three models which vary according to the orientation of the epicycle. The third model, the one ultimately accepted by him, is similar to that known from the *De motu octave spere*, in that the epicycle is in the tangent plane to the sphere. The numerical details are somewhat different, however. At one point in his analysis he gives numerical details of the model as they relate to observations of α Leo made by Hipparchus, Ptolemy, al-Battānī and himself. The variables used in the model include the angle of rotation around the epicycle (θ), the declination of the point on the epicycle from the equator (δ), the accession or recession (E), and the radius of the epicycle (r). These are related as follows

$$r = \sin \delta / \sin \theta$$
$$\sin E = \sin \delta / \sin \epsilon$$

where ϵ is the obliquity. These relations may be inferred from the account of the geometry, but are also given explicitly later in the tract[43]. Although I have had to make some slight emendations to the Hebrew text we may confidently list the the following details; the first three columns represent al-Zarqāla's information, while the remaining columns give my calculations.

observer	θ	E	δ	δ (calc)	ϵ (calc)	JD (calc)	date (calc)
Hipparchus	-67;27,48	-9;28,30	(absent)	-3;48,59.5	23;50,59	1668143	-145 Feb 17
Ptolemy	-40;58,30	-6;42,45	-2;42	-2;42,30.7	23;50,45	1772256	140 Mar 4
al-Battānī	28;10	4;52,56	1;56,58a	1;56,57.9	23;33,33	2044017	884 Mar 19
al-Zarqāla	46;0,40b	7;27,15	2;58,20	2;58,19.4	23;33,24	2114154	1076 Mar 28

In the manuscript the readings are as follows:
a) fol. 16v, l. 23: '1;36,18'.
b) fol. 16v, l. 25-6: '46;40,30';
fol. 17v, l. 9: '46 deg 40 sec 46;0,40'.

For the time of Hipparchus no value of δ is provided. Emendations of two of these values are plainly required. In case a) there has evidently been the usual confusion in the reading of the Arabic letters ي(10), ل(30), ن(50), where so often the dots were omitted. In case b), the confusion in the value of θ for the time of al-Zarqāla can be confidently put right, for the value is repeated inconsistently, and provides us with the emendation needed: 46;0,40[44].

The calculations are performed with the use of al-Zarqāla's value of r = 0;4,19,26[45]. With this value of r a value of δ can be found for each value of θ, and from that value of δ,and al-Zarqāla's value of E, the value of ϵ is found. He gives also his rule for θ as a function of time, as follows,

---

[43] Millàs Vallicrosa (1950), p. 335 seq.

[44] In his Spanish translation Millàs has wrongly put 46°40 in the second passage; Millàs-Vallicrosa (1950), p. 322.

[45] This is the value as quoted in text, and is clearly meant to used in the formula as shown. The related angle *arcsin(r)* = 4;7,58 is employed later by Zarqāla.

which caused the editor Millàs Vallicrosa much trouble: *In predicto autem libro nostro quem prefatus Azarchellus super hunc motum ordinavit exponens ita Albaten in hoc motu ita eum descripsit.* From this point the tract continues by giving an alternative Latin version of the *De motu octave spere*, including the numerical parameters in full agreement with the latter.

Millàs Vallicrosa is troubled by that connecting passage because, indeed, neither al-Zarqāla nor a book of his have been mentioned *expressly* up to this point. He is sure that it refers to a book about accession and recession, but he cannot believe that al-Zarqāla can have written a tract on accession and recession apart from the one known to us on the motion of the fixed stars (preserved in Hebrew), which disagrees in all numerical details from the *De motu octave spere*. He would rather continue to accept the traditional attribution to Thābit.

Certainly the passage seems to be corrupted or at least confused in some way, but I take it to mean, in effect, 1) that al-Zarqāla wrote the *Toledan Tables*, from which Johannes Hispanus has been quoting, 2) that al-Battānī explained the zig-zag model, which he did in chapter 52 of his *Zīj*, 3) that al-Zarqāla wrote the tract which immediately follows, whose content is the same as that of the *De motu octave spere*. For us, there is no contradiction in supposing that al-Zarqāla wrote two studies of accession and recession, one depending on the times of equinox and solstice, and the other on the longitude of α Leonis.

We have seen that accession and recession according to the *De motu octave spere* was intimately linked with the *Toledan Tables*, and this is quite consistent with a simple reading of Johannes Hispanus, that al-Zarqāla was the author of both. After all he comes across as a researcher in depth who would hardly have been permanently satisfied with the simple adoption of the synodic differences from the *Zīj* of al-Battānī, and that presumably marked an early stage in his work. Therefore the *De motu octave spere* and the *Toledan Tables* form together that early stage, to be followed by a later stage represented by the tract on the motion of the fixed stars and by the material extant in the work of Ibn Kammād, based expressly on that of al-Zarqāla, and also the so-called *Almanac* of al-Zarqāla.

In his Tract on the Motion of the Fixed Stars, al-Zarqāla readily acknowledges Thābit's information about the position of α Leonis, taken from Thābit's work on the solar year, he makes no mention of any other work of his.

The history of science *Ṭabaqāt al-umam* written by the Qāḍī Ṣā'id al-Andalūsī in 460/1068 is a work in which we would rather expect to find more information than we do on these points, particularly since Ṣā'id claims to have studied the subject of accession and recession, prompted by the work of Ibn al-Adamī and his disciple al-Qāsim b. Muḥammad b. Hishām al-Madā'inī al-'Alawī, which appeared under the title *Naẓm al-'iqd* in 338/949[29]. This led him to compose a work on 'the restoration of the motion of the stars'. His extended remarks on Thābit b. Qurra include nothing of concern to precession. The grandson Ibrāhīm is not mentioned. He also remarked on the visit to India by Muḥammad. b. Ismā'īl al-Tanūkhī, who returned it seems with some knowledge of their views on accession and recession. Al-Zarqāla, who also mentions him, says that he described a model with a period of 3125 years and an unspecified amplitude[30]. On al-Zarqāla Ṣā'id is rather vague, indicating only the breadth and stature of his researches, without naming any of his works. He does not mention Ibn Kammād.

In his justly famous *Système du Monde*, Pierre Duhem gave his arguments in favour of rejecting the attribution of the *De motu octave spere* to Thābit ibn Qurra, preferring instead al-Zarqāla[31]. To be sure these were arguments and not a conclusive proof, and in an earlier work I followed the traditional attribution in spite of Duhem's arguments[32]. It is clear now however that his conclusion was after all correct. We may summarise his principle points as follows.

1. Al-Bīrūnī in his *Chronology* referred to the *Book of the Motions of the Sun* (*K. Ḥarakāt al-Shams*) of Ibrahīm b. Sinān b. Thābit, and the *Zīj al-Ṣafā'iḥ* of Abū Ja'far al-Khāzin for accounts of accession and

---

[29] Blachère (1935), p. 114.

[30] Millàs Vallicrosa (1950), p. 320.

[31] See especially Duhem (1914), vol. 2, p. 246-259.

[32] Mercier (1977).

another year near to 1075. At t_1 and t_2 the sidereal mean Sun differs from 180 by equal and opposite amounts, *i.e.* $L = 180 \pm A$. Let t_0 be the time of the Autumn equinox in -146, when $L = 180 \pm A$. When linear precession is added to L the sum will equal 180 at the times t_1 and t_2, but this will not be the case of course at t_0. That is the reason why the linear precession is then replaced by a sinusoidal form

$$A = \arcsin(\alpha \sin \beta(t - t_T)).$$

The problem is to fix values for α and β such that the new tropical longitude will equal 180 also at $t = t_0$. This amounts to solving the two equations for α and β:

$$A_0 = \arcsin(\alpha \sin \beta(t_0 - t_T)).$$
$$A_1 = \arcsin(\alpha \sin \beta(t_1 - t_T)).$$

Take the sine of each side, and then the ratio,

$$\frac{\sin A_1}{\sin A_0} = \frac{\sin\beta(t_0 - t_T)}{\sin\beta(t_0 - t_T)}.$$

This is an equation to be solved for β. Essentially it is of the form

$$\frac{\sin rx}{\sin x} = y$$

where $x = \beta(t_0 - t_T)$, and $r = (t_1 - t_T)/(t_0 - tT)$.

This may be solved for x in the range -90,90. Since $|r| < 1$, there is a solution if $y > r$, a condition which is satisfied here. While it is easy for us to use modern numerical procedures to find a solution to any degree of precision, the medieval calculator would have proceeded perhaps through two or three iterations before exhausting the available precision of the sine function. In fact for the purposes of this reconstruction the greater problem is to establish just the way in which these equinoctial data were handled. There is also the problem, as noted above, of fixing the midpoint. The midpoint between the equinoxes of 132 and 1075 differs by some months from that actually used in the tables of accession and recession.

The equinoctial data which have to be fitted are these,

-146 Sep 26/27 midnight
132 Sep 25, 2 pm.

From these we find the period $360/\beta$ and the amplitude arcsin α for both these times of true equinox and the corresponding mean equinox, and also for the meridians, Alexandria and Toledo, 40;20 West of Alexandria. The values of A are read from the sidereal longitudes; in the first entry A = 12.95298. Recall that the correct values are period = 1481785.4 and 10;45.

	Alexandria MST	40;20 West
true A.E. -146	1668000.5	1668000.388
	192.95298	191.94259
true A.E. 132	1769539.083	1769538.971
	189.36145	189.25106
period	1242525	1249668
amplitude	12;15	12;10

From these it is obvious that the true position was not used.

	Alexandria MST	40;20 West
true A.E. -146	1667998.3	1667998.188
	189.88464	189.77425
true A.E. 132	1769536.873	1769536.761
	187.18326	187.07287
period	1242525	1249668
amplitude	10;46,9	10;43,31

Clearly, the dates of mean equinoxes yield values much closer to those used for accession. Instead of such carefully precise calculations, we might consider instead the combination of rounded data:

	Alexandria MST	40;20 West
true A.E. -146	1668000	1668000
	189;53	189;46
true A.E. 132	1769536	1769536
	187;11	187;4
period	1477315	1499184
amplitude	10;45,56	10;43,8

One could of course choose a meridian, perhaps a few degrees West of Alexandria for which the amplitude would be correct, and the period nearer the correct value, but this would not prove very much. Instead it best to observe that we have essentially the correct explanation of the parameters, even though there is not an altogether exact numerical agreement. Exact numerical agreement cannot be expected as long as we are ignorant of so many of the original details, including the method of obtaining the approximate solution of the equation $\sin rx/\sin x = y$. Moreover the amplitude 10;45 appears to be a round figure.

Authorship

In his recent collection of the astronomical works of Thābit ibn Qurra Morelon gives a list which is as authoritative as anything we could hope for, and this includes nothing corresponding to our tract on accession and recession. In his view the attribution is 'trés suspecte'[22].

As to the question whether al-Zarqāla is himself the source of the model of accession in the *De motu octave spere*, it must be said that while the work clearly made use of observational material gathered at the very time when he was active, there is not a fully satisfactory documentary link with him. In the tract of his which we do possess he argued from considerations of the longitude of α Leonis rather than the time of equinoxes. Of course his work on the motion of the Sun, which is sadly unavailable, may have included an approach such as we have tried to reconstruct here.

Al-Bīrūnī offers a rather negative view of Thābit's work. He is

[22] Morelon (1987), p. xix.

mentioned only insofar as a work on the solar year is 'attributed to him'[23], while it is said that Thābit merely polished up the work of others, such as the Banū Mūsā. No observations are attributed to Thābit, and those which he gives in his work are dated *ca* 830, around the time of his birth. Further to this, in his *Qanūn*[24], al-Bīrūnī notes an observation from the *Book on the Solar Year* of the Banū Mūsā 'often attributed to Thābit b. Qurra', an observation of the Summer Solstice in A.Y. 201 (A.D. 832 Jun 17)[25]. Since observations noted in the work are in the years 830-2, about the time when Thābit was born, and are always cited as made by the author, we must surely have here a work edited, and not composed, by Thābit[26].

While Ibn Yūnis however has no doubt that Thābit 'wrote' a book on the solar year, he remarks that the position of α Leonis given there is one of several found by Ḥabash, again leading one to the view that the work was in some sense edited by Thābit.

The Latin translator of the early 12th century Johannes Avendaut Hispanus has left a tract of great interest in this connection[27]. He begins by explaining the conversion between the Hijra date and the Spanish Era, and other conversions, and continues with remarks on the mean motions of Saturn and the Moon, where he cites the radices and motions in 30 lunar years which agree exactly with those of the *Toledan Tables*. He then describes the zig-zag model of accession and recession which goes back to Theon of Alexandria, in which the motion is 1° per 80 years. For this he gives an instance which agrees with one known from Abū Ma'shar, that the accession equals 5;12,45 in the year A.Y. 265[28]. He says that if this accession is added to the results of the *Zīj* of al-Khwārizmī there will be agreement with that of al-Battānī. He then has a connecting passage

[23] Sachau (1878), pp. 52.10; Sachau (1879) 61.45.

[24] Hyderabad (1956), p. 654.

[25] In the edition of the Arabic text, Morelon (1987), p. 33.

[26] Morelon discusses this question at length; Morelon (1987), p. xlviii.

[27] Millàs Vallicrosa (1936). On the indentity of this John, see Burnett (1994).

[28] In fact this text gives in error 5;12,15.

recession (*muqbila wa mudbira*) with an amplitude of 8 degrees[33]. This is just the occasion where a reference to Thābit would have been expected had he been the author of the *De motu octave spere*. Al-Bīrūnī does in fact refer to Thābit's work on the solar year[34].

2. Al-Biṭrūjī is cited according to the Latin version of Calo Calonymos[35]. Al-Biṭrūjī refers more than once to al-Zarqāla's tract on accession and recession (*maqāla fī ḥaraka al-iqbāl wa'l-idbār*)[36], and never attributes such a work to Thābit.

3. Ibn Ezra in his *Liber coniunctionum* remarks that since the time of al-Ṣūfī there was no comparable observer until al-Zarqāla, who agreed with him regarding the Sun. He makes no mention of Thābit.

4. Ibn Rushd ('Averroes') in his *Epitome of Metaphysics* attributes the theory of accession and recession to al-Zarqāla, contrasting it only with the uniform motion advocated by Ptolemy[37].

5. Abū'l-Ḥasan, al-Marrakūshī, writing in the 13th century, devotes a significant part of his *Jāmiᶜ al-mabādi' wa'l-ghāyāt* to the topic of accession and recession, which he attributes simply to al-Zarqāla, without ever mentioning Thābit. Duhem knew this work from the French version, Sédillot (1835).

For these and other good reasons Duhem became convinced of the mistake in attributing the work to Thābit.

Ibrāhīm ibn Sinān ibn Thābit ibn Qurra

This author is, according to his name, the grandson of Thābit ibn Qurra, his date of death being A.H. 335/A.D. 946-7. Less prolific than

---

33 Sachau (1878), p. 326; Sachau (1879), p. 322.

34 Sachau (1878), p. 52; Sachau (1879), p. 62.

35 Since then we have also an edition of Michael Scot's version, and also of the Hebrew version.

36 Calonymos (1531), fol. 6, 12; Carmody (1952), p. 85, 95; Goldstein (1971), vol. 1, p. 80, 91.

37 van den Bergh (1924), p. 112. Writing after Duhem, van den Bergh was one of the few to take Duhem's proposal seriously.

his forebear, his work as a theoretical astronomer was nevertheless distinguished and influential. He wrote a tract entitled *The Motions of the Sun* (*Ḥarakāt al-Shams*), known uniquely from the manuscript Bankipur 2468[38]. The work was published originally by the Osmania Publications Bureau, Hyderabad, Deccan, but in an altogether defective way[39]. Subsequently the whole work was transcribed carefully by Sa'idan (1983), although he did not control the diagrams or their lettering. An analysis of the contents of the whole manuscript, with a reordering of the dislocated pages, has been given by Hogendijk (1982).

The work is of great interest in the present discussion of accession and recession. It is referred to by al-Bīrūnī under the name *Kitāb Ḥarakāt al-Shams*[40], where he refers the reader to it for an account of the accession and recession with the amplitude 8. In his *Taḥdīd*, he refers to it again[41], noting that variations in the obliquity

> cannot be attributed to the rotation of the poles about a point, as was imagined by Sinān and Abū Ja'far al-Khāzin, because that rotation is irregular.

The only other reference known to me is by the very late commentator on the *Tadhkira* of Naṣīr al-Dīn al-Ṭūsī, Shams al-Dīn al-Ḥafarī [42], whose date is unknown, but which must be of the time of Ulug Beg or later.

The work is concerned with inequalities in the tropical year, which are linked with variations of the solar apogee and of the obliquity. While Ibn Yūnis, for example, was content to conclude that Ptolemy's

---

[38] Khuda Baksh Oriental Public Library, Patna, fols. 118-124, 323, 1.

[39] Thus the bulk of the work is contained in Hyderabad (1947), pp. 2-34, while the final paragraphs appear within, and entirely undistinguished from, one of al-Bīrūnī's tracts, in Hyderabad (1948), pp. 209 line 10 to 214 line 10. Even then the final page of the work (fol. 1[a]), dislocated from the rest in the copy, was not printed.

[40] Sachau (1878), pp. 326.5; Sachau (1879) 322.28.

[41] Ali (1967), p. 70.

[42] India Office Library, London, Loth 747, fol. 69.

$$\theta = \beta_1(t - t_H) + \beta_0$$

where $t_H = 1948438$

$\beta_0 = 3;51,11$

$\beta_1 = 360/B$

$B = 1415108.9$ days $\cong$ 3874 Julian years.

From this rule, then, and each value of θ, a date is derived. Al-Zarqāla actually provides three tabulations of θ for the Hijra, Rūmī and Persian calendars. These are not rigourously consistent with each other, but I have chosen the Hijri because for that calendar the radix and the table of motion are confirmed by al-Marrakūshī and Ibn Kammād[46]. The period according to al-Zarqāla is 3874 'solar years', and this is at least the exact period, 3874 Julian years, of the motion in the Rūmī calendar, but for the other calendars the period is slightly different[47].

Clearly the dates derived in this way are not exactly those of known observations for any of these observers. Al-Battānī relates, for example, that he carried out observations of α Leo in several years, but placed great confidence in one made in 1191 of Alexander, that is Sep/Oct 879-880[48]. Al-Zarqāla states in a number of places that he carried out an observation in AH 467 of the position of α Leo, that is in the period 1074 Aug 27 to 1075 Aug 15. Indeed his value 3;51,11 is supposed to have been derived on the assumption that the value of θ in A.H.467 was 40;0,40.

It is important to note that evidently with each observation he has assumed a value of the obliquity which he considers appropriate for that time. He does not say so however until the following section of the tract, where he gives a model of the variable obliquity, and states that the value of ϵ derived from it is to be used in the formula $\sin E = \sin \delta / \sin \epsilon$.

The model of variable obliquity requires that an epicycle be imagined centred on the pole of the ecliptic. The point moving around the epicycle is connected by a great circle to the pole of the equator, its length

---

[46] Sédillot (1835), p. 150. Ibn Kammād is considered below.

[47] This is in contrast to the tables of motion for the epicycle governing the variable ecliptic, where in each case the period is quite rigourously 1850 Julian years.

[48] Nallino (1903), Vol. 1, p. 124.

being the variable obliquity.

In modern terms we may summarise the trigonometry as follows, where r = 0;10 is the radius of the circle, ϵ_0 = 23;43 is the mean obliquity, u and δ are ancillary quantities, and ϕ is the rotation of the epicycle,

$$\sin u = \sin r \sin \phi$$
$$\tan \delta = \cos \phi \tan r$$
$$\cos \epsilon = \cos u \cos(\epsilon_0 - \delta).$$

The obliquity ϵ varies between the limits 23;33 and 23;53. To a first approximation we have

$$\cos \epsilon \cong \cos \epsilon_0 + r \cos \phi,$$

but al-Zarqāla's tabulation of ϵ as a function of ϕ shows that he calculated more precisely than this. The argument ϕ, which is represented in tables for the three calendars, and is listed in the text for multiples (1, 10, 100) of the various years, has the period 1850 Julian years exactly in every case. However the radices are not at all clear. The radix for the Hijri calendar, 295;14,33, at least fits the only dated requirement laid down by al-Zarqāla, namely that the obliquity should be minimum in '1266 Dhū'l Qarnayn', that is AD 954-5. The radices for the other calendars, 304;25,55 for Dhū'l Qarnayn, and 297;16,36 for Yazdajird, are inconsistent with that. These should be 113;43,41 and 297;10,26, respectively[49]. The graph of this variable obliquity is shown in Fig. 3.

Ibn Kammād

Ibn Kammād, some time after al-Zarqāla, composed treatises in

[49] The translation is inadequate here, since a section of the Hebrew text is missed in the quotation of the radices. The Spanish is equivalent to 'and at the beginning of the Hijra it would be 295;16,36', but should be 'and at the beginning of the Hijra it would be 295;*14,33 and at the beginning of the Era of Yazdajird it would be 297;*16,36'. Moreover the radix for Dhū'l Qarnayn is given in the text as '304 deg 25 min 55 sec 64 25 55', a confusion between 64 and 304 evidently resulting from the Arabic original, which must have had س (60) in place of ش (300). In the table of motion the radices are given as Hijri 295;14,33, Rūmī 304;25,55, Yazdajird 297;16,36.

Arabic which are known, at least in a summary form, from a Latin version. His dates are unknown, but he appears to be an immediate follower - a disciple - of al-Zarqāla[50].

The tables in the extant Latin work are interesting in the present connection because they present models of the solar apogee, solar equation, and precession which all include an oscillatory component, and besides his list of star coordinates shows the use of the sidereal longitude measured from the 'Indian' origin point, making the longitudes equal to Ptolemy's + 6;38. In contrast to al-Zarqāla he says nothing of the variable obliquity, at least in the extant Latin material. The following details may be determined from the Latin text[51].

The modification to the Sun amounts to a further epicycle added to the eccentric. The longitude of the apogee is the sum of a uniformly increasing part,

$$\Gamma = \Gamma_1 (t - t_H) + \Gamma_0$$

and an oscillatory part

$$q(c) = \arctan\left(\frac{\mu_0 \sin c}{\kappa_0 + \mu_0 \cos c}\right).$$

The argument *c*, *motus centri circuli exeuntis*, measures the rotation of the additional epicycle,

$$c = c_1(t - t_H) + c_0.$$

---

[50] Millàs Vallicrosa draws attention to the marginal note in a copy of John of Sicily's Canons for the Toledan Tables, 'Has tabulas composuit Abensahet iudex regis Maymon Toleto, eius discipulus Arzachel et alii cum eo, sed Arzachel regebat instrumenta et dirigebat considerationes. Postquam a Toleto dictus Rex fuit expulsus a Xristianis, Arzachel ruit Cordube et ibi fecit et invenit novas considerationes et fecit tractatum de motu solis et stellarum fixarum. Post venit Alcamet discipulus eius et fecit tabulas persianas et 4 principia tabularum, post ipsum venit Albubalet cordubensis qui consideravit apud Murciam et ipse complevit tabulas <...> et tabulas coniunctionis solis.' Cf Toomer (1969), p. 314.

[51] The folio references for the various parameters are Γ_1: 28v, Γ_0: 35r, θ: 28r, c: 55r,55v,56r.

The numerical details are as follows,

$\Gamma_0 = 76;45,21$ $\Gamma_1 = 0;18,3/(90\text{x}354;22)$
$c_0 = 83;40,31$ $c_1 = 10;26,55/(100\text{x}354;22)$
$\kappa_0 = 2;4,15$ $\mu_0 = 0;18,38.$

The tabulation of the argument c shows various inconsistencies, and the value c_1 used here is based on the table of motions in units of 100 years, since we are more interested in the long-term behaviour. Even then the radix c_0 is inconsistent with the subsequent rows of the table, which imply the unlikely value 83;40,1. This is a small matter however which does not affect the arguments here.

The mean motion of the apogee is one degree in 299.2 Arabic years, while the period of the added epicycle is 3445.4 Arabic years. In his *Tract on the motion of the fixed stars* al-Zarqāla states that he had explained in his *Tract on the Sun* that the motion of the apogee was one degree in 299 years. The period of the epicycle, 3445 years, is similar in magnitude to values derived by al-Zarqāla for his various models of accession and recession, but not identical to any of them, nor to that used in the *De motu octave spere.* The apogee and its linear part are both plotted in Fig. 5.

The distance from the centre of the world to the centre of the Sun's eccentric is the vector sum of the two radii,

$$\kappa(c)=\sqrt{\kappa_0^2+\mu_0^2+2\kappa_0\mu_0\cos c}\ ,$$

and this quantity is the maximum equation of the Sun at the time for which c is calculated. This determines in turn the equation of the Sun

$$\eta(\theta,c)=-\arctan\left(\frac{\kappa(c)\sin\theta}{1+\kappa(c)\cos\theta}\right)$$

where θ is the argument of the Sun (*centri solis*),

$$\theta = \theta_1(t - t_H) + \theta_0,$$

$\theta_0 = 36;35,9$; $\theta_1 = 87^r\ 113;42,16/(90\text{x}354;22)$.

The position of the true Sun is then

$$\Gamma_0+q(c)+\theta+\eta(\theta,c).$$

The maximum equation of the Sun is given by the value κ(c) for that date. Ibn Kammād provides a table of the equation of the Sun in which the maximum equation is 1;52,43[52]. From the above numerical details we can calculate that the date for which $\kappa(c)$ = 1;52,43 is A.D. 1068 June 15, A.H. 460 Shaban 19 (JD 2110891). However in the table for the equation of the Sun and the proportional minutes of the variable equation[53], the latter are only tabulated to the nearest minute, and in the range of the table for dates in the period of activity of al-Zarqāla, these differ by one from the calculated values. When κ(c) = 1;52,43, the proportional minutes are 11.5, which corresponds in this table to $c \cong 132.5$, for which the date is 468 Rabi' II /1075 Nov. In any case it falls well within the period of activity of al-Zarqāla, so that we can regard Ibn Kammād as contemporaneous.

In addition Ibn Kammād provides a model of accession and recession. The argument θ (*adventio puncti capitis arietis*)[54] is given by,

$$\theta = \beta_1(t - t_H) + \beta_0 = \beta_1(t - t_T)$$

where

t_T = 1933292
β_0 = 3;51,11
β_1 = 360/B = 81;8,3/(900×354;22)
B = 1415123 days.

[52] Fol. 35r; the largest value in the table is 1;52,44, but a figure of 1;52,43 seems to be a somewhat better fit for the table as a whole.

[53] Fol. 55v-56r.

[54] The folio references for the various parameters are A: 35v, θ: 28v.

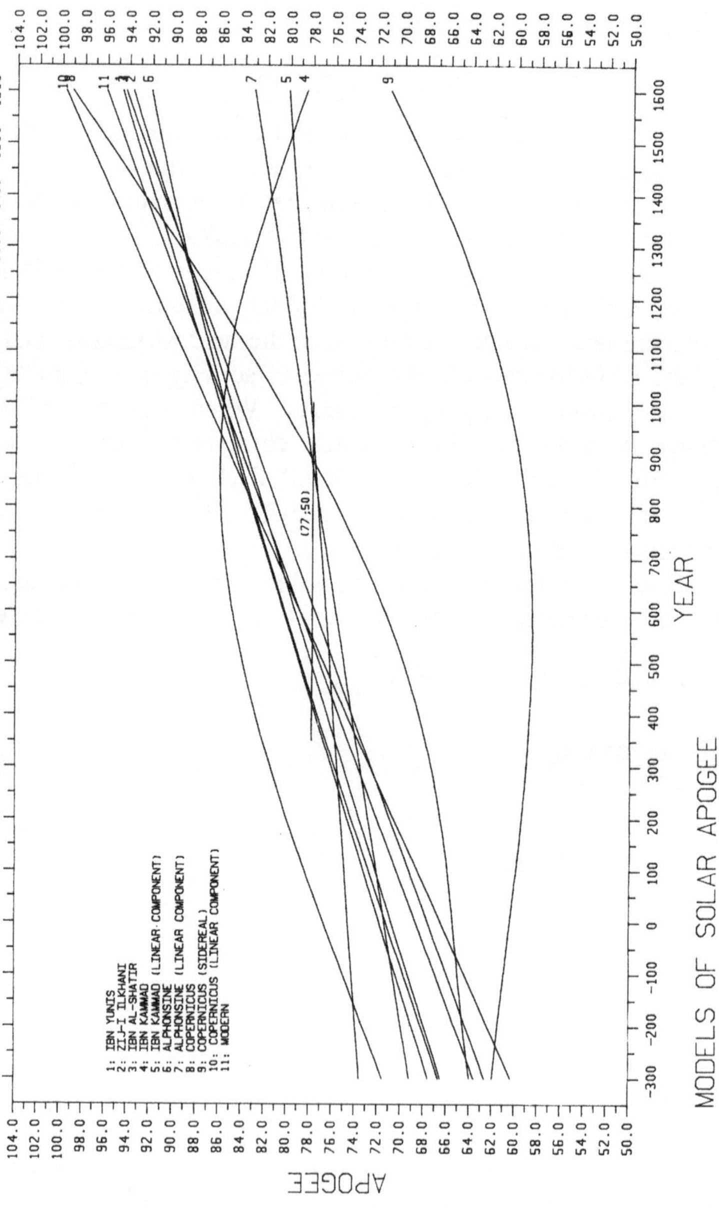

Fig. 5
Models of Apogee.

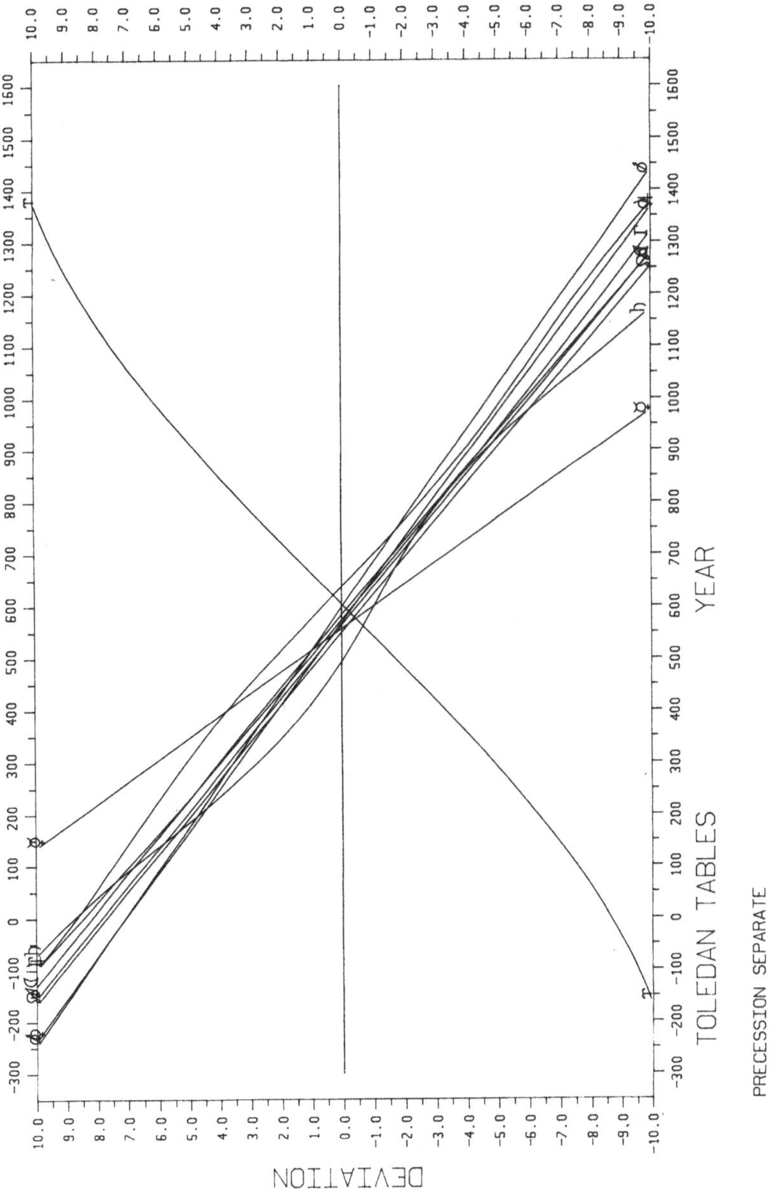

Fig. 6
Deviation curves, Toledan Tables.

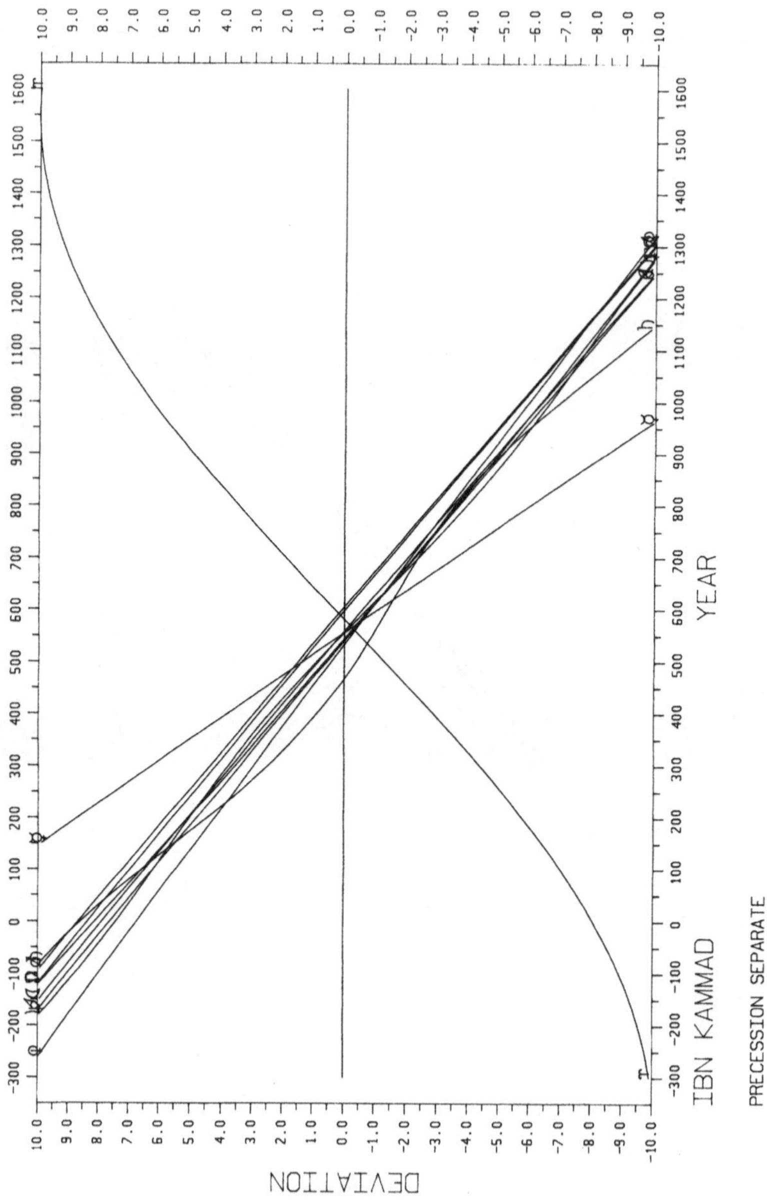

Fig. 7
Deviation curves, Ibn Kammād.

The date corresponding to t_T is AD 581 Jan 24. The period equals 3874.4 Julian years, 3993.4 Arabic years.

The amplitude (*directio adventionis capitis arietis*) as a function of the argument θ is given very approximately by the following expression.

$$A = \arcsin(\sin(10) \sin \theta).$$

This tabulation corresponds only roughly to al-Zarqāla's proposal, for the maximum amplitude was there some 10;16, when the variable obliquity was included. Ibn Kammād makes no reference to the variation of the obliquity.

We may finally observe that the mean motions employed by Ibn Kammād are for the most part intimately related to those of al-Battānī and the *Toledan Tables*. The rates of motion and the radices differ only slightly, and in a way that is demonstrated more effectively by the display of the deviation curves than by a listing of the parameters[55]. These are shown in Figs. 6, 7. It is evident that while the Toledan parameters are closely dependent on those of al-Battānī, those of Ibn Kammād have been adjusted somehow to agree better with observations, especially in the period from the ninth century to his own time.

Abū 'l-Ḥusain al-Ṣūfī

Although the star catalogue of al-Ṣūfī makes no reference to accession and recession he is credited with an interest in the matter by Abraham Zacut (ca. 1475), who records in chapter 9 of his *Great Composition*[56]:

> If you wish to ascertain the positions of the seven planets in the 8th sphere, as they were arranged in the 9th, subtract 13;52 from their positions. So for every 66 years you add one degree that the 8th sphere moves forward, and 0;0,54,30 for each year. It results thus

---

[55] The seconds part of the motions have in most cases to be estimated from a tabulation which seems to have been rounded down from a more exact tabulation elswehere.

[56] Cantera Burgos (1931), p. 249. The Hebrew version is rather confused here; Cohn (1918), p. 19.

> that Creation was in the 6th degree of Gemini at the Spring Equinox. When they left Egypt the Israelites used the head of Taurus as the Spring Equinox, and then counted the signs Taurus, Gemini, while at present we begin with Aries. So it appears from the words of the Muslim Abū'l-Ḥusain <al-Ṣūfī> in his book that they reckoned this way in antiquity. In the year 563 of the Christians, which corresponds to 4323 of Creation, the head of Aries was at the Spring Equinox, when the 8th and 9th spheres coincided.

The accession 13;52 is appropriate for the time of composition of Zacut's *Great Composition*, 563 + 13;52×66 = 1478. The 'book' of al-Ṣūfī may be a *zīj* which he wrote, which appears to be lost, and for which I have attempted to recover the parameters[57]. On present evidence, however, one should be very cautious about accepting Zacut's attribution. Perhaps, in the Spanish-Islamic context long after al-Ṣūfī, the year 563 was obtained from the date of zero accession 604 by subtracting the interval 41 years. For it was assumed by al-Ṣūfī that Ptolemy's star coordinates applied in fact not to his own time, but to that of Meneleus, 41 years earlier. To be sure, this is not perfectly logical, since it ignores the increment of 0;25 which was assumed by al-Ṣūfī to have been applied by Ptolemy. Nevertheless it is persuasively simple.

It is also curious that in this year 563 the Toledan true sidereal Sun is most accurate when compared with the modern Ephemeris, as indeed is that of the otherwise unrelated *Zīj* of al-Khwārizmī[58].

Star Longitudes in the Sidereal Ecliptic

In connection with accession and recession one finds various Arabic and Latin star lists which have evidently been computed for a year of zero accession, so that for another date the accession had only to be added to the longitudes given in the list. In these lists the increment over Ptolemy's longitudes is 6;38, or in a few cases 6;40. Among these we

---

[57] Mercier (1991).

[58] Mercier (1977), Part I, p. 205.

may mention the list of 240 stars given by Ibn Kammād[59] and al-Marrakūshī [60], and a number of lists in the MS Bibl. de l'Arsenal (Paris) 1036. Al-Zarqāla assumed, for example, at the beginning of his trials with various models that the value of accession at the time of Ptolemy was 6;38, although he varied this by small amounts for the other models. In the third model, as we have seen above, he assumed 6;42,45 at the time of Ptolemy. The tropical longitudes corrected in this way give the sidereal longitudes, measured from a point -6;38 relative to the tropical origin used by Ptolemy. In an earlier analysis of models of precession I showed that this sidereal origin is that used commonly in Indian astronomy, and it is therefore not surprising that it is found again in the present context of Spanish-Islamic research which was oriented from the outset along the lines determined by the *Zīj* of al-Khwārizmī.

In that earlier work I showed that this choice of sidereal ecliptic could be readily explained as a consequence of the star-mapping techniques which must have gone back to the time of Hipparchus[61]. Those techniques depended in the alignments found amongst the stars as a means of fixing their positions. In the star catalogue in Ptolemy's *Almagest*, one primary alignment stands out, namely that great circle which passes through the stars α Ari, β Ari, α Vir, ζ Psc. If we draw the great circle passing through α and β Ari it passes within 2.0 arc minutes of α Vir, and intersects the ecliptic at the longitude -6;38,43. At the point where this circle of alignment intersects the ecliptic it also passes within 2.2 arc minutes of the star ζ Psc, a star which, as is well known from studies of Indian astronomy, has been taken to mark the origin of sidereal cooordinates in that context. This circle moreover is the horizon for the latitude 35.6 degrees, so that at that latitude, essentially that used by Hipparchus on the Island of Rhodes, α, β Ari and ζ Psc rise simultaneously when α Vir sets. Finally we may argue from the fact that Ptolemy was evidently unaware of this fundamental alignment, that he did not know how the star catalogue was obtained, and hence that he was not the author of it. Undoubtedly Greek techniques which involved this

---

[59] The details are in the section below on this author.

[60] Sédillot (1835), p. 140-9.

[61] Mercier (1977), Part II, p. 34.

sidereal ecliptic passed to India, but when and how remains a mystery.

I should like to draw attention to three articles by Julio Samsó (1994 a-c), who treats, quite independently, problems that are intimately related to certain parts of this essay, especially the discussion of al-Zarqāla's third model of accession and recession, and the tables of Ibn Kammād, as far as they reflect on the model of the solar apogee apparently proposed by al-Zarqāla.

Appendix: Parameters of Zījes

In this section one will find in tabular form all the parameters of the *zīj* of al-Battānī, the *Toledan Tables* and Ibn Kammād.

AL-BATTĀNĪ

Sun & Moon	radix at 1947572	motion in 700×365.25 days
Sun	340;23,37	700^r 6;30,57
Moon	228;58, 5	9357^r 355;44,23
argument	312;14,10	9278^r 309;29,33
apogee	96;43,55	79^r 46;14,50
node	-187;53,56	-37^r 218;13,39
Planets	**radix at 1947572**	**motion in 600×365.25 days**
Mercury	243;23,37	2491^r 79;27, 6
Venus	211;34,37	975^r 117; 5, 6
Mars	117;58	319^r 11; 5
Jupiter	260; 4	50^r 218;47
Saturn	163;47,41	20^r 141;21

Apogees reduced to epoch T = 1947572 (A.D. 620 Feb 29) from that used by al-Battānī, T = 2042356 (A.D. 879 Sep 1).

	T_0	T_1
Sun	82;14	78;18
Mercury	201;28	197;32
Venus	82;14	78;18
Mars	126;58	123;2
Jupiter	164;28	160;32
Saturn	244;28	240;32

Parameters of equations

	Equation of argument		Equation of centre	
	M	maximum	*S*	maximum
Sun			0; 2, 5,48	2; 0,10
Moon	0;10,19,29	13; 9	0; 5, 6,27,27	4;53
Mercury	0; 3,50,27	3;40	0;22,49,55	22;22
Venus	0; 2, 4,38	1;59	0;43, 8,53	45;59
Mars	0;11,59,43	11;25	0;40,15,52	42; 9
Jupiter	0; 5,47,56	5;32	0;11,30	11; 3
Saturn	0; 6, 4,44	5;48	0; 6,29,50	6;13

TOLEDAN TABLES

Mean parameters

	radix at 1947572	motion in 600×354;22 days
Sun	113;41,11	582^r 40; 7, 1
apogee	77;50	0
Moon	120;58,18	7782^r 36;43,52
argument	108; 8,39	7716^r 116;39,26
apogee	192;49,39	65^r 280; 4,26
node	125;50,05	-31^r 107;35,21
Mercury	187;27,29	2416^r 347; 7,21
apogee	197;30	0
Venus	159; 9,48	946^r 85;43,11
apogee	77;50	0
Mars	211;24,59	309^r 181; 0,20
apogee	121;50	0
Jupiter	331;39 37	49^r 26;46,30
apogee	164;30	0
Saturn	115;51,15	19^r 273;30,19
apogee	240; 5	0

Parameters of equations

	Equation of argument		Equation of centre	
	M	maximum	S	maximum
Sun			0; 2, 5,48	2; 0,10
Moon	0;10,19,29	13; 9	0; 5, 6,27,27	4;53
Mercury	0; 3,50,27	3;40	0;22,49,55	22;22
Venus	0; 2, 4,38	1;59	0;43, 8,53	45;59
Mars	0;11,59,43	11;25	0;40,15,52	42; 9
Jupiter	0; 5,47,56	5;32	0;11,30	11; 3
Saturn	0; 6, 4,44	5;48	0; 6,29,50	6;13

IBN KAMMĀD

Mean parameters

		radix at 1948438	motion in 90×354;22 days
Sun	arg	36;35,9	87^r 113;42,16
Moon	centre	120;34,42	1167^r 113;30,18
	arg	108;11	1157^r 161;30,6
	dbl elong	14;33	2159^r 358;53,42
	node	126;30	-4^r 250;8,0
Mercury	arg	74;1	275^r 82;0,12
	centre	274;58	87^r 113;42,16
Venus [a]	arg	45;21	54^r 222;50,12
Mars	arg	91;46	46r 152;38,18
	centre	262;14	40^r 321;4,0
Jupiter	arg	171;58	7^r 129;42,24
	centre	143;2	79^r 343;58,0
Saturn	arg	236;52	2^r 346;42,18
	centre	357;48	84^r 127;0,4

a: The centre of Venus is not given.

Parameters of equations

	Equation of argument		Equation of centre	
	M	maximum	S	maximum
Sun			0;1,58,2 [a]	1;52,43 [a]
Moon	0;10,19,29	13; 9	0; 5,14,48	5;1
Mercury	0; 3,10,38	3;2	0;22,30,32	22;2
Venus	0; 2, 4,38	1;59	0;43, 8,54	45;59
Mars	0;11,58,39	11;24	0;39,28,55	41; 9
Jupiter	0; 5,30,6	5;15	0;11,30	11; 3
Saturn	0; 6,49,54	6;31	0; 6,29,50	6;13

a: See the account above concerning the variable Solar equation.

References

Ali (1967). *The Determination of the Coordinates of Positions for the Correction of Distances between Cities*, a translation from the Arabic of al-Bīrūnī's *Kitāb Taḥdīd Nihāyāt al-Amākin li-taṣḥīḥ Masāfāt al-Masākin*, by Jamil Ali, Beirut, 1967.

Billard (1971). *L'astronomie Indienne*, R. Billard, École Française d'Extrême-Orient, Paris, 1971.

Burnett (1994). '*Magister Iohannes Hispanus*: towards the identity of a Toledan translator', Charles Burnett, *Comprendre et Maîtriser la Nature au Moyen Age, Mélange d'Histoire des Sciences offerts à Guy Beaujouan*, Paris: Droz, 1994.

Cantera Burgos (1931). *El Judio Salmantino Abraham Zacut*, F. Cantera Burgos, Madrid, 1931.

Calonymos (1531). *Alpetragii arabi theorica planetarum nuperime latinis mandata literis a Calo Calonymos Hebreo Neapolitano, ubi nititur salvare apparentias in motibus planetarum absque eccentricis et epicyclis*. Venice 1531.

Carmody (1952). *Al-Biṭrūjī de Motibus Celorum*. Critical edition of the Latin translation of Michael Scot. Francis J. Carmody. California, 1952.

Caussin (1804). 'Le livre de la grande table Hakémite', J.J.A. Caussin de

Perceval, *Notices et Extraits de la Bibl. Nat.* (Paris, An XII, 1804).

Cohn (1918). *Der Almanach perpetuum des Abraham Zacuto*, Berthold Cohn, Strassburg, 1918.

Dobrzycki (1965). 'Teoria preceszji w astronomii średniowiecznej', *Studia i Materialy z dziejow nauki polskiej*, seria C, Z.11 (1965), 3-47.

Goldberg (1848). *Liber Jesod Olam seu Fundamentum Mundi, opus astronomicum celeberrimum auctore R. Isaac Israeli hispano*, edited with notes, and summary translation, by B. Goldberg and L. Rosenkrantz, Berlin 1848.

Goldstein (1964). 'On the theory of trepidation', Bernard R. Goldstein, *Centaurus* 10 (1964) 232-247.

Goldstein (1971). *Al-Biṭrūjī: On the Principles of Astronomy. An edition of the Arabic and Hebrew versions with translation, analysis, and an Arabic-Hebrew-English glossary*. Bernard R. Goldstein, 2 vols. Yale University Press, 1971.

Hogendijk (1982). 'Rearranging the Arabic mathematical astronomical manuscript Bankipore 2468', J.P. Hogendijk, *J. History of Arabic Science* 6 (1982) 133-159.

Hyderabad (1947). *Rasā'il ibn Sinān l'il-'alāmata b. Ibrāhīm b. Sinān b. Thābit b. Qurra al-mutawaffā fī sana khamsi wa thalāthīn wa thalāthami'a [A.H. 335 / A.D. 946-7] bi-Baghdād*. Osmania Publications Bureau, Hyderabad, Deccan, 1366/1947.

Hyderabad (1948). *Rasā'il al-Bīrūnī, Abū Rayḥan Muḥammad b. Aḥmad al-Bīrūnī*, four tracts based on Bankipore MS 2468 (42, 36, 38, 37). Osmania Publications Bureau, Hyderabad, Deccan, 1367/1948.

Hyderabad (1956). *Al-Qanūn al-Mas'ūdī*, of Abū Rayḥan Muḥammad b. Aḥmad al-Bīrūnī, Osmania Oriental Publications Bureau, Hyderabad, Deccan, 1956.

Ibn Ezra (1507). *Liber coniunctionum planetarum et revolutionum annorum mundi qui dicitur de mundo vel seculo*, Abraham Ibn Ezra. Translated by Henry Bate, included in composite volume, Liechtenstein: Venice, 1507.

Kunitzsch (1965). 'Sufi Latinus', P. Kunitzsch, *ZDMG* 115 (1965), pp. 65-74.

Meeus (1991). *Astronomical Algorithms*, Jean Meeus, Willmann-Bell, 1991.

Mercier (1977). 'Studies in the medieval conception of precession', Raymond Mercier, *Archives Internationales d'Histoire des Sciences*, 25 (1976), pp. 197-220, 26 (1977), pp. 33-71.

Mercier (1987a). 'Astronomical Tables in the Twelfth Century', by Raymond Mercier, in *Adelard of Bath, An English Scientist and Arabist of the early twelfth century*, edited by Charles Burnett. The Warburg Institute, London, 1987, pp. 87-118.

Mercier (1987b). 'The meridians of reference of Indian astronomical canons', Raymond Mercier, *History of Oriental Astronomy*, ed. G. Swarup, A.K. Bag, K.S. Shukla, Cambridge University Press, 1987; pp. 97-108.

Mercier (1991). 'The lost *Zīj* of al-Ṣūfī in the twelfth century tables for London and Pisa', Raymond Mercier, in *Lectures from the Conference on al-Ṣūfī and Ibn al-Nafis*, (Amman, Jordan, Oct. 1987), Beirut, 1991.

Millàs Vallicrosa (1936). 'Una obra astronómica desconocida de Johannes Avendaut Hispanus', J. Millàs Vallicrosa, *Osiris* 1 (1936) 451-475.

Millàs Vallicrosa (1947). *El libro de los fundamentos de las Tablas astronomicas de R. Abraham Ibn Ezra*, edicion critica, con introduccion y notas por José Ma. Millàs Vallicrosa, Madrid-Barcelona, 1947.

Millàs Vallicrosa (1950). *Estudios sobre Azarquiel*, José Ma. Millàs Vallicrosa, Madrid-Barcelona, 1943-1950.

Morelon (1987). Thābit ibn Qurra, *Oeuvres d'Astronomie*, Texte établi et traduit par R. Morelon. Paris, 1987.

North (1967). 'Medieval Star Catalogues and the movement of the eighth sphere', J.D. North, *Archives Internationales d'Histoire des Sciences*, for 1967, pp 71-83.

Regiomontanus (1514). *Tabulae Eclypsium Magistri Georgii Peurbachii. Tabula Primi Mobilis Ioannis Monteregio*. [Edited by G. Tannstetter Collinitus, Vienna], 1514.

Regiomontanus (1557). *Fundamenta operationum, quae fiunt per tabulam generalem: Vel, Apodixes et demonstrationes eorum, quae in tabulis primi mobilis, cum tabulis eclipsium Purbachii praeceptoris editis a Tanstetero* [Venice 1514], *praecepit, in communem omnium Mathematum studiosorum utilitatem nunc primum editae. Johannes Mueller, Regiomontanus. Neuburgi ad*

Danubium, 1557.

Ricci (1513). *Augustini Ricci de motu octavae sphaerae opus mathematica atque philosophia plenum*, Augustinus Ricci, Trent, 1513.

Sachau (1878). *Chronologie orientalischer Völker von Alberuni*, herausgegeben von C.E. Sachau, Leipzig, 1923.

Sachau (1879). *The Chronology of ancient Nations, an English version of the Arabic text of the Athār al-Bāqiya of al-Bīrūnī*, translated by C.E. Sachau, London, 1879.

Sa'idan (1960). 'The rasā'īl of Bīrūnī and Ibn Sinān, a rearrangement', A.S. Saidan, *Islamic Culture* 34 (1960, Hyderabad, Deccan) 173-5.

Sa'idan (1983). *Risā'īl ibn Sinān, transcribed from the MS preserved in the Library Bankipur/Patna under the number 2468, and in the Books Risa'il Biruni and Ibn Sinan*, Hyderabad, 1948, by Ahmad Salim Sa'idan, Kuwait, 1983.

Samsó (1994a). 'Trepidation in al-Andalus in the 11th Century', Julio Samsó, *Islamic Astronomy and Medieval Spain*, §VIII, Aldershot: Variorum, 1994.

Samsó (1994b). 'Sobre el modelo de Azarquiel para determinar la oblicuidad de la eclíptica', Julio Samsó, *Islamic Astronomy and Medieval Spain*, §IX, Aldershot: Variorum, 1994.

Samsó (1994c). 'Ibn al-Bannā', Ibn Isḥāq and Ibn al-Zarqālluh's Theory', Julio Samsó and Eduardo Millàs, *Islamic Astronomy and Medieval Spain*, §X, Aldershot: Variorum, 1994.

Schmeidler (1972). *Joannis Regiomontanus Opera Collectanea*. Faksimiledrucke von neun Schriften Regiomontans und einer von ihm gedruckten Schrift seines Lehrers Purbach, Felix Schmeidler. Osnabrück, 1972.

Sédillot (1835). *Traité des Instruments astronomiques des arabes, composé au treizième siècle par Aboul Hhasan Ali, de Maroc, intitulé Jāmiᶜ al-mabādī wa'l-ghāyāt (collection des commencements et des fins)* traduit de l'arabe sur le manuscrit 1142 de la Bibliotheque Royale par J.J. Sédillot et publié par L.Am. Sédillot. 2 vols, Paris 1834-5.

Sédillot (1844). 'Mémoire sur les instruments astronomiques des arabes', L.Am. Sédillot. *Mémoires présentés par divers savants à l'Académie Royale des Inscriptions et Belles-Lettres* 1 (1844) 1-229.

Steinschneider (1887). 'Études sur Zarkali, astronome arabe du XIe siècle et ses ouvrages', Maurice Steinschneider, *Bulletino di Bibliografia e di Studia di Scienze Matematiche e Fisice* 14 (1881) 171-182 (SS 1-8); 16 (1882) 493-504 (SS 9-10); 17 (1884) 765-794 (SS 11-16); 18 (1885) 343-360 (SS 17-18); 20 (1887) 1-36 (SS 19-24), 575-602, b-X (Appendice).

Toomer (1968). 'A Survey of the Toledan Tables', G.J. Toomer, *Osiris* 15 (1968) 5-174.

Toomer (1969) 'The Solar Theory of az-Zarqāl: A History of Errors', G.J. Toomer, *Centaurus* 14 (1969) 306-336.

Van den Bergh (1924). *Die Epitome der Metaphysik des Averroes*, übersetzt und mit einer Einleitung und Erlauterungen versehen, S. van den Bergh, Brill 1924.

IV

From Tantra to Zīj

It is a pleasure to recall the times when Paul Kunitzsch came to my assistance in linguistic matters, saving me from embarrassing pitfalls, and so in return I offer an essay essentially about one of the common, but misunderstood, terms from medieval astronomy. The Arabic word *zīj* is the generic name for those astronomical handbooks which consist of groups of tables representing all the functions required in calculations of planetary longitudes and other quantities of interest. The tables are accompanied by a series of chapters which explain how the tables are to be used. There may also be further text giving more background information to the subject, but the dominant impression is created by the large number of tables. This is reflected in Nallino's etymology which saw the word *zīj* as a reference to the tabular format. He wrote as follows, expressing a view which has been accepted ever since,

> La parola zīǵ deriva della lingua pehlevica usata in Persia all'epoca dei re Sāsānidi. In questa lingua zīk significa l'ordito su cui viene tessuta la trama; i Persiani applicarono questa parola alle tavole numeriche, perchè le lore righe verticali somigliano ai fili dell'ordito.[1]
>
> The word *zīj* derives from the Pahlavi language used in Persia at the time of the Sasanid kings. In this language *zīk* means the warp on which came to be woven the weft; the Persians applied this word to the numerical tables because their vertical lines resembled the threads of the warp.

This explanation in terms of the lines defining the tables did not originate with Nallino, however, for it is found in Persian texts. John Greaves[2] in *Astronomica quaedam ex traditione Shah Cholgii Persae: una cum Hypothesibus Planetarum* (1652) quoted at length from the *Zīj-i Jāmiʿ* of Maḥmūd Shāh Khaljī (died 1469). In the opening of that work we have an explanation of the word, as follows,

> زیج معرب زیك است و آن عبارت است از ریسمانی چند که نقش بندان نقش جامها بر آن بندند و آن قانونیست جامه بف را در بافتن جامها ملون چنانچه زیج قانون است منجم را در استخراج تقاویم و اعمال موالید و خطوط و جداول زیج شبیه است بریسمانها زیك که در طول و عرض بهم کشیدة باشد
>
> Zīj is Arabic, from zīk; this expression is from the cord used by embroiderers by which they weave clothes of various colours, and also the guide (qānūn) for the weaver, used to arrange the variety. Thus a zīj is a guide (qānūn) for astronomers in deriving true

1 Nallino (1944), p. 120.

2 Greaves (1652). On Greaves' work as an Arabist and astronomer, see Mercier (1994).

positions and in making nativities; the lines and tables of a zīj are similar to the threads of the zīk, that are drawn in latitude and longitude.

No doubt it is through this work of Greaves that this interpretation came to be current in Europe.

There is however a compelling reason to reject this explanation. In a Pahlavi text, the *Epistles of Manuščihr* (2.ii.9), which was indeed known to Nallino[3], there is a reference to *zīg ī hindūg*, *zīg ī šahriyārān* and *zīg ī ptlmyws* (Πτολεμαίος). It has become a habit to render zīg as 'astronomical tables', as we find not only in Nallino's reading, but everywhere else[4]. The term *zīg* is therefore used indifferently of Indian, Persian and Greek astronomical works. The problem with regard to the Sanskrit sources is that these works contain no tables.[5] They are in verse form, and all the numerical data and numerical procedures are provided within these verses, as mere lists. The prose commentaries likewise never employ the tabular format. It is necessary therefore to find another explanation for the choice of the word *zīg*.

In any case, he makes a conscious distinction between Indian and other sources. It is assumed here that the word was introduced in the first place when the Indian material was translated into Persian. Manuščihr writes in the ninth century, and so it might be argued that he is not close enough to the period when the Sanskrit material was in the course of transmission, but at a time when Pahlavi was no longer a living language. It may be argued, however, that he used the older Sasanid expressions. Nallino[6] argued that because the Pahlavi transcription of Πτολεμαίος is *ptlmyws*, it must have been derived from the Arabic form *Btlmyws*, not directly from the Greek, because the short vowels *o*, *e* would have appeared in the Pahlavi transcription. On the other hand, this Pahlavi form could also have come from the pre-Islamic Syriac form *ptlmyws*, although admittedly this is only one of numerous variations in the transcription of this word.[7] Moreover, in Pahlavi the letters *p* and *b* are clearly distinguished, so the fact that Manuščihr uses *p*-, rather than *b*-, argues that he did not work from an Arabic source.[8]

In the extant Pahlavi texts the word *zīg* occurs a few times. The primary meaning is 'tether, tow-rope', that is a cord used for *binding*.[9] In one of these texts, the *Škand Gumānīk Vičār*, the context is astronomical, where the planets are said to be held by this 'tether' to the Sun and Moon. The word also occurs with the meaning 'astronomical text' in two passages, in the *Epistles of Manuščihr*, as noted above,

3 Nallino (1922), p. 350.

4 Bailey (1943), p. 80; de Menasce (1945), p. 60; de Menasce (1958), p. 61. Note that we use the modern transcription *zīg*, except when quoting an earlier passage.

5 In commentaries of the tenth century and later, if one is to judge by arrangements adopted in modern editions, there is some very limited listing of parameters, sometimes placing degrees in one row, and minutes in the next row. In the much later Moghul period, certainly, tables in the usual sense are found.

6 Nallino (1922), p. 350.

7 Payne Smith (1910), p. 3089.

8 I owe this last point to François de Blois.

9 The Manichaean 'Book of the Giants': Henning (1946), pp. 59, 64; Skand Gumānīk Vičār: de Menasce (1945), pp. 52–3, and Zaehner (1955), p. 159. The Manichaean Šābuhragān: MacKenzie (1972b), line 327. The word is not included in MacKenzie (1971a); he gives only the closely related zīh, 'bowstring'.

and probably also in the *Dinkard*, Book III, Ch. 419. This last, however, is a most difficult passage which has been the subject of various translations.[10] MacKenzie, amending somewhat the version of de Menasce, suggests[11]

> 'The beginning (?) of the years, months and days (is (calculated?)) according to the *zīg* of the stars, <and> the birth of men, the year-count of kings, how many years have gone (by) from the creation (is) in the hands of the calculators.'

It is not absolutely clear to me whether *zīg* here means simply the 'binding' of the stars (to the Sun ?), or refers to some more abstract 'text' or 'rule'. The whole chapter is concerned with numerical details of the calendar and intercalation, and with the responsibility of calculators to maintain the (calculated – *ošmurtīk* –) year in step with the seasons, and to keep Nō Rōz at the Spring Equinox. Certainly the rule depends on the motion of the Sun, such as would be recorded in the usual astronomical text, Indian or otherwise, but the mention of *zīg* here may refer to nothing more than the need to follow the motion of the Sun in relation to the stars, which indeed is the sense of Nyberg's version, *Der Eintritt der neuen Jahre, Monate und Tage geschieht jedesmal gemäss der jeweiligen Stellung der Tierkreisbilder.* However elsewhere when *zīg* is a binding within the heavens, it is a case of planets bound to the Sun, not the Sun to the stars.[12] In any event, the translation of *zīg* by de Menasce as 'tables' certainly presumes too much.

Al-Bīrūnī, in his *Qānūn al-Mas'ūdī*, Book III, Ch. 1, offers his own explanation of the term *zīj*, which he bases on the Persian *zeh.*[13] He writes,

> ان هذه الصناعة اذا اريد اخراجها الى الفعل بمزاولة الحساب فيها فالاعداد مفتقرة الى معرفة اوتار قسى الدوائر ، فلذلك سمى اهلها كتبها العلمية زيجات من الزيق الذى هو بالفارسية زه اعنى الوتر، وسموا انصاف الاوتار جيوبا وان كان اسم الوتر بالهندية جيبا ونصفه جيبارد، ولكن الهند اذا لم يستعملوا غير انصاف الاوتار اوقعوا اسم الكل على النصف تخفيفا فى اللفظ،

> This art, if one wishes to put it into practice on the basis of careful calculation and preparation, requires knowledge of the chords of arcs of circles, and for this reason the practitioners of this art called their scientific books *zījāt*, from the [older] Persian *zīq*, which in [modern] Persian is *zih*, meaning 'bowstring/arc' (*watar*), and they called the halves of the arcs *juyūb*, this despite the fact that in the Indian language the name for the (whole) bowstring/arc is *jīva* and for its half *jīvārdha*, but the Indians, since they only used the halves of the arcs (in their calculations), transferred the name of the whole to the half as a way of simplifying the pronunciation.

---

10 The text is available in a facsimile of one manuscript, Dresden (1966), p. 518. Nyberg (1934), pp. 34, 68, gives a transcription, translation and various notes. He recalled Nallino's theory, but rejected it, adding that he regarded zīg simply as a technical term in astronomy. De Menasce (1972), p. 375, gives a translation and notes, but for this chapter indeed he acknowledges Mary Boyce as the source of the translation with its notes. Apart from these I am indebted to Prof MacKenzie for correspondance about this passage.

11 Private correspondance.

12 In his discussion of the 'binding' of the planets to the Sun and Moon by means of zīges, Zaehner brings together a number of passages from the Škand Gumānīk Vičār, the Bundahišn, and elsewhere.

13 Bērūnī (1954), Vol. 1, 271.

Much of this is correct, since we know that *zih*, as in the earlier Pahlavi, means bowstring or chord, that the Arabic *juyūb* is the plural of *jīb*, the Arabic transcription of the Sanskrit *jīva*, chord, and that *jīvārdha* is the Sanskrit for half-chord, that is 'sine'. Clearly *zīq* is his Arabic orthography for the Persian *zīg*. However, in saying 'from the [older] Persian *zīq*, which in [modern] Persian is *zih*', he has somewhat simplified the position, for while in Pahlavi *zīg* meant a stretched cord, *zih* meant a bowstring; while the two words are very closely related, there was a difference in meaning. In any case, the point of al-Bīrūnī's argument seems to be just that the work came to be called *zīj* because it contained an account of sines and chords. Incidentally, he was not concerned that there should be a *table* of sines.[14] However important the calculation of sines may be in these astronomical works, it seems to be stretching a point to argue that they are named as class after the word for 'sine'.

In early modern Persian, one finds *zīj* and *zīč* (not *zīg*, at least in the printed edition) in the Persian Epic, *Shāhnāmeh* of Firdausī, in the sense of an astronomical text. This is qualified occasionally as *kuhan* ('ancient'), *hindī* or *rūmī*[15]. Mohl always translated the word as 'table astronomique'.

The word *zīj* is glossed according to a Persian source in the late thirteenth century, as recorded in the Greek version of the *Zīj al-ʿAlāʾī*.[16] Although this was an Arabic *zīj* it came into Greek use with the aid of the Persian Shams Bukharī who supplied oral instructions. Here the sense of 'weaving', 'loom' was apparently provided by this Persian informant.

> Ζῆζι οὖν λέγεται τὸ βιβλίον τοῦτο κατὰ Πέρσας ἐπειδὴ δικὴν ἵστου ὑφαίνονται οἱ ἀστέρες ἐν τούτῳ καὶ τάττονται, δι΄ ὧν ὑφασμάτων παντοῖα εἴδη δείκνυνται ἐκ μεταφορᾶς τῶν ἱστῶν· ὃν γὰρ τρόπον ἐν τούτοις μετὰ τὸ ὕφασμα ἡ ποικιλία τῶν ἀσκουμένων διάφορος δείκνυται, τὸν αὐτὸν τρόπον καὶ ἀπὸ τοῦ τοιούτου ἐπιστημονικοῦ βιβλίου κατὰ μέθοδον κρείττονα ὑφανθέντος ἡ ποικιλία καὶ ἐναρμόνιος πλοκή τε καὶ τάξις ἀναφαίνεται τῶν ἀστέρων.
>
> This book then is called a Ζῆζι (*zīj*) according to the Persians because, like a loom, the stars are woven on it and arranged in the weaving, and thanks to such weavings all sorts of forms are shown, according to the metaphor of the looms. Indeed, from the manner in which, in the latter, after weaving, the variety of worked materials appears different, in the same way also, thanks to such a scientific book, woven in the best way, the variety and the harmonious twisting and arranging of the stars becomes evident.

The entry zīg in the Persian dictionary *Burhān-i Qāṭiʿ*, written 1062/1636 by Muḥ. Ḥusain b. Khalaf al-Tabrīzī, known as Burhān, dedicated to the Shīʿī King of Gol-

14 For example, he translated the Sanskrit work Karaṇa Tilaka into Arabic, giving it the name Ghurrāt al-zījāt, although as a typical Sanskrit astronomical work, it contained no tables, only verses; Rizvi (1965).

15 The detailed references to the passages in Mohl's edition are as follows. These are based on Wolff (1934), s. v. zīj, zīč.
Zīj: 2 p. 232, ll. 450, 453; 4 p. 34, l. 369, p. 252, l. 2923, p. 564, ll. 2437, 2440, 5 p. 284, l. 225.
Zījhā-i Kuhan: 7 p. 112, l. 1303.
Zīj-i Hindī: 4 p. 232, l. 2686, p. 704, l. 4105.
Zīč: 1 p. 328, l. 1404; 7 p. 278, l. 3290.
Zīčhā-i Kuhan: 7 p. 112, l. 1306.
Zīč-i Rūmī: 5 p. 494, ll. 38, 41; 6 p. 346, l. 2243

16 Pingree (1985), Part 1, p. 36.

conda, Abdullāh Quṭb Shāh, is a remarkably close echo of the *Zīj-i Jāmi'* of Shāh Khaljī, and the foregoing Greek text.[17]

تارهایی باشد که استادان نقشبند نقش جامهایکه بافند بدان بندند. وکتابیکه منجمان احوال و اوضاع نجوم و افلاك را از جداول آن معلوم کنند و همچنان که آن قانونی است جامه بافان را در بافتن نقشهای جامه این کتاب نیز دستوری است منجمان را در شناختن احوال و اوضاع فلکی، وهمچنانکه کیفیت نقوش جامها از آن تارها پیدا می شود، کمیات و حرکات کواکب از جدولهای این کتاب ظاهر میگردد، و معرب آن زیج است

> (*Zīg*) may be the cords to which the masters of design bind the design of a garment which they are weaving. It is a book from which are made known to astronomers the states and conditions of the stars and spheres from its tables. Just as the former is a guide for the weavers in weaving the design of a garment, so this book also is a guide for astronomers in recognizing the states and conditions of a sphere. And as the manner of the design of a garment is made known from those cords, the forms and motions of the stars are made apparent from the tables of this book. In Arabic it is *zīj*.[18]

The cords (tārhā') in question would seem to be the warp of the loom on which the weaving is done. The *Burhān-i Qāṭi'* does not name a source, but evidently it was an explanation which had circulated since the 13th century. Without a quotation from a text the sense 'loom' cannot of course be considered as absolutely certain, and it may after all be a ghost.

We may summarise the position then, by saying that *zīj* has been explained in Arabic and Persian sources as arising

1) from the fact that the knowledge of sines and chords is central to the work (al-Bīrūnī);
2) from a resemblance between weaving a pattern on a loom according to a fixed rule, and the calculations of astronomical positions from a set of rules (*Zīj al-'Alā'ī* and the *Burhān-i Qāṭi'*);
3) from a resemblance between the warp and weft of a loom and the lines defining an astronomical table (Shāh Khaljī).

The last interpretation finally invokes the actual shape of the table, with its horizontal and vertical divisions, the only argument which has circulated among European scholars down to Nallino.

Sanskrit origins

Since the word *zīj* arose in connection with an astronomical text at the time when Indian works came to be known in Iran, that is in the late Sasanid period, it may

17 Vullers (1855), s. v. zīg. The Burhān-i Qāṭi' is a fundamental work which has appeared in several editions, both European and Persian, for example by Th. Roebuck, Calcutta, 1818, 1834, and Tehran, Mo'īn (1962). It is much quoted by Vullers and Desmaisons (1910). For an account of the work see the Introductory volume of the Loghat-Nama, Mo'īn (1959), vol. 1, pp. 199 seq.

18 The passage continues as follows, 'Name of a small bird, smaller than a sparrow, ash-gray, red underneath, with very good and melancholy song. Name of a tribe of Kurds inhabiting the Gīlūye mountain (Arabic Jīlūya, Turkish Cilu Dagh)'.

well have been as a literal translation of a Sanskrit word. In the sense 'cord' it may have rendered *sūtra*, and in the sense of stretched cord, *tantra*. Both these words have strong literary associations, and the latter is particularly relevant to astronomical texts of the sixth century.

Sūtra is literally 'thread', or 'cord' but came to be used in the sense of a guiding thread through a subject, a collection of aphorisms, and can sometimes be rendered 'guide'. Sūtra texts, such as the work of Pāṇini, are collections of concise formulae, suited to memorising. In the astronomical context we have Āryabhaṭa's *Daśagītikāsūtra*, the opening part of the work which we commonly know under the title *Āryabhaṭīya*.

Tantra is a derivative from the root *tan*.

tan: to extend, continue, stretch, weave, prepare the way, accomplish, compose;
tantra: loom, warp, leading part, model, system, doctrine; class of works dealing with magic.

Among related derivatives,

tanti: cord, line, weaver
tantu: thread, cord; continuity.

Although *tantra* is most commonly used now in the context of Tantric practices, including forms of black magic, associated with the worship of Śakti, that sense is relatively late, perhaps later than the seventh century.[19] In the Ṛgveda the word simply means 'loom', while for Pāṇini and his early commentators it meant also a 'collection of procedures'. In the astronomical context, in the sixth century, we know that both works of Āryabhaṭa (i. e. the so-called sunrise and midnight systems) were referred to as *tantras*.[20] In the following I give a list of the uses of *tantra*, both referring to Āryabhaṭa, and in a generic sense.

Bhāskara[21], *Mahābhāskarīya* –

I,3: *sphuṭatantramāśmakam*, the perfect treatise of the Āśmaka (Āryabhaṭa)
I,21: *laghutantra*, short treatise (of Āryabhaṭa)
II,8: *tantrajña*, knowing the *tantra* (composed by Āryabhaṭa)
V,78: *sarvatantravit*, knowing all the treatises
VII,33: *tantrāntara*, the other treatise (of Āryabhaṭa, i. e. the midnight system)

Varāhamihira, *Bṛhatsamhitā* (I.9)[22] –

jyotiḥ śāstramanekabhedaviṣayam skandhatrayādhiṣṭhitaṃ tatkārtsnyopanayasya nāma munibhiḥ saṃkīrtyate saṃhitā
skandhe'smin gaṇitena yā grahagatistantrābhidhānastvasau horā'nyo'ṅgaviniścayaśca kathitaḥ skandhastṛtīyo'paraḥ

The Jyotiḥśāstra, treating of several subjects, is contained in three branches. A complete course of it is termed by the Seers Saṃhitā. In one branch are to be found the motions of the heavenly bodies, as determined by calculation: this is called Tantra, or

19 Kane (1962), p. 1031 seq.
20 Billard (1971), pp. 6, 25, 80, 91, 110, 122.
21 Shukla (1960).
22 Kern (1865), (1913).

doctrinal part. The second branch is Horoscopy, or the casting of the horoscope. There is a third branch, different again.

Brahmagupta, *Brāhmasphuṭasiddhānta* –
XI, (entitled) *tantraparīkṣā*, examination of treatises, referring especially to Āryabhaṭa.

Haridatta (A.D. 683), *Grahacāranibandhana* –
III,46: Bhaṭatantra, referring to Āryabhaṭa's work.[23]

It is clear that for Āryabhaṭa and the period immediately following him the general name of an astronomical treatise was *tantra*. It was only in later centuries that this came to be replaced by *siddhānta*.

Al-Bīrūnī, in his *India*, has a note on the words *karaṇa*, *tantra*, and *siddhānta*.[24] He remarks that works of the first two types do not reach the standard of the *siddhānta*, and that there are two famous *tantras* by Āryabhaṭa and Balabhadra.[25]

> In the course of the transmission of Sanskrit works to Iran in the sixth and seventh centuries it was natural therefore to search for a Persian equivalent of the word *tantra*, and the proposal here is that *zīg* was chosen for this purpose.

The alternative would have been a transcription, but I can see no indication that *tantra* was ever transcribed. In Tibetan translations from Sanskrit *tantra* and *tantu* are both translated by *rgyud*, from the verb *rgyud-pa*, 'to file on a string', another instance of a literal translation.[26] In Soghdian Buddhist literature, on the other hand, the word *sūtra* was transcribed as *swtr*, *swttr*.[27]

The words *zīg* and *tantra* are linked at an elementary level in the sense of 'stretched cord', and also, if we accept the Burhān-i Qāṭiʿ, with one of the principal meanings of *tantra*, 'loom'.

While ranging beyond the Persian and Sanskrit texts of immediate interest, we may note that the metaphor between 'something woven' and 'something written' is by no means confined to these languages. After all, the English words 'textile' and 'text' have a common origin in the Latin 'texere', meaning 'to weave'; consider the common expression 'to weave a tale'. In Latin itself 'textum' meant both 'textile' and 'text'. In Post Biblical Hebrew 'Maseket' מסכת meant not only a 'web on the loom', but 'tract' (as in the Talmud), both derived from נסך to 'weave'.[28] In Chinese, jīng 經, also has this ambivalence, 'That which runs lengthwise; the warp in a loom. ... The "Canon" of Confucianism. The *sūtras* of Buddhism and Taoism', (Giles' *Dictionary*, no 2122).[29]

23 Sarma (1954), p. 20.
24 Sachau (1910), I, pp. 155–6.
25 Balabhadra was a ninth-century commentator on the works of Varāhamihira and Brahmagupta, much used by al-Bīrūnī, and known to us through commentaries by Pṛthūdaka.
26 Jäschke (1881), pp. 111–2.
27 Benveniste (1940), p. 271.
28 I am obliged to Jill Butterworth for this example.
29 It remains unclear to me whether the metaphor refers to the mental act of arranging one's thoughts in order, or to the physical act of writing out lines of text.

Implications

If we accept this proposal, then, it follows that works in Persian or Arabic of the *zīj* type, especially those of Sanskrit origin, need not necessarily contain material in tabular format. This observation would pertain especially to the *Zīg-ī šahriyārān*, which would seem to have been the first stage in the translation from Sanskrit to Persian. This is apparently the antecedent of the later Arabic work *Zīj al-Shāh*. These works are not extant, but we have some information about them from al-Bīrūnī and others.[30] It is altogether likely that these were simply texts arranged in verse like the Sanskrit originals. We know at least of imitations of such verses by al-Fazarī, quoted by al-Bīrūnī.[31] Other works of *zīj* type may also have been without any tabular format, for example the lost *zījes* of Abū Maʿshar, and Māshā'allāh.

Further, if *zīj* was used to render *tantra*, this would support the view that the *Zīg-ī šahriyārān* was based in fact on *the* Tantra, the work of Āryabhaṭa. Finally since al-Bīrūnī remarks[32] that in the *Zīg-ī šahriyārān* the day was measured from midnight, not sunrise, it would follow that the Persian translation is based not on the *Āryabhaṭīya*, but on the lost *tantra* by Āryabhaṭa, which used the midnight system. However we know enough about the parameters of the original midnight system and about the *Zīg ī šahriyārān* to be sure that changes had taken place, at least in the details of the planetary equations.[33] It is interesting to note Manuščihr's remarks on the three *zījes*, *šahriyārān*, *hindug*, and *ptlmyws* (Πτολεμαίος), since he reports that these were accurate for the Sun and Moon, for Saturn, and for Mars respectively, so that he certainly distinguished between the Indian original and the Persian version.[34]

Acknowledgements: This article has undergone many revisions since its original draft in 1994, and I am indebted to a number of colleagues who have kindly offered criticism and assistance: François de Blois, Jill Butterworth, Craig Jamieson, E. S. Kennedy, D. N. MacKenzie, Anne Tihon, and Sergei Tourkin. They are not of course responsible for the views finally expressed here.[35]

30 Kennedy (1958), further work being reported in Kennedy (1959) and Kennedy (1976).

31 Kennedy (1976), vol. 1, p. 190, vol. 2, p. 118.

32 Sachau (1879), p. 6.

33 Apart from Kennedy (1958), one should consult Kennedy (1959) for the details of the Zīj al-Shāh as known to al-Bīrūnī. The magnitudes of some of the equations differ from those known from the Sanskrit sources of Āryabhaṭa's midnight system, such as the Sūrya Siddhānta of Varāhamihira.

34 West (1892), p. xlvii. Although this is an old translation, notable indeed as a correction to an earlier translation of West's, there is not likely to be any quarrel with this reading from it.

35 A recent study by Panaino (1998), which came to my notice when this paper was completed, surveys discursively the role of *zīg* as a 'planetary binding', exploring numerous texts, mainly Iranian. He evidently regards this sense of *zīg* as sufficient reason for its use in the sense of 'astronomical handbook'.

Bibliography

BAILEY (1943). *Zoroastrian Problems in the Ninth-century Books*, H. W. Bailey, Oxford, 1943.

BENVENISTE (1940). *Textes Soghdiens*, E. Benveniste, Paris, 1940.

BILLARD (1971). *L'Astronomie Indienne*, R. Billard, Paris, 1971.

BĒRŪNĪ (1954). *Al-Qānūnu'l - Mas'ūdī*, Abū al-Rayhān Muḥammad b. Aḥmad al-Bērūnī, Dāiratu'l-Ma'ārif-il-Oṣmānia, Deccan, 1954.

DESMAISONS (1910). *Dictionnaire Persan-Français* par Jean-Jacques Desmaisons, Paris: Maisonneuve, 1910.

DRESDEN (1966). *Dēnkart, a Pahlavi text; facsimile edition of the manuscript B of the K. R. Cama Oriental Institute*, Bombay, ed. M. J. Dresden, Wiesbaden: Harrassowitz, 1966.

GILES (1912). *A Chinese-English Dictionary*, by Herbert A. Giles, Shanghai, 1912.

GREAVES (1652). *Astronomica quaedam ex traditione Shah Cholgii Persae: una cum Hypothesibus Planetarum.* Studio et opera Johannes Gravii nunc primum publicata. London, 1652.

HENNING (1946). 'The Book of the Giants', W. B. Henning, BSOAS 11 (1943–6) 52–74.

JÄSCHKE (1972). *A Tibetan-English Dictionary*, H. A. Jäschke, London, 1881 (repr. 1972).

KANE (1962). *History of Dharmaśāstra*, P. V. Kane, Vol. V (part II), Poona, 1962.

KENNEDY (1958). 'The Sasanian astronomical handbook Zīj-i Shāh and the astrological doctrine of "transit" (mamarr)', E. S. Kennedy, *Journal of the American Oriental Society* 78, pp. 246–262, 1958.

KENNEDY (1959). *Al-Bīrūnī on transits*, E. S. Kennedy, Beirut: American University of Beirut, 1959.

KENNEDY (1976). *The Exhaustive Treatise on Shadows, by Abū al-Rayhān Muḥammad b. Aḥmad al-Bīrūnī*, E. S. Kennedy, 2 vols., Aleppo: University of Aleppo, 1976.

KERN (1865). *The Bṛhat Saṃhitā of Varāha-Mihira*, edited by H. Kern, Calcutta, 1865.

KERN (1875). *The Bṛhat Saṃhitā of Varāhamihira*, translated by H. Kern, Journal R. Asiatic Society (1870–5), Verspreide Geschriften vol. I, The Hague, 1913.

MACKENZIE (1971a). *A Concise Pahlavi Dictionary*, D. N. Mackenzie, Oxford, 1971.

MACKENZIE (1971b). 'Mani's Šābuhragān', D. N. Mackenzie, BSOAS 42 (1979) 500–534.

DE MENASCE (1945). *Škand Gumānīk Vičār*, J.-P. de Menasce, Fribourg, 1945.

DE MENASCE (1958). *Une Encyclopédie Mazdéenne. Le Denkart*, J.-P. de Menasce, Paris: PUF, 1958.

DE MENASCE (1972). *Le troisième livre du Dēnkart*, traduit du Pehlevi par J.-P. de Menasce, Paris: Klincksieck, 1972.

MERCIER (1994). 'English Orientalists and Mathematical Astronomy', in *The 'Arabick' interest of the natural philosophers in seventeenth-century England*, edited by G. A. Russell, Leiden: Brill, 1994, pp. 158–214.

MOHL (1878). *Le Livre des Rois*, Abou'l Kasim Firdousi, edition and translation Jules Mohl, Paris 1838–1878.

MOʿĪN (1959). *Loghat-Nama* fondé par Alī Akbar Dehkhodā, sous la direction de Moḥammed Moʿīn, Tehran (1959–).

MOʿĪN (1962). *Burhān-i Qāṭiʿ*, ed. Moḥammed Moʿīn, Tehran, H. Shamsī 1330.

NALLINO (1922). 'Tracce di opere greche giunte agli Arabi per trafila Pehlevica', C. A. Nallino, in *Volume of Oriental Studies presented to E. G. Browne*, Cambridge, 1922, pp. 345–363.

NALLINO (1944). 'Storia dell'astronomia presso gli Arabi nel Medio Evo', C. A. Nallino, in *Raccolta di Scritti editi e inediti*, vol. V, Roma, 1944, pp. 88–329. [Based on Arabic lectures given in the University of Cairo, 1909–1910.]

NYBERG (1934). *Texte zum mazdayasnischen Kalendar*, H. S. Nyberg, Uppsala Universitets Årsskrift, 1934.

PANAINO (1998). *Tessere il Cielo*, Antonio Panaino, Serie Orientale Roma lxxix, Roma: Istituto Italiano per l'Africa e l'Oriente, 1998.

PINGREE (1985). *The astronomical works of Gregory Chioniades*, Volume I *The Zīj al-ʿAlāʾī*, David Pingree, [Corpus des Astronomes Byzantins II], Amsterdam: Gieben, 1985.

RIZVI (1965). 'A Unique and Unknown Book of al-Beruni, Ghurrat-uz-Zijat or Karana Tilaka', *Islamic Culture* (Haiderabad, India) 1963 (Apr., July, Oct.) 1964 (Jan., July), 1965 (Jan., Apr).

SACHAU (1879). *The Chronology of Ancient Nations ... of Albīrūnī*, E. C. Sachau, London, 1879.

SACHAU (1910). *Alberuni's India* (English edition), E. C. Sachau, London, 1910.

SARMA (1954). *Grahacāranibandhana by Haridatta*, K. V. Sarma, Madras, Kuppuswami Sastri Research Institute, 1954.

SHUKLA (1960). *Bhāskara I and his works. Mahā-Bhāskarīya*, K. S. Shukla, Lucknow, 1960.

VULLERS (1855). *Lexicon Persico-Latinum* Ioannis Augusti Vullers, Bonn, 1855; repr. Graz, 1962.

WEST (1892). *Pahlavi Texts, Part IV, Contents of the Nasks*, Sacred Books of the East, Vol. XXXVI, Oxford, 1892.

WOLFF (1934). *Glossar zu Firdosis Schahname*, Fritz Wolff, Berlin: Reichsdruckerei, 1934.

ZAEHNER (1955). *Zurvan, a Zoroastrian dilemma*, R. C. Zaehner, Oxford, 1955.

V

THE MERIDIANS OF REFERENCE OF INDIAN ASTRONOMICAL CANONS

ABSTRACT The canons of Sanskrit astronomy depend on mean motions which are normally postulated to refer to the central meridian of Ujjain. The present work is a statistical analysis of these mean motions designed to discover the optimum position of the meridian, by comparison with modern mean motions. This follows earlier work done by Billard in determining the optimum year. The results confirm that from the time of Āryabhaṭa all the canons were referred to meridians lying well within India, and in many cases clearly identifiable with Ujjain within the statistical bounds.

1. INTRODUCTION

The general idea underlying the research which is summarised in the present paper is hardly original, and indeed is a direct development of that employed so successfully in Roger Billard's l'Astronomie Indienne (Billard (1971)). The medieval mean longitudes of Sun, Moon and planets are compared directly with the corresponding modern means, and the differences are treated by the method of least squares in order to determine values of the year and of the meridian to which the medieval longitudes are referred.

In Billard's researches the meridian was always assumed to be that passing through Ujjain, the central meridian repeatedly referred to in Sanskrit sources. In this way he was able to fix, within narrow limits as a rule, the year when the Indian mean longitudes best agreed with the true configuration.

In the present work not only the year, but also the meridian are allowed to vary in the search for the best fit. There are also improvements in the modern parameters, taking advantage of quite recent developments, and besides the statistical control follows more closely the formal method of least squares.

Apart from a natural confirmation of Billard's results for the optimum year, we have a series of results for the optimum meridian showing that it always lies in India, indeed generally near to Ujjain (longitude 75;46 East of Greenwich). This reinforces very well Billard's general conclusion that with the Sanskrit canons we are presented with a continuous millenium of observational astronomy in India.

All of the canons cited by Billard have been analysed in this way, as well as a number of others, but here there is room only to present a few particularly important ones.

2. MEAN LONGITUDES AND DEVIATIONS

Let the modern mean longitudes be denoted $L_i(t)$, where the suffix runs from 1 to 9:

1. Sun
2. Moon
3. Lunar apogee
4. Lunar node
5. Mercury
6. Venus
7. Mars
8. Jupiter
9. Saturn

All these are tropical longitudes. The precise numerical expressions are those now to be employed in the national ephemerides, following an IAU recommendation (Francou, e.a., 1983). These expressions are functions of Terrestrial Dynamical Time (TDT) which differs from Universal Time by the quantity ΔT, which allows for the changes in the rate of rotation of the earth; this includes both a steady rate of decrease, and fluctuations. A satisfactory expression for the steady part of ΔT depends necessarily on ancient eclipse records, and this calculation is in no way as accurate as that determining L_i. For many years a formula for ΔT which was determined by Spencer Jones (1939) has been used, although there are theoretical objections (van der Waerden (1961)) to its derivation. Many new formulae have been derived, but they do not agree especially well among themselves, nor do they meet the objections brought by van der Waerden against the older formula. Naturally the uncertainty in the value of ΔT appears directly in the meridian which we determine, but it is unlikely that further revisions of ΔT would lead to alterations in the meridian of more than a few minutes of arc.

In this paper the expressions for $L_i(t)$ include terms as far as t^3 (as taken from references listed by Francou, e.a., (1983)), and in addition a large number of trigonometrical terms expressing the various perturbations (1). These are included for Venus, the Sun, Mars, Jupiter and Saturn, but are effectively important only for the last two, which are affected by the well known resonance.

The medieval mean longitudes $\lambda_i(t)$, where t is simply the Universal Time, are linear expressions. In most cases, the canons which we analyse are already included in Billard's survey, and there one will find the numerical details. As a general rule the Sanskrit canons define sidereal longitudes.

3. THE METHOD OF LEAST SQUARES

We now change the notation slightly, so that t denotes the Universal Time, and TDT as required for L_i will be $t + \Delta t$, with Δt provided by the Spencer Jones formula, *faute de mieux*. The deviation between the modern and the medieval mean longitudes is defined as

$$D_i(t,\phi) = \lambda_i(t-\phi/360) - L_i(t+\Delta t)$$

where ϕ is the longitude of the meridian East of Greenwich.

The method of least squares will be used to determine jointly the optimum estimates t_0 and ϕ_0. For this purpose we must postulate 'true' values of $D_i(t,\phi)$, which may be done in two ways. If λ_i are tropical longitudes, such as we may calculate from the Sanskrit canon when we are also given a model of precession, then the 'true' value of D_i is zero, and then we would seek the minimum of the sum

$$Q = \sum_I D_i^2$$

where I is a selection of values of i. In this case we say that the deviations are 'absolute'. On the other hand, if λ_i is sidereal, the 'true' value of D_i would depend on the rate of precession together with an unknown constant. In this case therefore, which we refer to as 'relative', we can only take the mean of the set D_i as the true value, and calculate the minimum of

$$Q = \sum_I (D_i - (\sum_I D_i)/N)^2$$

where N is the number of deviations included. When the deviations are relative, the number of random variables is then N - 1. Let I be indicated by a sequence of 1's or 0's, indicating whether a particular value of i is included or not: thus (1101 00000) would indicate that the Sun, Moon and lunar node only are included. Further the set I will be augmented by an initial 1 or 0 to indicate whether we are concerned with absolute or relative deviations, respectively. With each determination of the year and the meridian, then, we associate a statistic I, such as (0 1101 0000). This is the same symbol as used by Billard.

The method of least squares has been explained in a suitably general and clear way by van der Waerden (1967). In the following only the briefest summary of the application is possible.

If we write in the neighbourhood of the minimum

$$Q = h_{11}(t-t_0)^2 + 2h_{12}(t-t_0)(\phi-\phi_0) + h_{22}(\phi-\phi_0)^2 + Q_0,$$

then the estimate of the variances of t_0 and ϕ_0 are s_t^2 and s_ϕ^2:

$$s_t^2 = \frac{Q_0}{n-r} \frac{h_{22}}{h_{11}h_{22}-h_{12}^2}$$

$$s_\phi^2 = \frac{Q_0}{n-r} \frac{h_{11}}{h_{11}h_{22}-h_{12}^2}$$

where n, the number of random variables, is one less than the number of 1's in the symbol I. Moreover if $\tilde{t}$ and $\tilde{\phi}$ and are the true values, then

$$(t_0-\tilde{t})/s_t, \quad (\phi_0-\tilde{\phi})/s_\phi$$

follow the Student t-distribution with n-r degrees of freedom. We may thereby assign confidence limits to the estimates t_0 and ϕ_0, as obtained from the minimum value Q_0. If n-r were large, then t_0 would have a probability of 0.682 of lying within one standard deviation of the true value. We use the Student distribution in this way to find the equivalent range for a probability of 68%.

If σ is the standard deviation of individual deviations D_i, then $s^2 = Q_0/(n-r)$ is an unbiased estimate of σ^2. Moreover Q_0/σ^2 has a χ^2 distribution with n-r degrees of freedom, so that one may find limits of confidence for s^2/σ^2 at a given level, in our case chosen to be 95%.

There are canons, such as that known from Lalla's Śiṣyadhīvṛddhidatantra, in which the year cannot be effectively determined. In that case it is more practical to fix in advance the value of t, and to apply the present methods to determine ϕ_0 alone, in which case $r = 1$, in the above expressions.

4. DISCUSSION OF THE RESULTS

4.1 General remarks

For each of the canons, and each of the statistics I, the results are given in the table below, Section 5.

These results for the determination of the meridian are satisfying in that they lie in every case in India. Nevertheless, they differ to a varying extent from the meridian of Ujjain. There are two obvious ways in which this may occur, for we may have either an error on the part of the author of the canon, or it may have been the case that the observations were established for some other meridian, without any attempt to reduce them to that of Ujjain. No doubt both these reasons have some degree of application.

A third possibility is that the discrepancy might arise from the equation of time. I have indicated elsewhere, however, (Mercier 1985), that while difficulties would arise because of the way in which the equation is defined in Greek, Arabic and Latin usage, in Sanskrit usage the equation takes positive and negative values symmetrically, so that there the Mean Solar Time would be the same as that used now. This would seem to be the case from the beginning of Sanskrit astronomy, since Varāhamihira, in his discussion of the Sūryasiddhānta (Pañcasiddhāntikā IX,9) indicates an approximate rule for the equation.

4.2 The Romaka Siddhānta

The earliest of the canons in this survey is the Romaka

Siddhānta, based on parameters given by Varāhamihira in the Pañcasiddhāntikā (VIII, 1-8). The deviations are limited in use, since the lunar apogee and node are not in good agreement with the Sun and Moon. Therefore only the statistic (1 1100 00000) is available, and we can only find a year and meridian such that $Q_0 = 0$, since $n = r$. Therefore no statistical bounds can be assigned to the year and meridian. Nevertheless the results are very interesting, especially in regard to the meridian 86;40. This is considerably too large for one to suppose that Indian observations were responsible for the canon, but if we recollect the longitude difference of 56;30 between Alexandria and Ujjain, according to Ptolemy's Geography (IV,5,9; VII,1,63)

Alexandria	60:30
Ujjain	117;0,

we see that the Romaka canon might very well have been transferred to Ujjain, by means of such presumed positions, from a Greek source at a meridian around 30;0, which is indeed the correct longitude of Alexandria. The name Romaka, and the use of the year 365.24666..., indicate clearly enough the dependence on a Greek source. The longitude difference 60;0 between Yavanapura (Alexandria) and Ujjain is entailed by a remark concerning two astronomers of the time, Lāṭācārya and Siṃhācārya (Pañcasiddhāntikā, XV, 17-20) (2).

4.3 Āryabhaṭa

In the list of results in Section 5 one naturally groups together the next three canons, those of the Āryabhaṭīya, and the two from Varāhamihira's Pañcasiddhāntikā: the Sūrya Siddhānta, and the emendation to the planetary parameters of that work defined in XVI, 10-11. We know that only the Āryabhaṭīya is accurate in the case of Jupiter, so that D_8 must be omitted from the statistic for the other two. That much is clear from Billard's analysis of the deviations, and the use of the most recent modern ephemerides does not alter the position. The use of the new ephemerides has however an extremely interesting consequence, for the optimum year is now very nearly equal to 499 A.D., incomplete. That year is singled out by Āryabhaṭa (Āryabhaṭīya III,10) although he contents himself with merely telling us that he was then aged 23, so that one has never been altogether clear as to its astronomical significance. This point is however distant 3600 years, quite precisely, from the Kaliyuga, so that it can hardly be doubted that it had, for Indian astronomy, some decisive importance. When the older ephemerides of Newcomb, Brown, etc., are used to obtain the year and the meridian, we obtain

(0 1111 01111) 507.6 ± 4.05 79;8 ± 1;24.

It is now apparent that although with the new parameters, no great change results, nevertheless the results point more clearly than before to the year A.D. 499 as that year to which Āryabhaṭa referred observational results which were collected around that time. It is not clear why exactly that year was chosen, although it may have been believed by

Āryabhaṭa that the precessional correction vanished then. In any case, how can one avoid the conclusion that the Kaliyuga, 3600 years earlier was defined and fixed as a direct consequence? The two versions of that epoch, Sunrise and Midnight, are fixed exactly, so that in terms of the related canons, Āryabhaṭīya, and the Sūryasiddhānta, respectively, 3600 sidereal years separate the epoch from Noon 499 March 21. The enduring use of the Kaliyuga in all the following centuries testifies to the decisive importance of the work of Āryabhaṭa in the history of Indian astronomy (3).

The statistical results for the emendation to the Sūryasiddhānta show remarkably that when some later astronomers established it, the effect was to improve the accuracy at the epoch year A.D. 499. We now have 498.1 for the Sun and Moon alone, and 499.1 with the planets included. It is as if those responsible had access to the observational data, and were able to make better use of it so as to obtain even more accurate longitudes for the selected epoch year. Is it not wonderful in any case, that the most modern and accurate parameters help us to discover and appreciate better than ever before the very accurate Indian observations made nearly 1500 years ago?

4.4 The canon of Lalla

The canon taken from a set of emendations given to us by Lalla is singularly interesting, because unlike most others, indeed all others preceding it, it is in good agreement with observations over a long period. This means that one cannot determine the year by the method of least squares, although other considerations such as the form of the emendations, lead one to associate it with the late ninth century. The determination of the meridian, however, is secure, and indeed safer than in some other cases, precisely because the optimum meridian is so insensitive to the choice of year. One is struck in this case by the close approximation of the optimum meridian to that of Ujjain, reinforcing the view that the canon was the product of outstandingly careful work. In the diagram there are shown the graphs of σ against ϕ for the selection of statistics given in Section 5.

4.5 The Dṛggaṇita

This work was composed in Śaka 1353 (A.D. 1431-2), by Parameśvara, an astronomer of Kerala, and constitutes the exposition of the Dṛk ('Observational') system (Sarma (1963)). It is produced by a set of emendations applied to that version of the Sūryasiddhānta on which Parameśvara wrote his commentary (Shukla (1957)), or equivalently, to the Karaṇa Tilaka of Vijayanandin, ca. 950 (Rizvi (1963)). The emendations of the solar and lunar parameters were derived from eclipses of Sun and Moon observed by Parameśvara in the period 1398 - 1432 inclusive, from which he determined mean longitudes referred to the date with ahargaṇa 1651700 (essentially 4522 sidereal years), at Sunrise, which is A.D. 1421 March 29. The text in which the dates and circumstantial details of the eclipses are given has been edited by Sarma (1966). For that date he found the following: Sun 0,13;0, Moon 10,4;6, Apogee 3,9;57, Node 4,23;55. These figures are most interesting

V

MAI016 AT SOTONVM AT 20:00:28 ON 16/02/1987

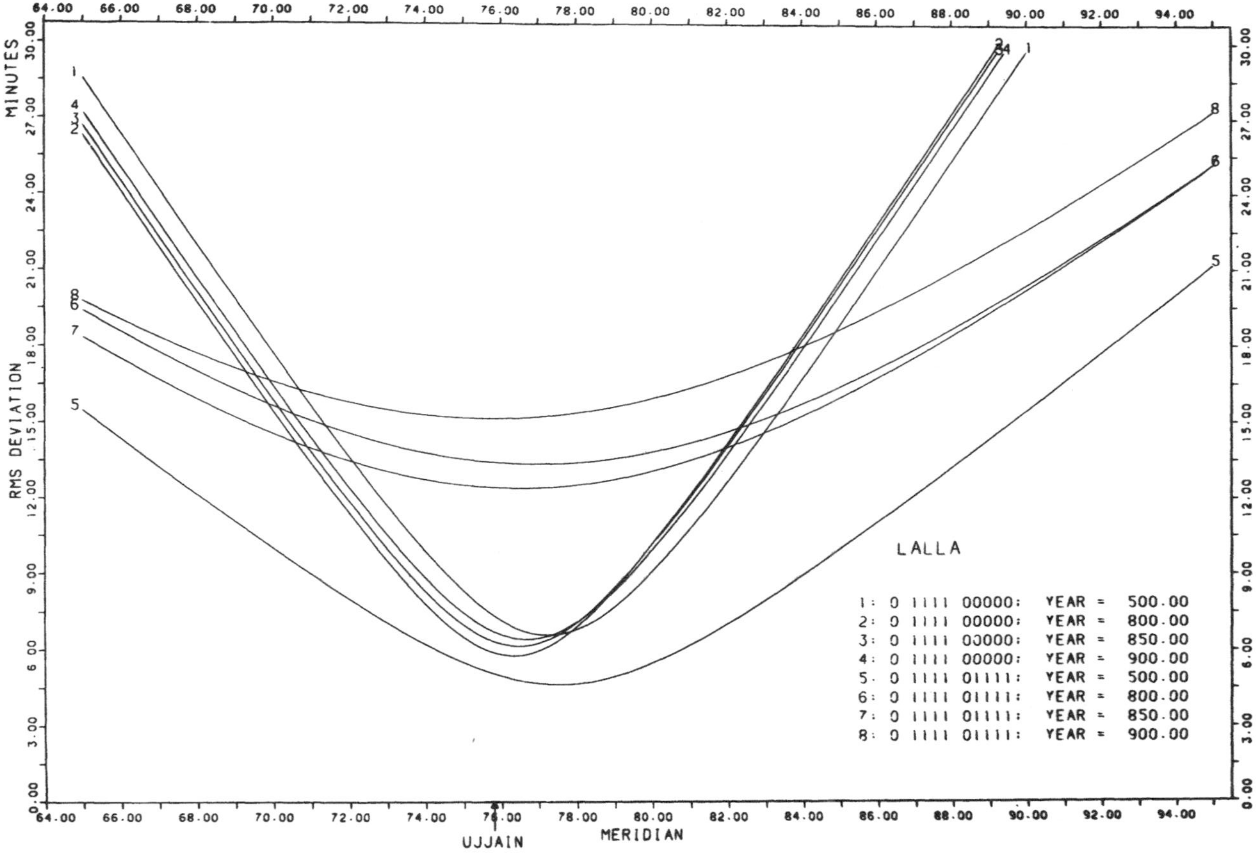

in that the longitude of the Sun is tropical, while the others are sidereal; that is the night-time observations are sidereal, the day-time tropical. Should we not see here an illustration of the paradigm of observation and tabular correction such as had been used throughout Indian astronomy, beginning with Āryabhaṭa?

The present methods applied to the eclipse deviations confirm well enough the documented circumstances, the year near 1421, and a meridian appropriate to Kerala (76;30) or perhaps Ujjain, for we cannot be certain whether Parameśvara reduced his parameters to the normal meridian.

4.6 Sphuṭanirṇayatantra

This work, by the Keralite astronomer Acyuta (1550-1620) was edited by Sarma (1974), and is of interest here because it is the one example known to me in Indian texts of a change of meridian. For, there is a short tract in another Malayalam MS, published as Appendix 5 by Sarma, in which the mean longitudes differ only by a shift of meridian, exactly 6.804 degrees Westward. Thus the respective meridians are approximately 80.5 and 74, neither of which however is suited to southern India. Moreover the accuracy is generally best in the thirteenth century, so one might infer that Acyuta obtained the canon from an earlier astronomer who lived further north.

5. TABLE OF RESULTS

The quantities given after the statistic are respectively, t_0, ϕ_0, with their standard deviations; σ, in minutes; the mean deviation at the point (t_0,ϕ_0); and the year and standard deviation for the meridian of Ujjain. In the case of those canons for which the year cannot be determined, the statistics concerning the meridian are given for certain preassigned values of the year. In all cases the year given is 'complete'.

Romaka Siddhānta

1 1100 00000				
400.0	86;40			

Sūrya Siddhānta

0 1111 00000				
498.1±31.67	77;19±6;25	10.67	0;14	493.8±14.7
0 1111 01101				
502.0± 5.41	77;28±1;54	5.00	0;11	501.3±5.23

Āryabhaṭa

0 1111 00000				
498.1±31.67	77;20±6;01	10.67	0;14	494.3±15.82
0 1111 01111				
502.1±5.10	77;38±1;45	4.70	0;12	501.7±5.13

Pañcasiddhāntikā

0 1111 00000				
498.1±31.67	77;19±6;25	10.67	0;14	493.8±14.7
0 1111 01101				
499.1±6.84	77;16±1.49	4.72	0;14	497.8±6.36

Lalla

0 1111 00000			
500	77;11±2;40	6.6	0;5
800	76;11±2;20	5.8	-4;47
850	76;23±2;30	6.1	-5;35
900	76;35±2;36	6.4	-6;24
0 1111 01111			
500	77;23±1;43	4.6	0;8
800	77;00±4;58	13.33	-4;44
850	76;23±4;37	12.38	-5;34
900	75;47±5;38	15.1	-6;24

Karaṇa Tilaka

0 11ȯ1 01011				
951.6±8.51	81;26±3;35	9.4	-6;59	950.7±10.43
1 1101 01011				
955.4±17.45	87;22±6;34	19.73	0;8	955.5±20.58

Dṛggaṇita

0 1111 00000				
1424.5±27.19	77;19±6;6	9.9	-14;40	1420.5±12.30

Sphuṭanirṇayatantra

0 11ȯ1 00ȯ10			
1250	80;47±1;45	4.64	-12;3
1350	81;35±1;53	5.00	-13;39
1550	81;47±5;39	14.99	-16;55

6. SELECTED CANONS

There are three canons which are not given by Billard, and which are not readily available elsewhere.

6.1 Romaka Siddhānta

Epoch A.D. 505 March 21 Sunrise (1905588.75)

Sun	(150t-65)/54787 revs
Moon	(38100t-1984)/1040953 revs
Lunar argument	(110t+664)/3031 revs
Lunar node	-(24t+56278)/163111 revs

The time t is measured in days from the epoch. These results are taken from unpublished work on R. Billard.

6.2 Sphuṭanirṇayatantra (a)
Sphuṭanirṇaya tulyagrahamadhyamānayanam (b)

Epoch 588465.75 (Kaliyuga Sunrise)
Period (Kalpa) 1577917517019 days

	radix (a)	radix (b)	yugabhagaṇa
Sun	0;0	0;1,7,4	4320000000
Moon	4;21,21,36	4;36,18,7	57753321009
Apogee	118;50,9,36	118;50,17,11	488123229
Node	201;18,43,12	201;18,39,36*	-232297832
Mercury	350;12,28,48	350;17,7,14	17937072112
Venus	35;6	35;7,49,1	7022270775
Mars	348;9,21,36	348;9,57,15	2296862959
Jupiter	15;15,50,24	15;15,56,3	364172296
Saturn	340;22,48	340;22,50,17	146626695

*The text as edited gives 204;18,39,36, which must be emended to 201;18,39,36.

The radices (a) are equal to 0.4569 bhagaṇa, which recalls the construction used in the Brāhmasphuṭasiddhānta, which has 0.4567 bhagana as a general formula for its radices.

Both the texts (a) and (b) were edited by Sarma (1974), who gives (b) in Appendix 5.

6.3 Karaṇa Tilaka

The parameters are identical to those of the modern Sūrya Siddhānta in the version given by Billard, except for two yugabhagaṇa:

Lunar apogee	488211	in place of	488203
Lunar node	-232234	in place of	-232238

See Rizvi (1963).

NOTES

(1) The trigonometrical terms are available from the Bureau des Longitudes, Paris.

(2) I am indebted to some important unpublished work on this canon by Roger Billard, who proposed the interpretation of the meridian 86;40.

(3) This proposal, that the Kaliyuga originates strictly with Āryabhaṭa, contradicts van der Waerden's conclusions (1978 and 1980). His argument begins by observing that Abū Maʿshar made use of the Kaliyuga as the date of the Deluge, and that he also made use of 'Persian' tables. If this only meant the Zīj-i Shāh in the mid-sixth century, there would be no problem, but van der Waerden argues for a Hellenistic dependence, via an earlier Persian system.

REFERENCES

Billard, R. (1971). L'Astronomie Indienne, Investigation des texts Sanskrits et des données numériques. Paris: École Française d'Extrême-Orient.

Francou, G., Bergeal, L., Chapront, J., and Morando, B. (1983). Nouvelles ephemerides du Soleil, de la Lune et des planetes. Astronomy and Astrophysics 128 124-139.

Mercier, R. (1985). Meridians of Reference in Pre-Copernican Tables. Vistas in Astronomy 28 23-7.

Rizvi, Sayyid Samad Husain (1963). A Unique and Unknown Book of al-Beruni, Ghurrat-uz-Zijat or Karana Tilakam. Islamic Culture (Haidarabad, India) 1963 (Apr., July, Oct.) 1964 (Jan., July) 1965 (Jan., Apr.).

Sarma, K.V. (1963). Dṛggaṇita of Parameśvara, critically edited with Introduction. Hoshiarpur, Punjab: VVRI.

Sarma, K.V. (1966). Grahaṇanyāyadīpikā of Parameśvara, critically edited with a translation. Hoshiarpur, Punjab: VVRI.

Sarma, K.V. (1974). Sphuṭanirṇaya-Tantra of Acyuta, with auto-commentary, critically edited with Introduction and ten appendices. Hoshiarpur, Punjab: VVRI.

Shukla, K.S. (1957). The Sūryasiddhānta edited with the commentary of Parameśvara. Lucknow: Lucknow University.

Spencer Jones, H. (1939). The rotations of the earth, and the secular accelerations of the Sun, Moon and planets. Mon. Not. R. astr. Soc. 99 541-558.

van der Waerden, B.L. (1961). Secular Terms and Fluctuations in the Motions of the Sun and Moon. Astronomical J. 66 138-147.

van der Waerden, B.L. (1967). Statistique Mathematique. Paris: Dunod.

van der Waerden, B.L. (1978). The Great Year in Greek, Persian and Hindu Astronomy. Archive for History of Exact Sciences 18 359-384.

van der Waerden, B.L. (1980). The Conjunction of 3102 B.C. Centaurus 24 117-131.

VI

THE PARAMETERS OF THE ZĪJ OF IBN AL-A'LAM

1. *Introduction*

The astronomer Ibn al-A'lam – Abū'l-Qāsim 'Alī ibn al-Ḥasan ibn al-A'lam al-Sharīf, al-Baghdādī (d. A.D. 985) – is mentioned in a number of sources including the Zīj al-Ḥākimī of Ibn Yūnis, who praises his work and cites the motion of the mean Sun and the rate of precession, apparently from his Zīj, now unfortunately lost. See the brief note by Sezgin (1978, 215), written before the appearance of the article by Kennedy (1977). However, parameters from it are cited in a number of sources in Arabic, Persian and Greek, as we know from Kennedy (1977), Mogenet (1962), and most recently in the accompanying paper of A. Tihon. The purpose of the present study is to establish from these sources the correct values of the parameters of mean motion and the associated equations.

Although many manuscript sources are explored by Kennedy, Mogenet and Tihon, essentially all the evidence which we need is contained in the following sources:

P1 Paris BN Supp. pers. 1488;

P2 Paris BN arabe 5968, f. 2-2^{v};

O1 Oxford Bodley *Selden supra* 17, f. 140-140^{v}.

P1 is the Zīj-i Ashrafī by Muḥ. b. abī 'Abdallāh Sanjar al-Kāmilī, Sayf-i Munajjim, which will be the principal source of the radices, the equations and apogees. P2 is a zīj whose title and author are unknown simply because of damage to the title page; in the catalogue of the Bibliothèque Nationale it is named as 'Dastūr al-Munajjimīn', but I do not see that name in the volume. The apogees and radices are given for the Era of Yazdajird. The accompanying paper by A. Tihon gives a full description of the Greek manuscript O1 (Tihon, 1989, 8-12), and the parameters which we need here are taken from the portions edited by her. O1 is the principal source of the rates of mean motion.

2. *The parameters of the Zīj of Ibn al-A'lam*

2.1 *Rates of mean motions*

The rates of motion that are given in P1 were the basis of Kennedy's list of daily motions according to the Zīj of Ibn al-A'lam. These were derived from the motions for 20 years (7305 days) listed on f. 234^{v}, and it is helpful to work from these figures, rather than the daily motions, in order to have the best control of the precision when making the comparison with the motions given by O1. From P1, then, we obtain the following table; the number of revolutions has been included, even though it is not given in the text.

Table of 20-year motions according to the Zīj-i Ashrafī

Sun	20r 0, 0; 9, 8
Moon	267r 4,13;33,40*
Double elongation	494r 8,26;49, 4
Moon: argument	265r 1, 9;41,47
Node	1r 0,26;55,59
Venus: argument	12r 6, 3;40,29
Mercury: argument	63r 0,14;26,11
Mars: mean	10r 7,18;20,51
argument	9r 4,11;48,17
Jupiter: mean	1r 8, 7;16,29
argument	18r 3,22;52,39
Saturn: mean	8, 4;41,59
argument	19r 3,25;27, 9
Apogees	0;17,10

* MS: 4,13;35,10

For each of the outer planets the mean and argument add up to equal 0; 9, 8 as required for consistency. The correction to the lunar motion is determined by consistency with the double elongation, and the decision as to which is to be emended – the mean or the elongation – is settled by comparison with the lunar motion in O1. The daily motion of the Moon given by Kennedy was taken from the uncorrected value.

In O1 the motions of all the longitudes are given for both 365 and 366 days. These are listed below, with remarks about the consistency of all three values. The motions in 365 days are calculated from the Zīj-i Ashrafī, working from the above table. These motions are not fully precise in the 3rds, whereas the figures from the Greek text O1, which go as far as the third, are here to be regarded as authoritative, although admittedly we do not know how they were obtained by the Greek author.

Table of motions from the Greek source and the Zīj-i Ashrafī

Greek version	Zīj-i Ashrafī	Remarks
Sun		
359;45,40,21*	359;45,40,19	full consistency
360;44,48,41		
Moon		
13r 129;23, 2,17	129;23, 2,15	full consistency
13r 142;33,37,18		
Moon: argument		
13r 88;43,–,–	88;43, 6,52	Ga=88;43, 3, 7 for
13r 101;46,57, 3		consistency with Gb.
		A larger by 4″.
Node		
19;19,43, 3	19;20, 0,17	Ga and Gb consistent.
19;22,53,42		A larger by 17″.
Mercury: argument		
3r 53;56,42,32	53;56,43, 0	full consistency
3r 57; 3, 6,39		
Venus: argument		
225;10,46,38	225; 1,46,35	Ga=225; 1,46,38 for
225;38,46, 7		full consistency
Mars: mean		
191;–,–,51	191;17,10,53	Ga=191;17,10,51 for
191;48,37,31		full consistency
Jupiter: mean		
30;20,34,40	30;20,34,38	Gb=30;25,33,56 for
30;25,33,53		full consistency
Saturn: mean		
12;13,35,48	12;13,35,48	Gb=12;15,36,24 for
12;15,16,24		full consistency
Apogees		
0;0,51,26	0;51,27,53	Ga and Gb consistent.
0;0,51,34 (or 33)		A differs only in 3rds

*Expressed as 360−0;14,20+0;0,0,21

In each case the motion in 366 days is set below that for 365. The two values from O1, and that of the Zīj-i Ashrafī are referred to as Ga, Gb, and A.

The only notable disagreement between the two sources is in the case of the lunar argument, a difference amounting to 4 seconds p.a.

In the case of the Sun, Ibn Yūnis (Caussin, 1804, 158) attributed to Ibn al-A'lam the motion 359;45,40,20 in 365 days.

In his recent edition of the anonymous twelfth-century Byzantine text "Methods of Computing Various Astronomical Hypotheses", Jones (1987, 15-16) has considered the comparison between the parameters of the zīj, as determined by Kennedy, and those available from Mogenet's text, and concluded that "of what we know from Ibn al-A'lam's lost zīj, only the solar mean motion and the rate of precession agree with the cited parameters of Alim". The discussion so far is a complete refutation of that conclusion, and it was not sound even when only the text given by Mogenet was known.

2.2 *Radices*

The radices of the mean motion are not given in the Greek sources and we must rely on the Zīj-i Ashrafī. In that work the radices are given on ff. 231ᵛ-232 for A.Y. 672, Khordād 13, for the meridian of Shīrāz (224 years complete in

Table of radices according to Zīj-i Ashrafī, A.Y. 672
Khordād 13 for the meridian of Shīrāz

Sun		357;58,41
Moon	−8;0	281; 9,44
Double elongation		222;22, 6*
Moon: argument	−14;0	282;49,17
Node		−78;20, 8
Mercury: mean	−30;0	327;58,41
argument	+4;0	247;41,19
Venus: mean	−50;0	307;58,41
argument	+2;0	321;37, 5
Mars: mean	−59;0	122;54, 1
argument	+12;0	188; 4,40
Jupiter: mean	−18;0	100;45,42
argument	+6;0	245;12,57
Saturn: mean	−13;0	150;47,41
argument	+6;0	200;11, 0

*MS: 222;26,58

the Jalālī Era, A.D. 1303 Mar. 13, J.D. 2197050). The radices from this work are listed in the above table, together with the displacements, which will presently be discussed.

The radices are also given in the zīj in P2, f. 2v, from which they were quoted by Kennedy (1977, 18). After a fresh detailed examination of this source, I have to reject Kennedy's readings, and also to reject this source as providing any real assistance in the determination of the radices. This will be explained below.

The displacements given in the table can only be determined by examining the tables of equations in the Zīj-i Ashrafī. For this purpose, we must observe how the Zīj-i Ashrafī is constructed. The author has given in his Book 3 all the equations for his own Zīj, ff. 60v-61, 62v-92, and then in the tables of Book 8, ff. 231v-251, he has indicated specifically those equations which differ from those of his own work, for each of twelve zījes. For the Zīj of Ibn al-A'lam he has given the equations of the Sun, both equations of Mercury, and the first equations of Jupiter and Saturn. We find the displacements then by considering all the equations of the Zīj-i Ashrafī, and then those specially for Ibn al-A'lam. They are listed in the following table. Here q(x) and p(x) are the usual first and second equations, supposed to be positive in the first quadrant, and without displacements. The displacements and various amplitudes, that is the maximum values of p and q can then be read from the table.

Table of equations showing displacements and amplitudes

		first	amplitude	second	amplitude
Zīj-i Ashrafī					
f. 60v	Sun	q(x)	1;59,10		
62v	Moon	14+q(x)	13; 8,30	8+p(x)	5; 1
87v	Mercury	4−q(x+30)	3; 2	26+p(x)	22; 1
82v	Venus	2−q(x+50)	1;59	48+p(x)	45;59
77v	Mars	12−q(x+59)	11;25,30	47+p(x)	41; 9
72v	Jupiter	6−q(x+18)	5;15	12+p(x)	11; 3
67v	Saturn	7−q(x+14)	6;30	7+p(x)	6;13
Ibn al-A'lam					
236	Sun	q(x)	2; 0,10		
237rv	Mercury	q(x)	3;40	p(x)	22;22
235	Jupiter	6−q(x+18)	5;32		
234v	Saturn	6−q(x+13)	5;48		

Evidently the only specific changes in the displacements are for Mercury and Saturn, because the changes in the Sun and Jupiter concern only the amplitudes of the equations. In applying this conclusion, however, there is a difficulty with Mercury, for the radices clearly demand the displacement 30, the difference between mean Mercury 10,27;58,41 and mean Sun 11,27;58,41. We must suppose that some error has corrupted the presentation here, and therefore we keep to the displacements of the equation of Mercury given for the Zīj-i Ashrafī, but adopt the amplitudes 3;40 and 22;22 given for Ibn al-A'lam.

Kennedy (1977) gave the details of the displaced equations in the case of Mars, where he compared the displaced equations of the Zīj-i Ashrafī, f. 77v-79, with non-displaced equations for that planet taken from the supplementary tables in Book 8, viz. that tabulated for the Zīj al-Kabīr of Muhya al-Dīn al-Maghribī, f. 247-8. Note that Kennedy gave the parameters of the displaced equation of Mars as 42; 9,48 and 60 instead of 41; 9,47 and 59.

Table of radices from the Zīj-i Ashrafī and P2

		P2 1952063	Ashrafī 2197050	from Ashrafī for 1952063
1.	Sun	87;17,53	357;58,41	87;14,26
2.	Moon	4;59,26	289; 9,44	3;49,11
3.	Arg	307;17,45	296;49,17	306;40,44
4.	Node	64; 7,16/56*	−78;20, 8	−65; 0,55
5.	Mercury (arg)	169;23,34	243;41,19	182;48,43
6.	Venus (arg)	122;11,57	319;37, 5	120;21,12
7.	Mars	312;56,47	181;54, 1	311; 1,41
8.	Jupiter	271;33, 2	118;45,42	272;38,53
9.	Saturn	239; 2,24	163;47,41	237;19,37

* This is an ambiguity between yā and nūn which cannot be removed

In P2, f. 2v, the author gives for A.Y.0 (A.D. 632 Jun. 14, J.D. 1952063) the radices from a number of sources, reduced to his own meridian, which may be inferred to be at 85;30 on the Ptolemaic scale, together with some of the values of the radices for the original meridians. The sources are named as Yaḥya Ibn Manṣur, Ḥabash, al-Khwārizmī, al-Battānī, Ibn al-A'lam and Abū Raiḥān. The writing is crabbed and scarcely legible in places, and conditions are made worse by a spillage of ink which has run the length of the page, by a portion of paper torn away in the centre, and finally by the loss of text at the bottom where the paper was cut short. In spite of this one can see enough of the author's figures to appreciate that while he often agrees exactly with the radices which we can compute from known sources, he also makes frequent mistakes. For example, in his reduction to his own meridian he has erred in the sign of the correction for both al-Khwārizmī and al-Battānī. It is to be inferred from his remarks on Ibn al-A'lam that the original meridian was at 80°, certainly meant to be Baghdād, stated to be 0;22^{h} (5;30) from that of the zīj of P2. These values are given in the table above, but in the discussion that follows, they will be rejected as unreliable.

Comment:

The method of analysing astronomical mean parameters by means of deviation curves was originated by Billard (1971), and further developed by me (Mercier, 1987). It is particularly useful here where we need to sort out possibly corrupted values of the radices, and also to confirm the displacements already derived.

In the following comment the notation is the following:

A,B radices as in the Ashrafī Zīj, and P2, respectively;
d altered by the addition or subtraction of a displacement;
y reduced to the Era of Yazdajird;
* emendation in the reading.

The comments are listed for the numbered rows of the last table.

1. Both A and B are acceptable in terms of the magnitude of the deviations, but they cannot both be correct, because they represent a separation of meridians of 21°, much too large for the interval between Shīrāz and Baghdād.

2. The deviation curves show that Ad is accurate, when the meridian of reference is taken to be Shīrāz (52;34 East of Greenwich), so confirming the displacement 8;0. B is too large to be accepted, and no plausible emendation suggests itself. The double elongation given in the Ashrafī Zīj is not quite consistent with the mean, nor would consistency be achieved by another choice of displacement, such as 7;40. The possibility that the mean, not the double elongation, should be emended, cannot be absolutely excluded.

3. The value of A is acceptable, confirming the displacement 14;0, but B is too large by about 0;20.

4. A is acceptable, while B is a full degree too small.

5. Wide experience shows that Mercury is always badly determined. In this case B gives a deviation which is clearly in line with that found in known 9th century sources. Nevertheless when we later examine two horoscopes which

appear on every other ground to be derived from the parameters of Ibn al-A'lam, we find that for Mercury A fits well, and B badly.

6.–9. In all cases A is acceptable in terms of deviations where B is not, as is evident from the complete set of deviations curves in figs 1, 2.

In figures 1 and 2 the deviation curves are shown for versions A and B; the displacements are, of course, included. The curves numbered 1 to 9 represent Sun, Moon, lunar apogee (Moon minus argument), lunar node, Mercury, Venus, Mars, Jupiter and Saturn; note that the mean longitudes of the inner planets are used, not their arguments.

While version A is in good agreement with the correct mean longitudes (except, as usual, for Mercury), version B is manifestly wrong, showing at a glance that the radices in P2 are to be rejected.

An inspection of the curves for version A suggests that agreement with observations is best during the ninth century. This is confirmed by a least-squares analysis of the deviation curves, a method which determines the optimum year and meridian; for details see Mercier (1987). In the following list the margins attached to the optimum year and meridian represent the limits for 68% probability:

statistic	year	meridian
0 1111 00000	797 ± 144	55;38 ± 6;56
1 1111 00000	775 ± 64	54;40 ± 2;51
1 1111 01100	831 ± 78	54;56 ± 4;11
1 1101 01000	966 ± 175	55;49 ± 3; 2
1 1101 01100	1094 ± 135	57;15 ± 3; 8
1 1100 01111	876 ± 75	55;19 ± 7;46
1 1100 00011	867 ± 121	55;10 ± 11;39

In the notation of the statistic, an initial 1 indicates that the deviations are compared with the zero level, while an initial 0 indicates that the deviations are compared only with the mean deviation. In the rest of the symbol the 0 or 1 indicates the choice of deviation, from Sun to Saturn.

The year is fixed only within broad limits, indicative of the fact that the bundle of deviation curves is near to zero over a long period. Nevertheless, the curves generally point to observations is the early or mid-ninth century. We know that Ibn al-A'lam was active in the late 10th century (d. A.D. 985), and so the indication is that he formed his tables by a review and consolidation of the earlier observations, rather than by means of fresh observations. This is not to deny that he made observations, at least to the limited extent noticed by Ibn Yūnis.

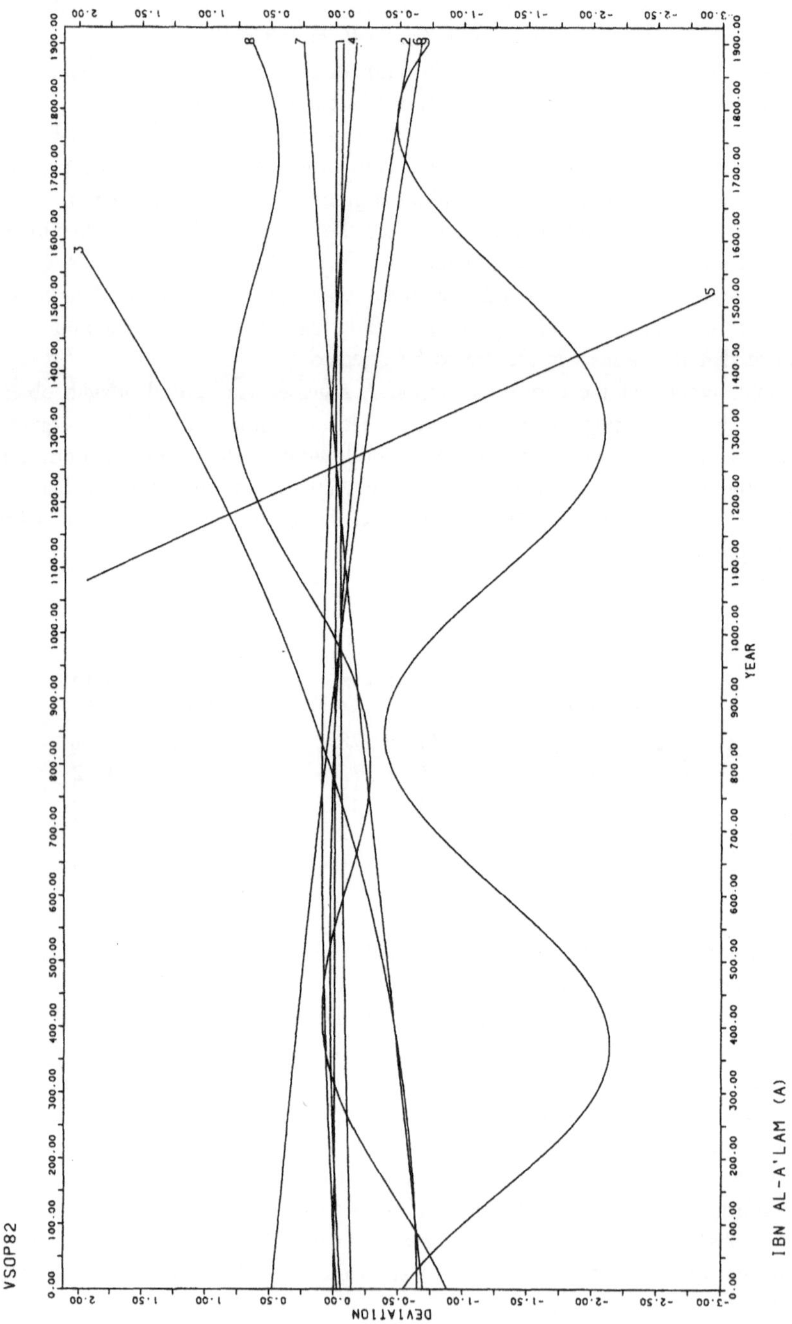

Fig. 1. – Deviation curves, version A

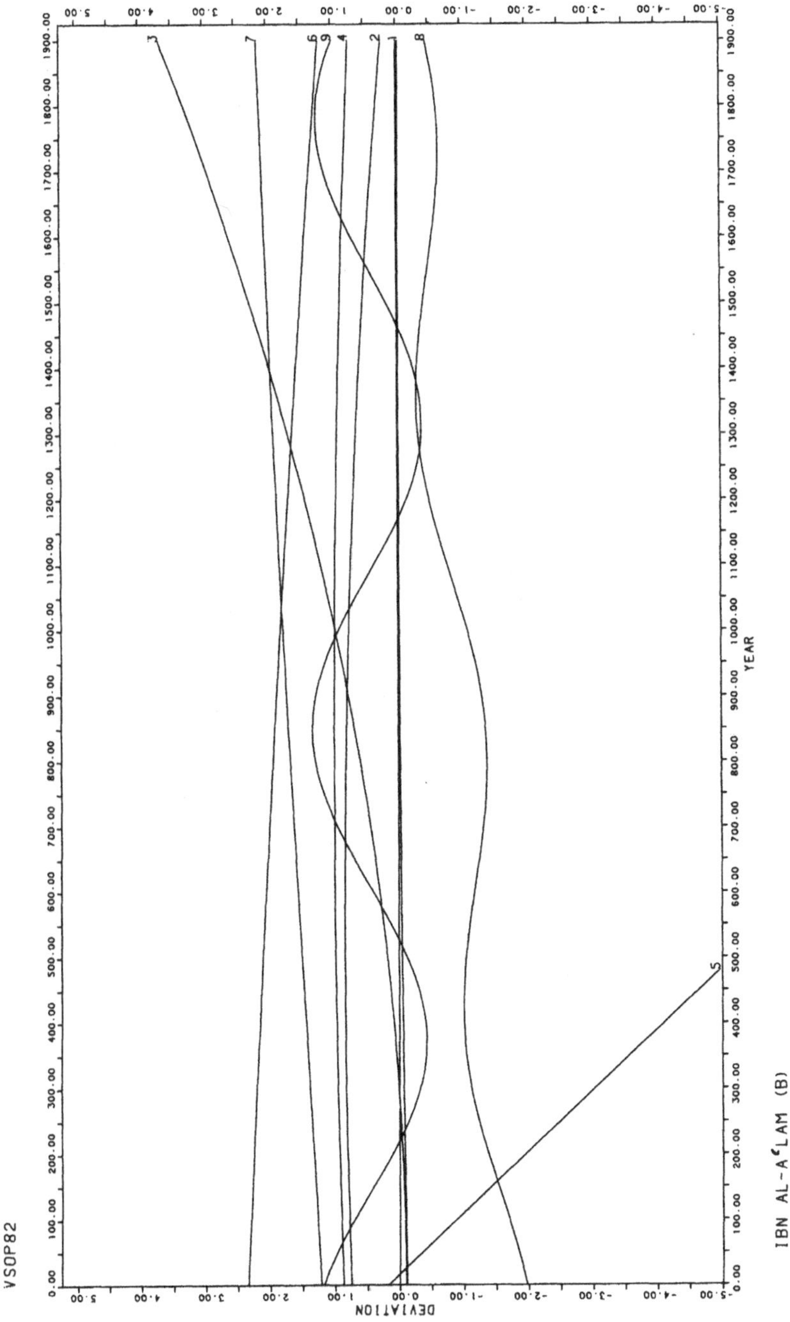

Fig. 2. – Deviation curves, version B

Since the year is not well determined, it is helpful to consider instead a statistical procedure in which the year is set in advance, while only the meridian is varied in the search for the minimum. The results are displayed in the form of a set of curves showing the mean square deviation as a function of meridian, for a given statistic and year; see fig. 3. In all the cases, the optimum meridian is near to 55, in a way that is not sensitively dependent on the year chosen. For the years 830 and 970, for example, we have

	A.D. 830	A.D. 970
1 1100 00000	55; 0 ± 2;25	54;47 ± 2; 2
1 1111 00000	54;47 ± 2;40	55;47 ± 6;55
1 1101 01000	54;47 ± 2;19	55;47 ± 1;54
1 1100 01111	55; 0 ± 5;58	56; 0 ± 8;54

This result shows that the version A is best suited to a meridian in the neighbourhood of the Greenwich longitude 55, confirming, surely, that the version A given in the Zīj-i Ashrafī was reduced to the meridian through Shīrāz (52;34 East of Greenwich) as used in Zīj-i Ashrafī itself. This is the place to remember that a difficulty arises here in connection with the equation of time, which disturbs the identification of the prime meridian (Mercier, 1985). I do not know the convention adopted in the Zīj-i Ashrafī, but it is certainly possible that the difference between the Greenwich longitude of Shīrāz and the position determined from the deviation curves is simply the result of the way in which the equation of time is defined.

We are not in a position to prove that the prime meridian originally employed in the Zīj of Ibn al-A'lam was Baghdād, although that is a most reasonable assumption, since his Zīj is referred to in the Zīj-i Ashrafī (f. 232), as Zīj-i Baghdādī.

From the foregoing analysis we can be sure that version A essentially preserves the original parameters of the Zīj of Ibn al-A'lam (apart from the change of era and prime meridian) even though it is taken from the early 14th century Zīj-i Ashrafī.

2.3 *Summary of mean parameters*

In the light of the above considerations we have the following set of mean parameters, in which all the displacements have been incorporated.

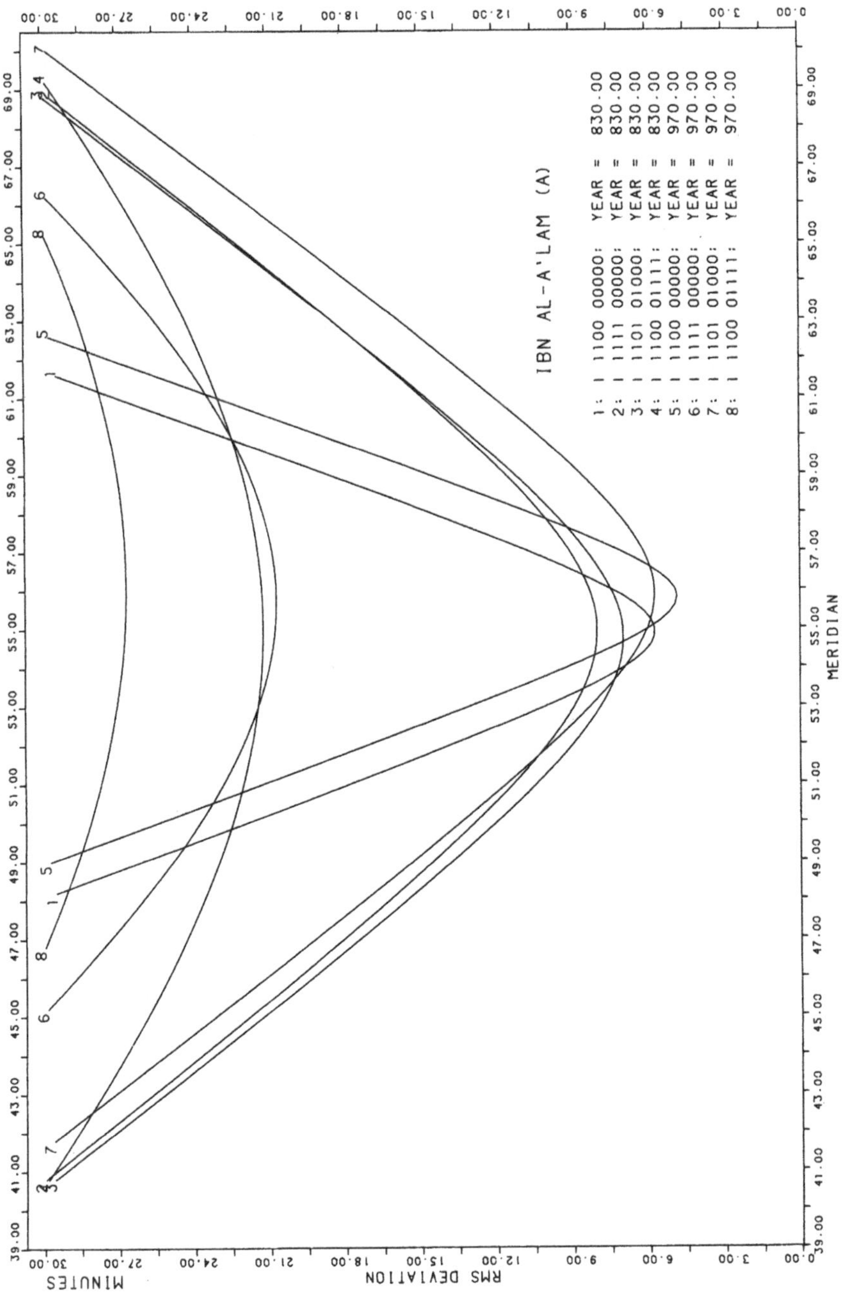

Fig. 3. — Variation of deviations with meridian

Table of mean motions and radices				
		motion in 365 d	radix(B)	radix(A)
date			1952063	2197050
meridian (from Greenwich)			45;0	55;0
Sun		359;45,40,21	87;17,53	357;58,41
Moon	13r	129;23, 2,17	4;59,26	289; 9,44
Arg	13r	88;43, 3, 7	307;17,45	296;49,17
Node		−19;19,43, 3	64; 7,16/56	−78;20, 8
Merc (arg)	3r	53;56,42,32	169;23,34	243;41,19
Venus (arg)		225; 1,46,38	122;11,57	319;37, 5
Mars		191;17,10,51	312;56,47	181;54, 1
Jupiter		30;20,34,40	271;33, 2	118;45,42
Saturn		12;13,35,48	239; 2,24	163;47,41

2.4 *Equations*

The amplitudes of the various equations have been given in section 2.2. The apogees are found also in the Zīj-i Ashrafī, f. 232^v, as well as in P2, and we list below the values quoted by Kennedy from the latter.

The rate of motion of the apogees is 1;0 in 70×365 days, expressed in the Greek source as 0; 0,51,26 per 365 days; more exactly, 0; 0,51,25,42,51, ... The apogees given by the Zīj-i Ashrafī end with ; 5,19, ;55,19 or, ;57,19. The occurrence of "5,18" for Mars is likely to be a scribal error. In the interval since the Era of Yazdajird, 244987 days, the precession is

$$244987/(365\times 70)=9;35,19,$$

so we can safely assume that the apogees have been calculated from values given originally for the Yazdajird epoch. The apogees listed in P2, as well as those listed in the Zīj-i Ashrafī for both eras, are given in the next table. The disagreement of the apogees from P2 is evidence of confusion which we have already seen in the calculations of that anonymous author. For the values ending in ..;.., 51 are certainly to be explained by the assumption that he calculated these for an epoch 20 years prior to the Era of Yazdajird. From that date (A.D. 612 Jun. 14, J.D. 1944758) to A.Y.0 (A.D. 632 Jun. 14) the precession is 0;17, 9 which, added to the apogees of P2, gives a series of values either equal to or differing by an integral number of minutes from those of the Zīj-i Ashrafī. Of course we accept the latter, rather than the very confused text P2.

Table of apogees			
	Zīj-i Ashrafī		P2
date	2197050	1952063	
Sun	89; 5,19	79;30	79;30
Mercury	226;55,19	217;20	217; 2,51
Venus	75;55,19	66;20	66; 4,51
Mars	136; 5,18	126;30	126;15,51
Jupiter	180;57,19	171;22	171; 4,51
Saturn	256;55,19	247;20	247; 5,51

The table on p. 36 gives the amplitudes of all the equations, together with the underlying parameters. In the Ptolemaic model the eccentricity and the radius of the epicycle, denoted M, and S, are calculated from the following formulae when the amplitudes are known, and these parameters are listed also.

First equation

Sun, Moon and the planets: $S = \sin(\max)$.

Second equation

Moon: $N = M/(1-M)$

$\tan(\max) = N/\sqrt{((1-N^2)(1-4N^2))}$,

planets: $M = 2\tan(\max/2)$.

3. *The Greek form of the tables of Alim*

The text edited by Mogenet, from *Vat. gr.* 1594, f. 2 seq. (Mogenet, 1962, 210-217; Tihon, 1989, 5-7) gives us some idea of the Greek version of the tables of Ibn al-A'lam. The essential features are picked out by A. Tihon (1989), and are listed here.

1. The starting point of the table was in the year '6540, end of February', when the mean Sun was 344;55,48,28. One can show that '6540, end of February' must in this case mean A.D. 1032 Mar. 1 (J.D. 2098056), for then the mean Sun according to version A was 344;49,28, 6.

2. Successive years, 6541 to 6544, are considered, for which the mean Sun is computed at intervals of 365, 365, 365, 366 days, in each case therefore for Mar. 1, in the years 1033-1036. Of these years, 1032, 1036 are leap years, as in

Table of parameters of equations			
Sun		max=	2; 0,10
		S=	0; 2, 5,48,45,25
Moon	first	max=	5; 1
		S=	0; 5,14,48,14
	second	max=	13; 8,30
		M=	0;10,19,13, 2
Mercury	first	max=	22;22
		S=	0;22,49,55, 0
	second	max=	3;40
		M=	0; 3,50,27,44
Venus	first	max=	45;59
		S=	0;43, 8,53,45
	second	max=	1;59
		M=	0; 2, 4,37,44
Mars	first	max=	41; 9
		S=	0;39,28,55, 3
	second	max=	11;25,30
		M=	0;12, 0,14,31
Jupiter	first	max=	11; 3
		S=	0;11,29,59,46
	second	max=	5;32
		M=	0; 5,47,56,24
Saturn	first	max=	6;13
		S=	0; 6,29,50,19
	second	max=	5;48
		M=	0; 6, 4,44, 1

the Latin calendar. There is no reference to tables of the mean Sun in which the years are grouped together by 20, or 24, for example, as is the case in Latin astronomical tables. The motion of the Sun in 365 days is 359;45,40,21, and in 366 days 360;44,48,41, in agreement with the data given above from the other Greek source, O1.

3. There is an indication that radices were provided, at least for the apogees, for the year zero in the Byzantine calendar. For the Sun and Saturn, the only

two for which figures are supplied, the radices of the apogees are 351;44 and 159.

4. The motion of the radices up to the years 6338 and 6540 is given, viz. 90 and 93 degrees respectively. The anomaly of the Sun in 6540 is 260.

5. For "each year" since the Creation (Byzantine year zero) the apogee and the mean Sun are given – no doubt not to be understood too literally, as if it meant every single year!

Since the mean Sun on 1032 Mar. 1 differs from that of version A, we may compute the difference of longitude between the meridians of reference of the two versions,

$$(344;55,48,28-344;49,28,\ 6)\times360\times365/359;45,40,21=38;35,23.$$

There are good reasons to expect that the meridian of the Byzantine version is that of Constantinople. The difference in longitude between that place and Shīrāz would be between 25 and 30 degrees, according to various Islamic sources. From the Zīj al-Sanjarī, 75;30−49;50=25;40, from the Qānūn of al-Bīrūnī, 78;35−49;50=28;45. We might well suppose that the interval 38; 35 arose in error from a misunderstanding about the longitude scale, a recurrent mistake among both Islamic and Byzantine calculators, which gave rise to an error of 10;0.

The Byzantine calendar is regularly taken to begin on Sep. 1, and so its origin point would be A.D. −5508 Sep. 1 (J.D. −290496). In the Greek tables of Alim, however the year is taken to begin on Mar. 1, although the years are still reckoned from "the Creation of the World", that is with essentially the same year numbers as in the regular Byzantine calendar. We might assume that in this reckoning 1032 Mar. 1 begins the 6540th year in this numbering, whereas in the regular reckoning the year 6540 begins on 1031 Sep. 1, 181 days earlier. Thus the new zero point is −290496+181 = −290315, A.D. −5507 Mar. 1.

The apogees of the Sun and Saturn are supposed to have been given for that date, and according to version A, we would have

$$(-290315-2197050)/(365\times70)=-97;21,10$$

to be added to the radices 89; 5,19 (Sun), and 256;55,19 (Saturn), giving for the Sun 351;44, 9, and for Saturn 159;34, 9. These results in the text agree exactly for the Sun (351;44), but only roughly for Saturn (159), showing that the last figure was incomplete in the Greek text. Incomplete also are the intervals from the origin point to the years 6338 and 6540, 90 and 93 degrees. If it is a question of the initial points of those years, we have for 6338, 2024275 (A.D. 830 Mar.1), and for 6540, 2098056, so that for those years, the motions of the apogee from the time of Creation are

$$2314590/(365\times70)=90;35,26$$
$$2388371/(365\times70)=93;28,42.$$

At the beginning of 6540 the Sun's apogee is 85;12,50,56, so that the Sun's anomaly is 259;42,57,44. The given figure of 260 is evidently rounded off.

4. *Analysis of two thematia*

This section contains analysis of the two thematia, or horoscopes, for the years 1153 and 1162, which have been presented by Tihon (1989, 12-20). After a theoretical account of the construction of the houses in a horoscope, and their dependence on the geographical latitude, the details of the horoscopes will be taken up.

4.1 *Theoretical analysis of the houses*

The diagram of the horoscope is divided into 12 houses surrounding a central square. The analysis of the construction of the houses given here is derived from the account by North (1986, 3 seq.), and follows his notation. Each house is assigned to a point of the ecliptic. These points λ_1 ... λ_{12} are the 'cusps' of the houses, that is, the longitudes at which the houses begin. λ_1 and λ_7 are the intersections of the ecliptic with the horizon in the East and West respectively, and λ_4, λ_{10} are the intersections of the ecliptic with the meridian in the North and South respectively.

The remaining cusps are determined according to one of several schemes. In the ordinary scheme one determines initially the right ascensions corresponding to the four cusps defined above, which we denote α_1, ... , α_{10}, as in fig. 4. Then let P denote the equal intervals

$$\alpha_4-\alpha_1, \quad \alpha_{10}-\alpha_7,$$

and let Q denote the equal intervals

$$\alpha_7-\alpha_4, \quad \alpha_1-\alpha_{10}.$$

The trisection of each of these intervals of magnitude P and Q determines the remaining right ascensions α_2, ... α_{12}, while the remaining cusps are the corresponding longitudes in the ecliptic. It is evident that houses are diametrically opposed, that is $\lambda_i = \lambda_{i+6} + 180$.

The Right Ascension α_i is expressed as α (λ_i), where

$$\tan\alpha(\lambda)=\tan\lambda\cos\varepsilon \qquad (1)$$

ε being the obliquity.

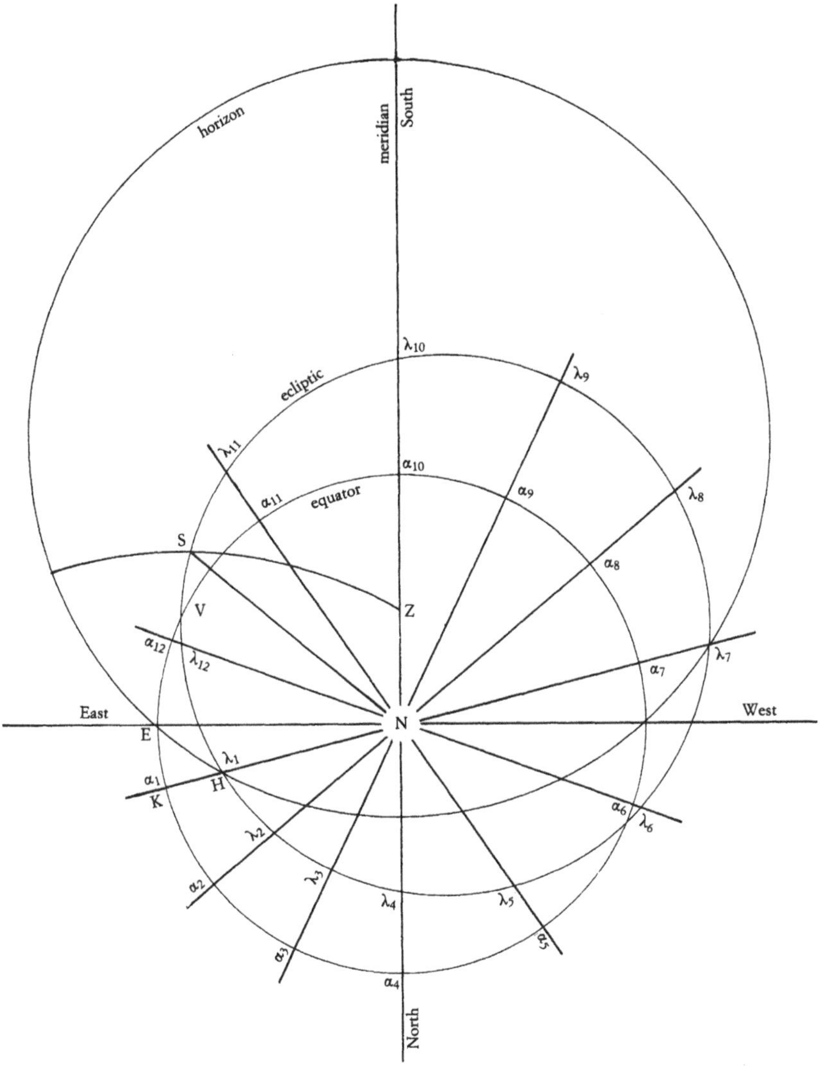

Fig. 4. – Construction of Houses

Since the horizon is involved in the definition of λ_1 and λ_7, the latitude ϕ may be related to the relative positions of λ_1 and λ_4, or α_1 and α_4. In the triangle EKH, the angle KEH is $90-\phi$, because the altitude of the pole N above the horizon EH is ϕ. The side KH is δ_1, the declination of H, while EK $= 90 - (\alpha_4 - \alpha_1)$. Since

$$\tan KH/\sin EK = \tan KEH,$$

we have

$$\tan \phi = \cos(\alpha_4 - \alpha_1)/\tan \delta_1. \qquad (2)$$

Moreover, in the triangle VKH,

$$\tan KVH = \tan KH/\tan VK,$$

that is,

$$\tan \delta_1 = \tan \varepsilon \sin \alpha_1,$$

$$\tan \phi = \frac{\cos(\alpha_4 - \alpha_1)}{\tan \varepsilon \sin \alpha_1} \qquad (3)$$

From a knowledge of λ_1 and λ_4, as well as ε, we may determine the latitude ϕ.

If λ is the longitude of the point of the ecliptic where it meets the horizon, such as the point H in fig. 6, then the portion of the equator similarly cut off is referred to as the "oblique ascension", such as VE in fig. 1, denoted $a(\lambda)$. Then

$$\begin{aligned} a &= VK - EK \\ &= \alpha - \arcsin(\tan \delta \tan \phi) \\ a(\lambda) &= \alpha - \arcsin(\sin \alpha \tan \varepsilon \tan \phi) \end{aligned} \qquad (4)$$

α being determined from λ according to (1).

In one of the thematia to be considered the 'horoscope' is given, that is the oblique ascension of the first cusp $a(\lambda_1)$. One should observe that this, together with the position of the cusp, determines a relation between the latitude and the obliquity, since in (4) we can express α in terms of ε and λ_1. This provides a check on the consistency of the calculations, since the positions of the cusps 1 and 4 also determine the relation between the latitude and the obliquity.

If the Sun is at that point of the ecliptic where its longitude is λ, then the duration of daylight in terms of uniform hours is 12D, where

$$\begin{aligned} D &= (180 + 2EK)/180 \\ &= 1 + \arcsin(\sin \alpha \tan \varepsilon \tan \phi)/90. \end{aligned}$$

On a day when the Sun's longitude at sunrise was λ, let the point of the ecliptic at some time during that day be on the south meridian at λ_{10}. Then the time, expressed in uniform hours relative to Noon, is

$$(\alpha(\lambda_{10}) - \alpha(\lambda))/15.$$

The period of daylight was also divided into 12 'seasonal' or non-uniform hours between sunrise and sunset. It may be shown that, at that moment, the time of day in terms of seasonal hours, relative to Sunrise, may be expressed either as

$$6+(\alpha(\lambda_{10})-\alpha(\lambda))/15D$$

or as

$$(a(\lambda_1)-a(\lambda))/15D.$$

In order to find the altitude of the Sun, we consider the triangle SZN, S being the position of the Sun, and Z being the zenith. The altitude of the Sun, ψ, is the complement of SZ, and besides,

$$NZ=90-\phi, \quad NS=90-\delta \quad \text{angle } ZNS=\alpha_{10}-\alpha.$$

From the cosine rule, which gives cos SZ in terms of the opposite sides and angle, we have

$$\sin\psi=\sin\phi\,\sin\delta+\cos\phi\,\cos\delta\,\cos(\alpha_{10}-\alpha).$$

A computer program is employed which calculates from cusps 1 and 4 the geographical latitude, and from cusps 1, 4, 7, 10, the remaining cusps. In this way we have a simple basis on which to test the consistency of the 12 cusps given in the horoscope. From cusp 1 and the position of the Sun we have the time of day, and finally from the latitude, the position of the Sun, and cusp 10, the altitude of the Sun. A value of the obliquity has to be assumed, and in the absence of any indication from the text, we can try the common values 23;30, 23;33 and 23;35, as well as the Ptolemaic value 23;51,20.

4.2 *The themation for 1153*

4.2.1 The houses and the latitude

The data according to the text is (without emendation) as follows:

1.	344;33 (ascendent)	7.	164;33
2.	19;53	8.	199;21
3.	51; 3	9.	231; 3
4.	82;59	10.	262;59
5.	107; 4	11.	287; 4
6.	164;33	12.	315;56

The position of the Sun,	323;50,31
The horoscope,	350;58
The time after sunrise,	0;50 hour
The altitude of the Sun,	7; 0

Cusp 6, and the other details in the 6th house, were repeated by dittography from the 7th house, although we may infer its correct position 135;56 from cusp 12. Cusps 2 and 8 are not diametrically opposed, and one has to choose apparently between two possible values of the number of minutes, 53 and 21.

When we calculate from these data, with the various values of the obliquity, we have the following results for latitude, time and altitude of the Sun.

obliquity	23;30	23;33	23;35	23;51,20
latitude	47; 0	46;55	46;51	46;21
time (unequal hours)	0;53	0;53	0;54	0;54
time (uniform hours)	−4;15	−4;15	−4;15	−4;16
altitude of Sun	6;55	6;56	6;56	7; 1

These results for the time and the Sun's altitude agree well with those of the text, but the latitude is surprisingly large.

As remarked in section 4.1, since we are given the horoscope, 350;58, we have another relation between the latitude and the obliquity, which depends on that number and the first cusp 344;33. The relation implies the following pairs of values,

obliquity	23;30	23;33	23;35	23;51,20
latitude	40;15	40;10	40; 7	39;39

Whatever value one takes for the obliquity, however, there is clearly no reconciliation between this determination of the latitude and the preceeding one.

If one tried to emend the cusps, particularly cusp 4, changing it from 82;59 to say 81;59, the latitude would be reduced to about 41;45 (when the obliquity is 23;33), which is more 'reasonable', but then the altitude of the Sun would be 8;40, and the time would be 0;58, spoiling the good agreement obtained without any emendation to the text.

The discussion of the radices of version A and of the Greek version strongly suggests that Costantinople was intended as the meridian of the Greek version. In an anonymous Greek text of the 11th century, which has quite recently been edited, *Methods of Computing Various Astronomical Hypotheses*, the latitude of Costantinople was assumed to be 41 (Jones, 1987, 36-37). This is essentially the correct value, since the modern figure is 41; 2, but it is surprising nevertheless, because in Ptolemy's *Geography* and all the Arabic sources the values of the latitude were considerably larger, viz., 43;12 from Ptolemy (*Geography* III,11,5), 43;10 from al-Battānī, and 45;0 from the many sources which followed al-Khwārizmī. One cannot escape the conclusion that some respectable local observations had provided this accurate value, one which never found its way into later Arabic works. Another clear example of the use of this correct latitude is found for the Brescia astrolabe, Dalton (1926, 139), a Greek instrument dated firmly to A.D. 1062, and carrying three plates, for Rhodes (36;0), Byzantium (41;0) and Hellespont (40;0). The last figure, like that for Byzantium, does not follow Ptolemy, who puts the cities of Hellespont in the range 41;0 to 41;12. The latitude determined from the horoscope and the first cusp, 40;0, is rather smaller than the Greek value, while the value 47;0 determined from the cusps is larger even than that from Arabic sources. When we come to the second themation, section 4.3.1, we will find however, that the cusps there fix the latitude as 41.

The remaining houses are shown in the following complete list.

	Text	Theory		Text	Theory
1.	343;33	343;33	7.	164;33	164;33
2.	19;53	19;29.2	8.	199;21	199;29.2
3.	51; 3	52;35.5	9.	231; 3	232;35.5
4.	82;59	82;59	10.	262;59	261;59
5.	107; 4	108;36.2	11.	284; 3	288;36.2
6.	135;56	135;29.2	12.	315;56	315;29.2

The agreement between text and theory is certainly not very good, so that, in particular, one cannot decide whether 53 or 21 is the correct number of minutes for cusps 2 and 8.

4.2.2 Positions of the Sun, Moon, and planets

According to the text which appears to accompany this themation, it was made according to a summary (περίληψις) of the *Canon* of Alim. By way of testing the possible sources, calculations of all the true longitudes have been made according to the reconstructed parameters of Alim (Ibn al-A'lam), and a number of other possible sources. We have seen that version B of Alim is not satisfactorily established, but it is included here nevertheless. The sources which have been tried are these:

A: Almagest	B: Zīj al-Ḥākimī	C: Zīj-i 'Alā' ī
D: Zīj al-Sanjarī	E: version B	F: version A

The *Almagest* was of course available and certainly used in this period. D was composed in the mid-twelfth, and C in the late twelfth century, although neither is known in Greek before the late thirteenth century. B is not known in Greek. In each case one has to assume a shift of meridian from that of the source. The shift will be chosen on the assumption that the calculations were intended for Constantinople, Greenwich longitude 28;57. In considering the sources, of course, we need to consider the medieval values of the longitude of the reference meridians. For the *Almagest*, Ptolemy's longitude 56;0 for Constantinople will be used, but for the others we use 59;50, or 49;50 the commonly found Islamic value on the Ptolemaic scale, or Khwarizmian scales respectively. Accordingly the differences from Constantinople are as follows:

A: Alexandria	60;30 – 56; 0 = 4;30
B: Fūsṭāṭ	63; 0 – 59;50 = 3;10
C: Shirwān	84; 0 – 59;50 = 24;10
D: Merv	86;30 – 49;50 = 36;40
E: Baghdād	80; 0 – 59;50 = 20;10
F: Shīrāz	38;30

The value 38;30 is taken in line with the earlier discussion about the conversion of the prime meridian in the Byzantine version of Alim.

The date is not fully specified in the text, being given only as 1153 February. The day may quickly be discovered, however, as the only one which suits all the given positions. The following table gives the true longitudes, then, for A.D. 1153 Feb. 5, –4;15 hours (from Noon), for the meridian of Constantinople.

	Themation	A	B	C	D	E	F
Sun	323;50,31	318;39.4	323;25.7	323;53.6	324; 6.2	323;27.2	323;26.7
Moon	-------	67; 3.0	72;27.0	73; 0.3	72;54.1	73;28.4	72;57.1
Node	-------	298;33.6	304;59.5	304;52.1	304;34.2	305;43.5	304;49.7
Mer	301;27,45	302;58.9	305;30.8	298; 1.5	308;15.5	307;37.0	300;32.4
Ven	324; 6,48	315;27.7	324;15.4	323;11.5	323;16.4	323;40.3	322;54.1
Mar	272;12	266;35.9	271;51.9	271;38.4	271;44.3	272;57.4	270;55.3
Jup	246;54,12	243; 7.0	247;15.6	246; 9.2	246;41.6	245;50.5	246;51.0
Sat	--------	129; 1.7	132;48.1	134;10.7	133;18.3	132;56.2	130;54.0

The positions are generally given clearly in the manuscript, but with two minor difficulties. The Sun appears twice, once as given above, and again in the 10th house, with the longitude 249;59 (or 240; 9,59), which I cannot interpret.

In the other case, Mars is placed at Capricorn 2 12, which could mean either 270; 2,12 or 272;12 (or even 272; 0,12). The calculations rather suggest that 272;12 was intended as the entry.

4.3 *The themation for 1162*

4.3.1 The houses and the latitude

According to the text, the houses and other details are as follows:

1.	62; 5 (ascendent)	7.	242; 8 (and 245;51)
2.	80; 6 (or 86)	8.	260; 6 (or 266)
3.	105;50	9.	285;50
4.	128;23	10.	308;23
5.	166; 5	11.	346; 5
6.	208; 7	12.	28; 7

The position of the Sun,	337;25
The horoscope,	–
The time after sunrise,	4th hour
The altitude of the Sun,	34; 0

The cusps 1 and 7 are not quite in diametral opposition. In the seventh house two positions are given, but one can rule out Sag 5;51 simply because the other position is much nearer to diametrical opposition. The need for other emendations as well is suggested by the following analysis. One also has to resolve the ambiguity regarding the 2nd and 8th cusps.

The formula for the geographical latitude at the end of section 4.1 gives us four different values when the two different values of the obliquity are combined with the two possible positions of cusps 1 and 7, cusp 7 being assumed to be opposite cusp 1

	cusp 1 = 62; 5			
obliquity	23;30	23;33	23;35	23;51,20
latitude	41; 5	41; 0	40;56	40;51
time (unequal hours)	3;56	3;56	3;56	3;57
time (uniform hours)	−1;53	−1;53	−1;53	−1;53
altitude of Sun	33;40	33;44	33;47	34; 7

	cusp 1 = 62; 8			
obliquity	23;30	23;33	23;35	23;51,20
latitude	41; 9	41; 3	41; 0	40;29
time (unequal hours)	3;56	3;56	3;56	3;56
time (uniform hours)	−1;53	−1;53	−1;53	−1;53
altitude of Sun	33;40	33;44	33;47	34; 7

It would appear that the latitude was taken to be 41, in agreement with that used by the author of the *Methods*, but that still leaves one free to choose between the pairs 62; 5, 23;33, and 62; 8, 23;35, and indeed, in view of the rather poor degree of precision which we have observed throughout the work, we cannot effectively rule out any of the other combinations. For the sake of argument, it will be assumed that 23;35 represents a reasonable choice, since the obliquity 23;35 was used in the *Methods* (Jones, 1987, 36-37). We also take the first cusp as 62; 5, although there is no argument against an emendation to 62; 8.

When the first cusp is 62; 5, the remaining cusps are as follows:

	Text	Theory		Text	Theory
1.	62; 5	62; 5	7.	62; 5	242; 5
2.	80; 6	84; 7.3	8.	260; 6	264; 7.3
3.	105;50	105;51.0	9.	285;50	285;51.0
4.	128;23	128;23	10.	308;23	308;23
5.	166; 5	166; 5.5	11.	346; 5	346; 5.5
6.	208; 7	205;28.6	12.	28; 7	25;28.6

These values for the remaining cusps suggest further emendations, viz.

2. 84; 6	8. 264; 6
6. 205; 7	12. 25; 7

Although the text allows an ambiguity in the positions of the 2nd and 8th, neither reading supports the required value.

The calculated time 3;56 agrees well with that given in the text, the 4th hour (from sunrise).

4.3.2 Positions of the Sun, Moon, and planets

The date would appear to be given in the right hand margin as "month February 2 day 6th Indiction 10th year 6670".

The year, A.D. 1162, and the month are clear, but the reference to the day is uncertain. However the given positions determine that the day is Monday 19th Feb., so that the '2' refers presumably to Monday, the second day of the week. The '6th' remains unexplained.

In considering the time used for the calculation, we might hesitate to use the figure 1;53 derived from the above analysis of the houses. Its apparent precision should not mislead us, and it is more than likely that the simpler figure of 2;0 uniform hours before Noon was used, and this is the time which will be used in the following calculations of the true longitudes.

The next table gives the true longitudes from the same sources A to F, as explained in section 4.2.2, for A.D. 1162 Feb. 19, 2 hours before Noon, for the meridian of Constantinople.

	Themation	A	B	C	D	E	F
Sun	337;25	332;26.0	337;22.3	337;50.2	338; 3.3	337;22.9	337;22.5
Moon	13;43	8; 4.1	13;31.1	14; 3.1	13;56.7	14;26.5	13;57.0
Node	-------	123;42.3	130;10.7	130; 4.1	129;46.1	130;54.9	130; 1.1
Mer	339;43	330;59.4	335;36.4	343;48.4	336;52.2	334;50.6	338;47.2
Ven	291;54	286;27.3	292;11.2	292;25.5	292;34.7	292;18.6	292; 2.3
Mar	-------	191; 1.3	197;39.4	196;15.2	195;26.8	199;56.9	195;10.7
Jup	158;23*	154;19.6	158;16.1	156;46.3	157;22.9	157; 2.2	158;16.3
Sat	245;51	241;21.8	245;27.3	246;32.1	246; 0.5	247; 1.5	245;28.6

* In the text the longitude of Jupiter is placed in the fourth house at "Vir 50;23", and this is emended as Vir 8;23, taking the ν as a misformed η

4.4 *The tables used for the thematia*

The decision as to which of the various sources provides the best fit to the data in the thematia can be decided by calculating the standard deviation

Standard deviations

Themation of:	1153	1162
Almagest	5.473	5.761
Zīj al-Ḥākimī	1.835	1.692
Zīj al-'Alā 'ī	1.646	1.842
Zīj al-Sanjarī	3.073	1.294
version B	2.826	2.146
version A	0.914	0.427

between the given and the calculated positions, that is, the square root of the average squared-difference between the positions. The smaller this standard deviation is, the better the fit.

From these values it is clear that the Zīj of Ibn al-A'lam, indeed in the version A, is the best fit, by a considerable margin.

Acknowledgement

The many helpful discussions with Dr Tihon are remembered with pleasure, and I am grateful also to E.S. Kennedy for a loan of the film of the MS of the Zīj-i Ashrafī.

Appendix

Mean Parameters of the Zījes Used in Section 4

The various sources are not discussed here. Except for Ptolemy, the data are drawn from various manuscripts. The daily motion is equal to that quoted in the 'motion' column divided by the interval given at the head of the table. The motion is given as so many revolutions and so many degrees. These various intervals are chosen in accordance with the original tables. Note that the lunar apogee is given, not the argument, and the mean longitudes of Mercury and Venus, not their arguments.

Ptolemy (*Almagest*)

epoch 1448638.0 interval 46656000000 days

	radix	motion	
Sun	330;45	127738332,	31;0
Moon	41;22	1707659135,	30;0
apogee	312;33	14437356,	151;0
node	317; 7	−6864754,	127;0
Mercury	352;40	530366528,	21;0
Venus	41;52	207637863,	359;0
Mars	3;31	67918138,	213;0
Jupiter	184;40	10772667,	271;0
Saturn	296;43	4340114,	291;0

Zīj al-Ḥākimī

	epoch 1948439.0	interval 10631 days
	radix	motion
Sun	115;24,12,56	29, 38;24,44,56, 9
Moon	113;28,44,15	389, 38;14,48,19,12
apogee	193;26,55,24	3, 104;24,50,17,36
node	−232;45,17,41	−1, 202;58,35,50,13
Mercury	187;14,52,19	120, 305;47,25,14,14
Venus	159;12,42,41	47, 112;56,10,59, 4
Mars	212;29,55,24	15, 171;24, 9,28, 1
Jupiter	331;51,48,55	2, 163;45,23,21, 3
Saturn	116;31,56,30	356; 6,51,13,21

Zīj-i ʻAlā ʼ ī

	epoch 2149163.0	interval 30 days
	radix	motion
Sun	318;49,41,19	29;34,10,17,43,30
Moon	31;50,29,46	1, 35;17,30,57,45,54
apogee	236;32,54, 1	3;20,32,48,49,54,30
node	−62;24,41,57	−1;35,18,47,44,39,30
Mercury	126;24,53,31	122;46,21,21,42
Venus	274;48,58,23	48; 3,54,39,13,18,49
Mars	285;40, 1,22	15;43,19,55,40,30
Jupiter	96;50, 2,20	2;29,37,49,50,30
Saturn	1;33,19	1; 0,18, 2,16,46,30

Zīj al-Sanjarī

	epoch 1948440.0	interval 1 day
	radix	motion
Sun	116; 0,25	0;59, 8,20,33,53, 4,29,40
Moon	145; 3,40	13;10,35, 2, 0,41,28,38,50
apogee	193;10, 5	0; 6,41, 5,48, 7,17,11,20
node	−233;52,34	−0; 3,10,37,47,51, 2,50,50
Mercury	192; 2,30	4; 5,32,27,43,32,36,15,30
Venus	161;53,20	1;36, 7,49,16,55,42, 8, 0
Mars	212; 8,31	0;31,26,39,36,34, 5,16,50
Jupiter	331; 7, 9	0; 4,59,16,19,53,47,11,20
Saturn	116;33,52	0; 2, 0,36, 4,43, 2, 8, 0

REFERENCES

Billard (1971) = R. Billard: *L'astronomie indienne*. Paris: Maisonneuve, EFEO, 1971.
Caussin (1804) = A.-P. Caussin de Perceval (ed.): "Le Livre de la grande table hakémite. Extraits du MS de Leyde", *Notices et extraits des MSS*, VII. Paris, 1804.
Dalton (1926) = O.M. Dalton: "The Byzantine Astrolabe at Brescia", *Proc. British Academy, 12* (1926), 133-146, 3 plates.
Jones (1987) = A. Jones: "An Eleventh-century Manual of Arabo-Byzantine Astronomy", *Corpus des Astronomes Byzantins*, III. Amsterdam: J.C. Gieben, 1987.
Kennedy (1977) = E.S. Kennedy: "The Astronomical Tables of Ibn al-A'lam", *J. for the History of Arabic Science, 1* (1977), 13-23.
Kennedy (1987) = E.S. and M.H. Kennedy: *Geographical Coordinates of Localities from Islamic Sources*. Frankfurt-am-Main: Institut für Geschichte der Arabisch-Islamischen Wissenschaften an der Johann Wolfgang Goethe-Universität, 1987.
Mercier (1985) = R. Mercier: "Meridians of Reference in Pre-Copernican Tables", *Vistas in Astronomy, 28* (1985), 23-27.
Mercier (1987) = R. Mercier: "The Meridians of Reference of Indian Astronomical Canons", in G. Swarup, A.K. Bag, K.S. Shukla (eds), *History of Oriental Astronomy*, IAU Colloquium 91 (Cambridge, 1987), 97-107.
Mogenet (1962) = J. Mogenet: "Une scolie inédite du *Vat. gr.* 1594 sur les rapports entre l'astronomie arabe et Byzance", *Osiris, 14* (1962), 198-221.
North (1986) = J.D. North: *Horoscopes and History*. London: The Warburg Institute, University of London, 1986.
Sezgin (1978) = F. Sezgin: *Geschichte des arabischen Schriftums*, Bd. VI. Leiden: Brill, 1978.
Tihon (1989) = A. Tihon: "Sur l'identité de l'astronome Alim", *Archives internationales d'histoire des sciences, 39* (1989), 3-21.

VII

ASTRONOMICAL TABLES IN THE TWELFTH CENTURY

Introduction

When the *Zīj*, or astronomical handbook, of al-Khwārizmī was translated by Adelard of Bath in the early twelfth century, the Latin world was totally unprepared for such a work. In its scientific and mathematical assumptions it far transcended anything found in the computus literature, and, as we shall see, a whole technical vocabulary had to be acquired and put to use. The process of assimilation and understanding was certainly very slow, and properly speaking one has to wait until Copernicus to find a European sufficiently equipped to meet as an equal the professional astronomers of old. In this short survey we confine our attention to the first response among those Latin scholars of the twelfth century, Adelard and others who followed in his wake, who began the accumulation of the translations of Arabic *zīj*es. Only tables of Sun, Moon and planets will be examined here, although lists of stars, corrected in various ways for precession, were also found in the new material, and are important for a full understanding of the tables.

This survey is divided into the following sections:

1.1 The *Zīj* of al-Khwārizmī

1.2 The mean motions

1.3 The equations of motion

1.4 Petrus Alfonsi

1.5 Robert of Chester

1.6 Relations among the sources of the Latin *Zīj*

1.7 Other versions of the *Zīj* of al-Khwārizmī

2 Walcher of Malvern

3 Abraham bar Ḥiyya (Savasorda)

4.1 The Toledan tables

4.2 Raymond of Marseilles and Roger of Hereford

5 Tables for Pisa and London

6 The *Zīj* of al-Battānī

7 Ptolemy's *Almagest*.

1.1. *The Zīj of al-Khwārizmī*

Adelard of Bath would appear to be the first Latin translator of an Arabic *zīj*, or astronomical handbook. Such handbooks were produced in some profusion throughout the Islamic culture area and were modelled on the *Handy Tables* of Ptolemy. That prototype served for Arabic *zīj*es even in the very early Abbasid period when parameters were drawn not from Ptolemy, but from Indian and Persian sources. Indeed, the word *zīj* is derived from the Persian *zīk*, '(guiding) thread', and so by extension, a 'ruled table', and there are pre-Islamic Persian astronomical tables which are referred to in this way, such as the Royal Tables, *Zīk-i Shahriyārān*.[1] The *Zīj* of al-Khwārizmī[2] was compiled during the Caliphate of al-Ma'mūn (A.D. 813–33) and represents a somewhat eclectic choice of Indian and Persian components. Apart from the style of the work, which, like every other *zīj*, follows that of the *Handy Tables*, there are only some slight indications of Ptolemaic parameters. The mean longitudes and other special features will be discussed in detail below, but here we give a brief sketch of the background to the work.

Apart from a number of calendrical tables in Maslama's version,[3] we have no copy of the Arabic text of the *Zīj* of al-Khwārizmī. This is the reason why Adelard's translation is so precious for the historian of medieval astronomy. Al-Khwārizmī's *Zīj* represented an early stage in an extraordinarily rapid development of Arabic astronomical science, and it was soon superseded by others depending on Ptolemaic theoretical models, rather than Indian, and in which the parameters of mean motion reflected fresh and accurate observations. As a result his work was soon forgotten in the Middle East. In the Islamic province of Andalusia, however, in that it was rather provincial and slowly developing in cultural matters, the early *Zīj* of al-Khwārizmī continued to be a matter of interest. The late tenth-century mathematician Abū'l-Qāsim Maslama b. Aḥmad al-Majrīṭī ('of Madrid')[4] reconstructed the work, recalculated the mean motions for the Hijra Era in place of the Yazdajird Era used by al-Khwārizmī, added various tables for the conversion of calendars, and various trigonometrical and eclipse tables, some now designed for the meridian of Cordova. The result was a work of a richly heterogeneous character, with layer upon layer of developments representing some thirteen hundred years of astronomical history,

[1] Remarks of the Pahlavi source of *zīk*, in the astronomical sense, viz., the *Epistles of Manuščihr*, are given by H. W. Bailey, *Zoroastrian Problems in the Ninth-Century Books*, second edition (Oxford, 1971), p. 80. The term *zīk* as 'guiding thread' is used in the *Škand-Gumānīk Vičār*, ed. ~~P.-J.~~ J.-P. de Menasce (Fribourg, 1945), pp. 52–3.

[2] It is convenient to refer to the article on al-Khwārizmī in F. Sezgin, *Geschichte des arabischen Schrifttums* (1978), VI, p. 140.

[3] These are found in MS Escorial, Real Biblioteca, arabe 908, fols 240v–243. They were printed by C. A. Nallino, *Al-Battani sive Albetenii Opus astronomicum*, Pubblicazioni del Reale Osservatorio di Brera in Milano, 40 (1899–1907), II, pp. 300–5, as *Tabulae spuriae*; al-Battānī's *Zīj* is found in the same manuscript. The tables are given in the edition of Adelard's version of al-Khwārizmī's tables as numbers 2, 2a, 2b, 3, 3a (H. Suter, *Die astronomischen Tafeln des Muḥammed ibn Mūsā al-Khwārizmī in der Bearbeitung des Maslama ibn Aḥmed al-Madjrīṭī und der latein. Übersetzung des Athelard von Bath auf Grund der Vorarbeiten von A. Bjørnbo und R. O. Besthorn*, Det Kongelige Danske Videnskabernes Selskab, Skrifter, 7 Række, hist.-filos. Afd. 3 (Copenhagen, 1914), pp. 110–14), but there is one Arabic table given by Nallino on p. 300, which is not included in the Latin versions. This gives the week-day index (*ʿalāma*) for Arabic years 0(1)219.

[4] Sezgin (n. 2 above), V, p. 334; id., VI, p. 226.

since we might suppose that the India planetary equations derived from pre-Ptolemaic Alexandrian methods. This is the work which, although lost in its Arabic original like so much of Andalusian science, is preserved for us in Adelard's version.

Al-Majrīṭī's work is known to us through various Arabic bibliographical sources, such as Ibn al-Qifṭī, and also through the impressive history of learning – *Classes of Peoples* – written by the contemporary *qāḍī*, Abū'l Qāsim b. Ṣāʿid al-Andalusī (1029–1079), who wrote of al-Majrīṭī:[5]

> Abū'l-Qāsim Maslama b. Aḥmad, known as al-Majrīṭī, was the leader of mathematics in al-Andalus in his time, and no one before him was more learned in astronomy. He was a capable observer of the stars and was filled with a desire to understand the book of Ptolemy known as the *Almagest*. He wrote an excellent book on the arithmetic of profits, which we know as *muʿāmalāt* (transactions) and a book in which he epitomized the equations in the *Zīj* of al-Battānī. He toiled over the *Zīj* of Muḥammad b. Mūsā al-Khwārizmī, in which he changed the calendar from the Persian to the Arabic, and established its mean motions for the beginning of the Hijra Era. He added to it excellent tables, although he followed the authority of it (the *Zīj*), and did not indicate the occasions of error in it. I have already pointed that out in a book composed on the correction of the motions of the stars and the exposition of the error of the observers. Abū'l-Qāsim Maslama b. Aḥmad died shortly before the outbreak of the period of affliction (*fitna*) in the year 398 (A.D. 1007–8). He produced distinguished students, such as no other expert in al-Andalus produced. Among the most famous are Ibn al-Samḥ, Ibn al-Ṣaffār, al-Zahrāwī, al-Kirmānī and Ibn Khaldūn.

He also wrote a revision (or gloss) of Ptolemy's *Planisphere*,[6] the theory of the plane astrolabe, a work preserved still in various Arabic manuscripts. This was translated by Hermann of Carinthia, whose preface will be cited in the following pages for precious evidence of Hermann's other translations and those of Robert of Chester.[7] His observations of the Sun and of stellar positions were cited by al-Zarqāla in his study of accession and recession.[8]

In the absence then of any Arabic text, our only acquaintance with al-Khwārizmī's *Zīj* is through the translation attributed to Adelard of Bath. This is known through nine manuscripts altogether,[9] although only two of these are both complete copies and reasonably close to Adelard's original. The others are either incomplete (or indeed mere fragments), or are attributed to Petrus Alfonsi or Robert of Chester.[10] The two primary manuscripts, both of the twelfth century,

[5] *Kitāb Ṭabaqāt al-Umam ou Les Catégories des Nations*, ed. L. Cheikho (Beirut, 1912), p. 69, trans. R. Blachère (Paris, 1935), p. 129.

[6] Sezgin (n. 2 above), v, p. 170, where the Arabic manuscripts are listed.

[7] C. H. Haskins, *Studies in the History of Mediaeval Science*, second edition (Cambridge, Mass., 1927), p. 47.

[8] J. M. Millás-Vallicrosa, *Estudios sobre Azarquiel* (Madrid-Granada, 1950), p. 310.

[9] *C* Chartres, Bibliothèque municipale, 214, fols 41–102; *O* Oxford, Bodleian Library, Auct. F.I.9, fols 99ᵛ–159ᵛ; *M* Paris, Bibliothèque Mazarine, 3642 (1258), fols 82–9ᵛ; *N* Madrid, Biblioteca nacional, 10016, fols 8–68ᵛ; *C′* Chartres, Bibliothèque municipale, 498, fols 190–237ᵛ, 241–6; *Cc* Oxford, Corpus Christi College, 283, fols 114–45; *L* London, Lambeth Palace, 67, fols 64–95; *LA* London, British Library, Arundel, 377, fol. 6ᵛ, 79ᵛ–81; *P* Paris, Bibliothèque nationale, lat. 16208, fol. 95, 131–2.

[10] *C* and *O* are the principal sources. *M* is a thirteenth-century extract from *C*, giving the first twenty-one chapters and nine tables. *N* is the revision due to Robert of Chester; see section 1.5. *Cc* and *L* are attributed to

will be denoted *C* and *O*, following Suter's sigla. He published a text which was prepared not by himself but almost completely by A. A. Bjørnbo on the basis of the four manuscripts – *C*, *M*, *N* and *O*. In an effort to prepare a definitive text, *C* was taken as the main exemplar, while readings from *O* or *N* (or occasionally *M*) were frequently selected. The difficulty is that we are dealing here not with a unique composition, but with a series of revisions following the first effort at a translation, so that it is somewhat illusory to choose 'better' readings from one manuscript or another, when each one seems to represent a distinct stage in the progress toward a Latin text freed of Arabic transcriptions.[11]

Since Suter's edition other sources have come to light. The manuscript *Cc*, discovered by Haskins, was reported very fully by Neugebauer.[12] That version is only a portion of the work, with tables of mean motion attributed to Petrus Alfonsi. Since then another copy, *L*, has been found. Petrus Alfonsi's tables are discussed in section 1.4 below, while Robert of Chester's version is discussed in section 1.5, and finally in section 1.6 some effort is made to discover the complex relations among these sources.

1.2. *The Mean Motions*

Insofar as the primary mathematical component of any *zīj* is the collection of mean longitudes, certainly the main source of the *Zīj* of al-Khwārizmī is the Sanskrit work *Brāhmasphuṭasiddhānta* of Brahmagupta (early seventh century).[13] Nevertheless it can be shown that the mean longitudes of the *Zīj* are not taken directly from the Sanskrit source, but from a revision proposed, or at least transmitted, by al-Fazārī,[14] who flourished under the early Abbāsid Caliph al-

Petrus Alfonsi; see section 1.4. *LA* is fragmentary: cap. xxii and table 57[a] from *C*; tables 78, 75, 76, and 73–4, with minor rearrangement. *P* is fragmentary: cap. viii from *O*; tables 78, 73–4, 76. *C'* is part of the *Heptateuchon* of Thierry of Chartres. This copy includes the tables 4–113 in Suter's enumeration (n. 3 above), but without any of the titles of the tables, nor any of the explanatory chapters. The various sections of the manuscript should be rearranged as follows: 198–213^{v}, 190–7^{v}, 230–7^{v}, 222–9^{v}, 214–21^{v}, 241–6. It is evident from the numerical entries in the tables and from the order of the tables that *C'* is copied from *C*. The other contents of this manuscript which, like *C*, was destroyed by fire in World War II, are listed in C. S. F. Burnett, 'The Contents and Affiliation of the Scientific Manuscripts Written at, or Brought to, Chartres in the Time of John of Salisbury', in *The World of John of Salisbury*, ed. M. Wilks, Studies in Church History, Subsidia 3 (Oxford, 1984), pp. 127–60 (142–3). I am grateful to Anne Tihon who made it possible for me to examine a film of *C'*. Note that in *O*, fol. 159 has become displaced from its correct position between fols 99 and 100. The first four sigla are those used by Suter (n. 3 above).

[11] At times the sense directs the choice. In cap. viii (ed. Suter, n. 3 above, p. 10, n. 3) 'secundae' from *O* is used in place of 'dakaice' ('minuta') from *CM*. However, this does not necessarily bring us close to the original translation; indeed the Arabic text itself may have contained the error. Elsewhere the choice is merely capricious. In cap. xxxiii (p. 27, n. 19 and passim) 'ex minutis residui' is chosen from *N* instead of 'dakaicarum elfudhul' from *C*, while a few lines later (n. 27) 'dakaicae' is chosen from *C* instead of 'minuta' from *N*. In *C* 'elwacat' is always used, while *O* has the translation 'medium' or 'medialitas'. Normally *C* is chosen, but in cap. xxxiii (p. 24, n. 2) *O* is chosen for no apparent reason. These examples could be multiplied.

[12] O. Neugebauer, *The Astronomical Tables of al-Khwārizmī, translation of the Latin version edited by H. Suter supplemented by Corpus Christi College MS 283*, Det Kongelige Danske Videnskabernes Selskab, histor.-filosof. Skrifter, 4, 2 (Copenhagen, 1962).

[13] The *Brāhmasphuṭasiddhānta* has not been translated, apart from the exposition of the algebra. The Sanskrit text has been edited most recently by A. R. S. Sharma, *Shri Brahmagupta Viracita Brāma-Sphuṭa Siddhānta, with Vāsanā, Vijñāna and Hindi Commentaries*, 4 vols (New Delhi, 1966). The revolution numbers are given in book I, verses 15–22. These data and the essentials of the computational procedure are presented in the most accessible form by R. Billard, *L'Astronomie indienne* (Paris, 1971).

[14] Muḥammad b. Ibrāhīm b. Habīb al-Fazārī. See Sezgin (n. 2 above), VI, p. 122, for further information.

Manṣūr (A.D. 754–75). Brahmagupta had adopted an *a priori* scheme in which a certain extremely remote epoch was constructed, at which all the mean longitudes were equal to zero, while all the rates of increase were expressed in terms of the number of revolutions in a *kalpa* (4,320,000,000 years). In his *India*,[15] al-Bīrūnī tells us that the Arabs knew the revolutions through the '*zīj*es of al-Fazārī and Yaʿqūb b. Ṭāriq', and that 'these differed somewhat from the primary statements of the Hindus'. Al-Hāshimī earlier had compiled an account of the various *zīj*es founded on Persian and Indian sources, and said that al-Khwārizmī selected rather uncritically from the *zīj*es of al-Fazārī and Yaʿqūb b. Ṭāriq.[16]

In the *Zīj* the revolution numbers as such are not quoted. In the manner of the Greek tables, the mean longitudes are given for a historical epoch, in this case the Hijra (Noon, 4 July 622), although there was no change of the meridian of reference from the Indian meridian through Ujjain as used by Brahmagupta. The values of the mean longitudes are given as regular intervals of thirty lunar years (10631 days) up to five hundred and seventy years after the Hijra, from which one can compute the rate of motion.

It is not difficult to compute from the parameters of the *Brāhmasphuṭasiddhānta*, or from the alternatives given by al-Fazārī for Saturn and the lunar node, the mean longitudes at the Hijra and the movement in five hundred and seventy lunar years. These results are displayed in Table I. The conclusion is clear enough: that the mean longitudes are founded on al-Fazārī's version of the Indian work.[17] This part of the *Zīj* therefore dates from ca. A.D. 775 at the latest. The *a priori* scheme used by Brahmagupta compels him to postulate minute movements of the apogees and nodes of the Sun and the planets, which are ignored in the *Zīj*. Brahmagupta rejected precession for reasons which would seem to us entirely unjustified, and there is no sign of it in the *Zīj*, although al-Fazārī apparently proposed the Ptolemaic rate, one degree per hundred years.

1.3. *The Equations of Motion in the Zīj*

Adelard's translation includes not only the mean motions, but also the tables of the various equations which must be added to the mean motions to obtain the true longitudes. In view of the Indo-Persian origin of the *Zīj* one is not surprised to find that the equations differ from those of Ptolemy. However, the *Brāhmasphuṭasiddhānta* is not the source either, although the procedure adopted here

[15] *Kitāb fi taḥqīq mā li'l-Hind* by Abū'l-Raiḥān Muḥammad b. Aḥmad al-Bīrūnī, trans. E. C. Sachau, *Alberuni's India* (London, 1910). The quotations from the *Zīj* of al-Fazārī are given in ch. 50.

[16] *The Book of the Reasons behind Astronomical Tables (Kitāb fi ʿilal al-zījāt), by ʿAlī ibn Sulaymān al-Hāshimī: A Facsimile Reproduction of the Unique Arabic Text contained in the Bodleian MS Arch. Seld. A 11*, trans. Fuad I. Haddad and E. S. Kennedy, with a commentary by D. Pingree and E. S. Kennedy (New York, 1981), p. 98.

[17] J. J. Burckhardt, 'Die mittleren Bewegungen der Planeten im Tafelwerk des Khwarizmi', *Vierteljahrsschrift der Naturforschenden Gesellschaft in Zürich*, 106 (1961), pp. 213–31, came to the same conclusion at the end of a somewhat involved discussion.

Table I

The Parameters of the *Zīj* of al-Khwārizmī in Comparison with Those of the Brāhmasphuṭasiddhānta (1), and of al-Fazārī (2)

		Epoch at Hijra		Motion in 570 lunar years	
		calculation	Zīj	calculation	Zīj
Mean					
Sun		113;25,49.2	113;25,48	553^r 1;4,6.4	553^r 1;4,4
Moon		117;45,16.8	117;45,17	7392^r 358;51,32.3	7392^r 358;51,30
Mercury		182;18,53.9	182;18,58	2296^r 41;8,22.2	2296^r 41;8,14
Venus		159;25,51.3	159;25,31	898^r 336;51,9.3	898^r 336;50,58
Mars		210;25,15.0	210;25,15	294^r 6;4,56.8	294^r 6;4,56
Jupiter		330;16,52.2	330;16,49	46^r 224;51,33.2	46^r 224;51,35
	(1)	115;7,43.1		18^r 274;20,45.2	
Saturn			117;58,37		18^r 274;26,19
	(2)	117;58,36.6		18^r 274;26,14.7	
Apogee					
Sun		77;54,32	77;50	0;1,20	0
Moon		193;9,16.7	193;9,16	62^r 173;39,56.4	62^r 173;39,56
Mercury		224;53,13	224;54	0;0,55	0
Venus		81;14,19	81;15	0;1,48	0
Mars		128;23,41	128;24	0;0,48	0
Jupiter		172;31,31	172;32	0;2,22	0
Saturn		260;54,17	244;55	0;0,7	0
Node					
	(1)	126; 8,39.6		29^r 265;43,49.6	
Moon			126;12,22		29^r 265;46,29
	(2)	126;12,12		29^r 265;46,30.5	
Mercury		21;11,11	21;10	0;1,26	0
Venus		59;48,25	59;27	0;2,28	0
Mars		21;54,47	21;54	0;0,44	0
Jupiter		82;1,28	82;1	0;0,10	0
Saturn		103;12,39	103;12	0;1,37	0

clearly falls within the class of procedures known from the many Sanskrit works on astronomy.

It is convenient to deal at first not with the Sun and Moon, but with the planets. The true longitudes of the planets are computed in a sequence of steps in which two basic functions are employed. The first $\sigma(\theta)$ is the *śīghra* correction of the Sanskrit source, here called *argumenti examinatio*, or *examinatio ipsius elheza*, and is defined by the expression

$$\sigma(\theta) = \arctan\left\{\frac{\sin \sigma_m \sin \theta}{1+\sin \sigma_m \cos \theta}\right\}$$

The second $\mu(\theta)$ is the *manda* correction of the Sanskrit source, here called *centri examinatio*, and is defined by the expression

$$\mu(\theta) = \arcsin(\sin \mu_m \sin \theta)$$

The procedure then is as follows:

Terminology	Procedure: Outer Planet	Procedure: Inner Planet
argument (*argumentum*)	$\alpha_1 = \lambda_S - \lambda$	$\alpha_1 = \lambda - \lambda_S$
corrected apogee (*sublimatio examinata*)	$s = \Gamma - {}^1/_2\sigma(\alpha_1)$	
centre (*centrum*)	$\kappa_2 = \lambda - s$	$\kappa_2 = \lambda_S - s$
corrected centre (*centrum examinatum*)	$\kappa_3 = \kappa_2 - \mu(\kappa_2)$	
corrected argument (*argumentum examinatum*)	$\alpha_2 = \alpha_1 + \mu(\kappa_2)$	
final centre (*centrum ultimum*)	$\kappa_4 = \kappa_3 + \sigma(\alpha_2)$	
true longitude (*locum*)	$L = s + \kappa_4$	

Here λ, λ_S are the mean longitudes of the planet and of the Sun, and Γ is the planet's apogee. The remaining notation follows Neugebauer's commentary.

The quantities which are tabulated in the *Zīj*, in the tables of equations, are $s(\alpha_1)$, $\sigma(\alpha_2)$, $\mu(\kappa_2)$. The procedures may be recast in a form which is more significant in view of the Sanskrit context. For the outer planets,

$$\beta = \Gamma - {}^1/_2\sigma(\lambda_S - \lambda) \qquad \gamma = \lambda - \mu(\lambda - \beta) \qquad L = \gamma + \sigma(\lambda_S - \gamma)$$

For the inner planets, λ_S and λ are interchanged. The very same steps are used in the *Áryabhaṭīya* (ca. A.D. 500) for the outer planets, and in two later works of the karaṇa type, the *Karaṇa Prakāśa* of Brahmadeva (A.D. 1092) and the *Grahalāghava* of Ganeśa Daivajña (A.D. 1520).[18] Generally speaking, however, the Sanskrit authors employed four rather than three steps in this computation, but in any case, the numerous variations proposed by those authors seem now rather capricious. In the *Zīj* the procedures are not merely presented in tabular form, something unknown in the Sanskrit works of the time, but the original

[18] Brahmadeva's work has been edited with a commentary by Sudhākara Dvivedī, *Karaṇaprakāśaḥ, Śrī brahmadeva viracitaḥ...Śrī sudhākaradvivedinā vāsanābhir āsannamānādyanekasiddhāntair vibhūṣya saṃśodhya ca mudritaḥ*, Chowkhamba Sanskrit Series, 5 (Benares, 1899), while the *Grahalāghava* was consulted in the recent edition by S. Śarma, *Śrī gaṇeśa daivajña viracitaṃ grahalāghavam sampādaka anuvādaka* (Delhi, 1958).

Indian procedures are recast considerably, as we have just seen. These are two illustrations of the considerable distance between the *Zīj* and its Indian ancestry.

In Sanskrit sources the equations for the Sun and Moon are *manda* corrections, expressed by means of the function given above. In the *Zīj*, however, another expression is used, as Kennedy and Muruwwa discovered.[19] For now a proportion of the function which expresses the solar declination is adapted to express either equation,

$$L = \lambda - (\mu_m/\varepsilon_o) \arcsin(\sin \varepsilon_o \sin(\lambda - \Gamma))$$

where ε_o is 23;51, the Ptolemaic value of the obliquity.

We can summarize in a table the values of the parameters which have appeared in the above formulae.

Table II

Apogees and Parameters of the Equations in the *Zīj* of al-Khwārizmī

	Γ	σ_m	μ_m
Sun	77;50	-	2;14
Moon	tabulated	-	4;56
Mercury	224;54	21;30	4;2
Venus	81;15	47;11	2;14
Mars	128;24	40;31	11;13
Jupiter	172;32	10;52	5;6
Saturn	244;55	5;44	8;36

Neugebauer's commentary on the equation procedures follows the traditional assumption that the model was constructed from uniform epicycles. However, the presence of the '$^1/_2$' in the above formula for β ensures that the final true longitude approximates significantly better to the equant model, as explained in Ptolemy's *Almagest*, than to a simple eccentric/epicycle model. This may be shown by means of a series expansion in powers of μ_m and σ_m. If, as is altogether likely, the Indian methods are related historically to pre-Ptolemaic Greek astronomy, then, as Van der Waerden has convincingly argued,[20] Hellenistic astronomers must at some point have discovered the equant hypothesis. He singled out Apollonius as the most likely figure, for he certainly created some of the essential elements of this theory, while Hipparchus is believed not to have produced a new model of planetary motion.

There is no space here to enter into the theory of planetary latitudes as represented by the tables in the *Zīj*, but one can take the matter beyond the stage reached in Neugebauer's commentary, as I will show elsewhere.

[19] Neugebauer (n. 12 above), p. 95; E. S. Kennedy and A. Muruwwa, 'Biruni on the Solar Equation', *Journal of Near Eastern Studies*, 17 (1958), pp. 112–21.

[20] B. L. Van der Waerden, 'Ausgleichpunkt, "Methode der Perser" und indische Planetenrechnung', *Archive for History of Exact Sciences*, 1 (1961), pp. 107–21.

1.4. *Petrus Alfonsi*

The Spaniard Petrus Alfonsi, formerly Moše Sefardi, was a converted Jew who on his baptism took the name of his godfather Alfonso I of Aragon. It is clear that he passed some time in England, and he is reported to have been the physician of Henry I (1100–35). Better known to medievalists for his apologetic works *Disciplina clericalis* and *Dialogi cum Iudaeo*, he was, however, also quite involved with the rapidly growing interest in Arabic astronomical sources. In three manuscripts,[21] all of which are English, we have a collection of tables and texts attributed to him.

Virtually all the astronomical material has been published,[22] but without taking into account the Lambeth manuscript *L*, which was found by J. H. L. Reuter,[23] and also by myself. It is difficult to give a satisfactory description of the material, in spite of Neugebauer's thorough analysis of the manuscript *Cc*.[24] On first inspection it appears to be an incomplete translation of the *Zīj* of al-Khwārizmī. It includes, for example, the same texts of capp. vii–xvii which are given in *O*. These chapters form a natural unity and cover the topics of mean and true motions, stations and retrogradations, declination and planetary latitude. There is then a long tract written by Petrus himself; it is obviously a preface of a rather didactic character to his translation of a set of tables, which he does not name, but which is evidently the tables in these same manuscripts *Cc* and *L*.

The long set of tables, which is found in the Corpus Christi and Lambeth manuscripts, stands in some relation to tables of the *Zīj*. These are tables of mean motion and equations for Sun, Moon and planets, and indeed those of equations coincide in all essentials with those of the *Zīj*. The tables of mean motion, however, are bizarre. They appear to be a version of the corresponding tables of the *Zīj*, but many differences emerge on a simple analysis. They are calculated for the Latin calendar, with radices for 1 October 1116, and at intervals of thirty years. Since thirty years is $10957^1/_2$ days, only the integer 10957 is kept. Nowhere else in the astronomical literature is the Latin calendar used in such an awkward fashion; normally the years would be grouped in multiples of four, such as twenty or twenty-eight. Further, all the entries in these tables of mean motion are throughout clumsily and erroneously calculated. For the Sun, the *Zīj* of al-Battānī is used for the rate of motion, but the initial value is incorrect. In all the other tables the source is clearly the *Zīj* of al-Khwārizmī, for although the calculations are generally crude, one discovers with some surprise that the radices for 1 October 1116 are all quite accurately calculated from the *Zīj* (a point not appreciated by Neugebauer). Evidently the same person could not have calculated both those precise initial values and also the remaining entries.

The epoch date 1 October 1116 is incorporated in another table, apart from those of mean motion. This is a small table which gives the interval from the

[21] Oxford, Corpus Christi College, 283, fols 114–145; London, Lambeth Palace, 67, fols 64–95 (these two were noted above and given the sigla *Cc* and *L*); London, British Library, Arundel 270, fols 40ᵛ–45.

[22] Neugebauer (see n. 12 above), pp. 132–234; J. M. Millás-Vallicrosa, 'La aportación astronómica de Pedro Alfonso', *Sefarad*, 3 (1943), pp. 65–105.

[23] 'Petrus Alfonsi: An Examination of his Works, Their Scientific Content and their Background', unpublished D.Phil. dissertation, University of Oxford, 1975, p. 149.

[24] See n. 12 above.

various eras up to the epoch date. This table is included in the tract *Liber ysagogarum Alchorismi in artem astronomicam a magistro A. compositus.*[25]

1.5. *Robert of Chester*

There is a version of the *Zīj* of al-Khwārizmī in the manuscript *N*. The text begins:

> Incipit liber ezeig id est chanonum Alghoarizmi per Adelardum Bathoniensem ex arabico sumptus et per Rodbertum Cestrensem ordine digestus.

In other words, this version is the *Zīj* of al-Khwārizmī as revised by Robert of Chester. The manuscript is apparently English and contains also the version of the Toledan Tables produced by Roger of Hereford for the meridian through Hereford. The manuscript is somewhat mutilated. At least one quire and some separate pages have been lost; another quire (fols 12–19) has been misplaced and belongs between the present folios 29 and 30. This is evident from an easy reconstruction of the tables and a comparison with the principal manuscripts of the *Zīj*. In this way it is clear that Robert intended to include essentially all the chapters and all the tables of the *Zīj*, as we know it in Adelard's version.

Robert's 'setting in order' involved the following changes:

1. The chapters are not collected together at the beginning, but are distributed among the various tables.

2. The tables of mean motions are not now collected into one section, but each table of mean motions is followed immediately by the corresponding tables of equations.

3. Some tables and chapters are found here, but not in the versions of *O* and *C*. These include calendrical tables for the Eras of Yazdajird and Diocletian, and a table which gives the conversion from Hijra to Latin dates. There are also chapters on the equation of time (fol. 42rb) and on solar eclipses (fol. 49rb). Unfortunately, none of these additional tables and chapters were printed by Suter. The table of the equation of time was given in *C* and *O*, but not in *N*. There is a reference to it at the end of cap. xxi.

4. There is given in a series of footnotes the corrections required for the displacement of the meridian of reference to London. Although only the corrections for the outer planets and the lunar node are extant, because of the loss of pages, it is clear that a shift of 55° westward from Ujjain was assumed. One can perhaps assume that a displacement of 8° to the East of the Toledo/Cordova meridian was intended, a difference approximately in line with Ptolemaic longitudes (Toledo 10;0, London 20;0). Neugebauer, in his commentary to the *Zīj*, thought that perhaps some computation involving the longitude of Baghdad had been made,[26] but he did not appreciate that the *Zīj* in Adelard's version still referred to the

[25] Consulted in MS Paris, Bibliothèque nationale, lat.16208, fols 67–71; see p. 38 above.

[26] Neugebauer (n. 12 above), p. 94.

meridian through Ujjain. These footnotes are all written in the same hand as the tables, but it does not necessarily follow that the corrections were calculated by Robert.

5. There is also a table for the motion of the lunar node for the meridian of Cordova, set at 63° from Ujjain. For this same meridian are calculated tables of conjunction and opposition, which are found in both versions of the *Zīj*.
6. Robert ignores almost all the Arabic transcriptions found in the earlier versions. The Appendix gives a selection of the Arabic terms and their various equivalents. We may divide Robert's innovations as follows:

   a. Arabic term retained: *kabīsa*: 'elkebice' ('leap year').

   b. Latin terms where none had been given earlier: *jaib*: 'sinus'.

   c. New Latin terms in place of those given earlier: *faḍl*: 'residuum'; *suṭūr al-ʿadad*: 'linee numeri'; *taʿdīl*: 'coequatio'.

Elsewhere he fixed the terminology by choosing one of the earlier Latin equivalents. His tables show a more advanced use of Latinization than does his text.

1.6. *Relations among the Sources of the Latin Zīj*

The relations among the four manuscripts *C, M, N, O* used by Suter are reported in his critical apparatus. Broadly speaking they all represent the same translation of the *Zīj*. The tables and text of Petrus Alfonsi in *Cc* were reported in detail by Neugebauer, while the other copy *L* has not yet been reported. However, I have examined it in detail in order to discover its relation to the others.

The version *L* seems on first impression to be quite inferior to *Cc*, if only because the tract by Petrus Alfonsi (beginning *Dixit Petrus*) is placed inconsequentially between capp. xii and xiii of the *Zīj*, while in *Cc* it comes at the end of the selection of capp. vii–xvii. When examined in detail, however, and compared also with *O*, it is seen to be in many ways closer to *O* than is *Cc*, and also it is often more correct than *Cc*. In the chapters, for example, the spelling *dakaicas* is found in *L* and *O*, but *dekaicas* in *Cc*. Further, Roman numerals are used in the chapters in *L* and *O*, but the names of the numbers are written out in *Cc*. In the tabular part, the monthly movements of the mean Moon are correct in *L*, but incorrect in *Cc*, from March to September. On the whole, perhaps the numerical entries are more often accurate in *Cc*, but where they differ, *O* agrees now with one, now with the other. Many of the differences between the tabular entries in *Cc* and *L* on the one hand, and *O* on the other, reveal significantly that the Arabic manuscripts were separately consulted for the two versions, so that we can be certain that *Cc* and *L* do not represent a mere extract from *O*. This was noted by Neugebauer and Toomer.[27] The column headings 'signs, degrees, minutes, seconds' are given in *Cc* and *L* as 'signa, gradus, puncta, minutie', but in *O* as 'signa, gradus, minuta, secunde'. Here *Cc* and *L* show the usage of Petrus Alfonsi, which one finds also in Walcher's tract *De dracone*, and provide a clear argument for Petrus's activity in this version of the tables.

The texts in *O* and *C* are nearly the same translation, but nevertheless there are very many differences of detail, and in capp. viii and xv the difference is so great as to argue for two different Arabic sources, and not merely two revisers.

In these versions some ninety to one hundred Arabic words and phrases are found (apart from fifty-six names of Sun, Moon, the planets, the zodiacal signs and months), and about one third of these are listed in the Appendix, so that one may judge the progress made towards the Latin version. The versions *O* and *C* do not differ much in the fund of vocabulary, giving much the same Latin equivalents. They differ rather in the independent and random decision as to whether an Arabic word or title is to be given along with the Latin, or in place of it.[28] The Arabic transcriptions differ slightly, in that in *C* 'c' is used for either the hard or soft fricative represented by 's' or 'z' in *O*, *Cc* and *L*. There are signs of a more enlightened hand at work in the tables, and in any case there are clear indications of systematic differences in the transcriptions and translations between text and tables. For example, the Sanskrit *buht* is translated in the text as *stupor* by the false assumption of an Arabic etymology (*b-h-t*, 'to wonder') while in the tables the correct meaning *motus* is found. A similar confusion is found over the term *sinus*, which was introduced on the assumption of an Arabic meaning *al-jaib*, 'pocket', 'hollow', while in fact that was only a misreading of an Arabic transcription of the Sanskrit *jīva*, 'bowstring', which was a good metaphor for a chord in a circle. Suter[29] drew attention to the interesting mistake over the translation of *mankūs*, meaning 'reversed', which, however, was understood as *manqūṣ*, 'diminished', which suggests that oral assistance was at hand. The term *coniunctio* occurs in the astronomical sense in the *De essentiis* of Hermann of Carinthia.[30] In the computus literature *accensio* or *coitus* are the usual terms. It is interesting that the word *spera* is nowhere used, although as a translation of the Arabic *kura* it occurs in the tract on the spherical astrolabe.[31]

27 Neugebauer (see n. 12 above), p. 97 and G. J. Toomer, review of Neugebauer's book in *Centaurus*, 10, pp. 203–12. A representative selection is as follows:

Mean Sun (22^h)	*L,Cc*: 0;54,12 (recte); *O*: 0;54,52
Obliquity (0^s 13)	*L,Cc*: 5;13,7 (recte); *O*: 5;13,50
Obliquity (1^s 8)	*L,Cc*: 14;24,54 (recte); *O*: 14;24,34
Obliquity (1^s 28)	*L,Cc*: 20;3,15 (recte); *O*: 20;3,55
Obliquity (3^s 20)	*L,Cc*: 22;19,58; *O*: 22;19,53 (recte)
Lunar eqn. (4^s 26)	*L,Cc*: 2;42,50; *O*: 2;42,7 (recte)
Lunar eqn. (5^s 6)	*L,Cc*: 1;17,32; *O*: 1;57,32 (recte).

28 In *O*, for example, tables 10–73 have with few exceptions only Arabic titles, while *C* gives mainly the Latin. Sometimes the intermingling of Latin and Arabic in *C* has odd results, as in one of the titles of table 54 (equation of Mercury), 'statio mukaam elprima mercurii avval autarid'. Here 'mukaam' = 'maqām, statio'; 'avvel' = 'awwal, prima'; 'autarid' = 'ʿutārid, Mercurius'.

29 Suter (n. 3 above), p. 69.

30 Ed. C. S. F. Burnett (Leiden, 1982), p. 212.

31 Consulted in MS *O*, fol. 79ra: 'incipit de horologio secundum alchoram id est speram rotundam'.

There are a number of entries in the tables in *C* and *O* which suffice to prove that those versions differ through independent consultations of the Arabic sources,[32] and indeed it has already been inferred from differences in the chapters that different Arabic texts were probably used for *C* and *O*.

In the case of version *N*, due to Robert of Chester, one finds that while the text of the chapters follows essentially *C*, except that the Arabic terms are eliminated, the entries in the tables often agree with *O* rather than *C*. It is therefore clear at least that *N* is not based directly on *C*. Bearing in mind that there must have been a common ancestor of *O* and *C*, it is possible to account for the variations between the tabular entries of *N* and *C* by supposing that *N* is derived from that ancestor, rather than by supposing that *N* is more directly dependent on *O*. However, there is no reason to suppose that *N* was formed with the aid of yet another check against an Arabic text. On the few occasions when *N* differs from both *O* and *C*, it is not such a variation as would arise from the usual ambiguities in the Arabic script. On the contrary, these variations, which sometimes given the best reading, show that *N* is closer than either *O* or *C* to their common ancestor.

In the versions *C*, *M* and *N*, at the end of cap. ii there is a paragraph in which the translator refers to a table which enables the user to convert from the Latin to Arabic calendar, and in which the Latin years are in steps of thirty, starting with the 'year of the translation'.[33] Although that table is not extant, the tables of mean motion in *Cc* and *L* depend in just that way on the Latin calendar. This may be taken as indicating that *C*, *M* and *N* descend from the work of Petrus Alfonsi. In the version *O* this paragraph is absent, but instead there is a reference at the end of cap. iv to the date 26 January 1126 (the start of A.H. 520), which would certainly appear to be copied from a paragraph in *Cc*, fol. 142.[34]

Proceeding now to a hypothetical construction, we may suppose to begin with that *Cc* and *L* represent the very first effort ca. 1116 to present to a Latin audience a work of this type. We can understand that its author, apparently Petrus Alfonsi, tried especially to interpret the work for Latin users, and indeed over-reached himself in his effort to convert the tables of mean motion to the

[32] The following examples will illustrate the difference between the tables in *O* and *C*:

Solar eqn. (3^s 8)	*O*: 2;12,37 (recte); *C*,N: 2;12,57
Solar eqn. (5^s 8)	*O*: 0;48,58 (recte); *C*,N: 0;58,58
Lunar eqn. (5^s 4)	*O*: 2; 6,43 (recte); *C*,N: 2;6,53
Obliquity (4^s 14)	*O*: 16;14,33; *C*,N: 16;54,33 (recte)

All these variations could arise from ambiguities in the reading of an Arabic script.

[33] The passage in Suter's edition (n. 3 above), pp. 3–4, is as follows: 'Quia vero ut supra dictum est liber iste status planetarum atque draconis secundum annos atque menses Arabum determinat, nos vero Latinorum studemus utilitati, ideo et in primis tabulam suppositam praenotavimus, per quam ab eo quidem anno quo hic liber in nostrum sermonem translatus est usque in tempora infinita ex annis quotlibet romanis et mensibus cum diebus annorum et mensium et dierum arabicorum aequalitas sumi queat. Huiusce ergo tabulae introitus talis est: Cum enim quotus sit ab elhigera annus vel mensis vel dies quaeris, per annos domini secundum trigenarium collectos in annos Arabum...'

[34] Neugebauer (n. 12 above), p. 207. The statement, 'Anno ab incarnatione domini 1126 die Januarii 26 prima fuit dies almuharram et feria tertia aliter annus arabicus 510 planus 10' clearly refers to the epoch year 510, although the remark is otherwise unconnected with the texts of *Cc*. When copied in *O* (Suter, n. 3 above, p. 5) it is altered to read '...annus arabicus DXX', and is placed at the end of a chapter on the determination of the feria at the start of the Arabic year. Because the passage is copied from *Cc*, which we take to antedate Adelard's version, there can be no real ground for taking A. D. 1126 as the date of that version.

Latin calendar. Once interest had been awakened, however, people such as Adelard went back to the Arabic text in order to present in Latin something much nearer to the original. The source of *Cc* and *L*, now lost, is the first hypothetical manuscript '*X*'.

Now the versions *O* and *C*, differing as they do at so many points, cannot have reached their separate forms as translations of the same Arabic manuscript, and so we may suppose that two Arabic texts were involved. Nevertheless the agreement between them was of such a degree that they must have been separate revisions of a common predecessor, which we shall refer to as '*Y*', no longer extant. That version *Y* would have been the first attempt to enlarge the early partial selection *X*, so as to include not only all of the texts and tables of the *Zīj*, but also the chronological table referred to at the end of cap. ii, which relates somehow to the first effort. As it happened, *C*, and even *N*, continued to include that reference to a table no longer kept, while the author of *O* correctly deleted it. The hypothetical version *Y* must have depended on the same Arabic text (A1) already used for *Cc* and *L*, because it serves as the intermediate stage to *O*, which has the same detailed character as *Cc* and *L*, and in particular the same versions of capp. viii and xv. There is finally a common ancestor of *C* and *N*, which we denote '*Z*', and which depends on the second hypothetical Arabic text, A2. These proposals are shown in the following diagram. These conclusions may be seen as a development of arguments put forward by Millás-Vallicrosa,[35] especially as far as they concern the importance attached to the contribution of Petrus Alfonsi.

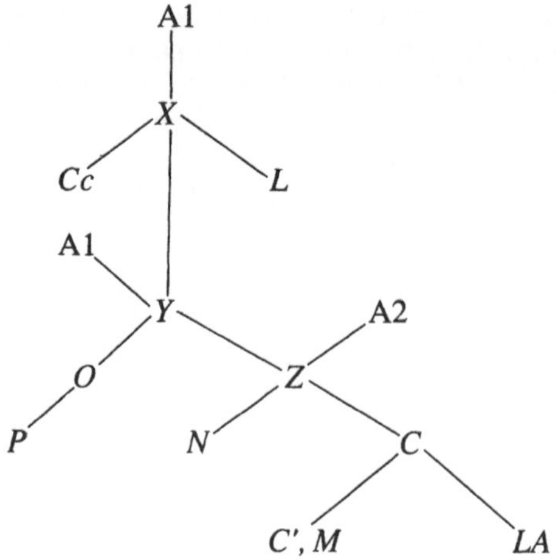

[35] See n. 22 above.

1.7. *Three Other Versions of the Zīj*

In the twelfth-century MS Paris, Bibliothèque nationale, lat. 16208, fols 2–3^v, 5–12^v, one finds a version of the *Zīj*, complete with tables of equations and texts which give the appropriate procedures. There is no textual relation between this version and those of Adelard or Robert. Here the radices are computed for dates in the Latin calendar from 1100 to 1200 in steps of twenty years, the year commencing at midnight at the end of December, as in the Toledan tables of Raymond of Marseilles. The meridian of reference is situated 61;30 from Ujjain, that is, 28;30 from the western limit. This would seem to indicate that the calculations were intended for Toledo.

The tables are written throughout in Roman numerals. The tabulation of the planetary equations is briefer than that known from Adelard's or Robert's versions, but is nevertheless complete. Although the computation of the equations produces the same result as in those other versions, the procedure is somewhat different, because only the *śīghra* and *manda* corrections are tabulated, i.e., $\sigma(\theta)$ and $\mu(\theta)$. The computation of s, the corrected apogee, is specified in the text and must be carried out by the user, instead of being provided in the table. Insofar as the tabulation is simpler, one might well take the view that these tables and the associated text are closer to the version originally provided by al-Khwārizmī, in contrast to the version prepared by al-Majrīṭī.

The texts give the procedures for the computation of the mean and true longitudes, the planetary latitudes, the solar declination, the construction of tables of mean motions, and the determination of planetary stations. There are no Arabic terms, but the Latin terms which are used differ from those of Robert. It has, for example, *rectitudo centri* for *coequatio centri*, and *equatio portionis* for *coequatio argumenti*. Unfortunately the chapters here correspond to just those absent from Robert's version, so that a full comparison is frustrated. It is of course possible that this is not a translation, but an original Latin composition.

Another version of the *Zīj* was produced for the meridian through Jaén in Spain (al-Jayyān). Its author was Abū ᶜAbdallāh Muḥammad b. Muᶜādh al-Jayyānī, who flourished in the eleventh century. A translation of these tables is listed among the works of Gerard of Cremona, although no manuscript of the work has yet been found. Nevertheless, the texts of the procedures, in a Latin version, were found to have been printed anonymously at Nuremberg in 1549, as an appendix to the *De elementis et orbibus coelestibus* of Messahalla.[36] The meridian is given expressly as Jaén, at a distance of 62° from Ujjain.

The procedures for the computation of the equations are very similar to those given in the *Zīj*. The final result is the same, but the steps are rearranged slightly so that $\kappa_2 + s$ is considered in place of κ_2, and is termed the 'second mean'. The corrected apogee, denoted s above, is evidently tabulated, so there is no likelihood of an identification of the missing tables for Jaén with those of MS Paris, Bibliothèque nationale, lat. 16208.

36 H. Hermelink, 'Tabulae Jahen', *Archive for the History of Exact Sciences*, 2 (1964), pp. 108–12, gave a summary description of the volume, but with few technical details.

We may remark, finally, that the eighteenth-century Samaritan tables for the Sun and Moon, as well as the lunar argument and node, quoted at some length by Robertson[37] are calculated from the same parameters as found in the *Zīj*, and for a meridian situated 30° to the west of Ujjain, as noted by Pingree.[38]

2. *Walcher of Malvern*

In the same Oxford manuscript which contains Adelard's *Zīj* there is a brief tract by Walcher, Prior of Malvern, in which he develops some teaching which he has received from Petrus Alfonsi.[39] Entitled *De dracone* ('On the Lunar Node'), it was appended to another tract of his on lunation cycles. It was edited by Millás-Vallicrosa from this manuscript and the Erfurt manuscript.[40] Its interest for us lies in the possibility that it might shed some light on the translation and use of the *Zīj*.

Walcher is very unsophisticated. When he opens his account by explaining that in the new astronomy one divides the circle into 360 degrees, not 365 days, he startles us into an awareness of the state of his education in scientific matters. One would imagine, of course, that he had met this technicality earlier when experimenting with the astrolabe.

Aiming eventually to discuss the occurrence of eclipses, and the periodic conjunction of the Sun and the lunar node, he begins by giving the longitudes of Sun, Moon and node on Thursday, 1 April 1120. This date marks the close of the year 513 of the Hijra and was presumably selected with a view to a simple consultation of tables drawn up in that calendar. His figures are:

Sun = 7;19,57 Moon = 23;30,51 node = 211;1,1

taking all the numbers as cardinal, although there is some doubt about that. If these longitudes are computed from the *Zīj*, we obtain:

Sun = 7;16,41 Moon = 20;35,9 node = 210;57,40

These cannot be reconciled with those given by Walcher by a simple shift of meridian. Longitudes computed from the tables of Petrus are, however, even further from those of Walcher, as one would expect from the inconsistent combination of al-Battānī's mean Sun with the lunar parameters of the *Zīj*. It might be concluded that the *Zīj* was the source, but that the computation from it was incompetent.

In the rest of the tract Walcher works with daily movements of Sun, Moon and node which differ substantially from those given in the *Zīj*, even from those of Petrus's tables. The solar movement which he gives is a shade less than 360 divided by $365^1/_4$ per day, and in commenting on this, he leads in a rather

37 'Notes and Extracts from the Semitic MSS in the John Rylands Library. vi. The Astronomical Tables and Calendars of the Samaritans', *Bulletin of the John Rylands Library*, 23 (1939), pp. 458–86.

38 'Al-Khwarizmi in Samaria', *Archives internationales d'histoire des sciences*, 33 (1983), pp. 15–21.

39 MS Oxford, Bodleian, Auct. F.I.9, fols 96ra–9ra. The tract on lunation cycles is on fols 86ra–96ra. *De dracone* is also found in MS Erfurt, Amplonian, Q 351, fols 17^v^–23^v^.

40 Millás-Vallicrosa (n. 22 above), pp. 87–96.

inconsequential way to a version of accession and recession evidently learned from Petrus. According to that the solar longitude gains and loses 7° alternately in intervals of nine hundred years, while the average movement is defined by the Julian year. In the interval when the Sun is gaining, this implies the reasonable value of $365^1/_4 - ^1/_{127}$ for the tropical year.

3. *The Tables of Abraham bar Ḥiyya (Savasorda)*

Abraham bar Ḥiyya flourished in the period 1110–35, and was associated with Barcelona. He wrote a number of important Hebrew works on mathematics, astronomy and the calendar, some of which found their way into Latin. His *Ḥibbur ha-Mešiḥah* of 1116, which became an authoritative source for arithmetic and mensuration, was translated by Plato of Tivoli in 1145 as the *Liber embadorum* ('Book of Areas'). Bar Ḥiyya, in his turn, appears to have been an active assistant to Plato in his translations from Arabic.

There is known in Hebrew, in a good number of manuscripts, a set of astronomical tables by him, the *Luḥōt ha-Nasī* ('Tables of the Prince'),[41] but there was no indication that these existed in Latin, until I found such a version in the twelfth-century manuscript, Cambridge, University Library, Hh.VI.8, vol. 1, fols 41–77. These tables are a curious blending of the Jewish calendar, the parameters of Ptolemy's *Almagest*, and Islamic observations of the ninth century. The epoch is 21 September 1104, the point in time which marks the completion of 256 nineteen-year cycles of the Jewish calendar, reckoned from the *Molad tohu* of 6 October, –3760. The year is 365 days, and the years are grouped in intervals of nineteen. Of course the Jewish calendar is structured around the nineteen-year cycle, but not with a year of 365 days! All the rates of motion are those of the *Almagest*, quite exactly, but these motions are not a direct continuation of them. Instead these have been adjusted so as to agree, more or less, with observations made in the middle of the ninth century. Indeed the dependence on al-Battānī is brought out by a comparison of the times of luni-solar mean conjunction with those of al-Battānī's tables. This is possible in exact terms, because the lunation periods are identical. Thus one discovers that the tables of Bar Ḥiyya are drawn up for the meridian used by al-Battānī – al-Raqqa.

In the same Latin manuscript, fols 13ᵛ–40ᵛ, there is another set of astronomical tables (fol. 13ᵛ ff.), explicitly attributed to 'Savasorda', the usual Latin name for Bar Ḥiyya. No Hebrew original is known, apparently. As in the first set of tables, Ptolemy's rates of motion are employed, but the motions are not continuous with the former set. The meridian of reference is stated in several tables to be Toulouse (Tolosa), and a calculation of the times of luni-solar conjunction determines the meridian to be situated at 30;15, relative to Baghdad at 80. The Latin calendar is now used, instead of the Jewish calendar, and the

41 *La obra Séfer Ḥešbón Mahlekot ha-Kokabim de R. Abraham bar Ḥiyya ha-Bargeloní*, ed. and trans. J. M. Millás-Vallicrosa (Barcelona, 1959). Millás-Vallicrosa includes in his edition and Spanish translation only those chapters which precede the tables, together with a brief indication of the contents of the tables themselves. These tables correspond to the first of the two sets of tables described here in the Latin manuscript. See also the review by B. R. Goldstein in 'The Survival of Arabic Astronomy in Hebrew', *Journal for the History of Arabic Science*, 3 (1979), pp. 31–9.

radices for collected years are at intervals of twenty years, beginning at 28 February 1110, which point is referred to here as '1109', that is 1109 complete years.

It is interesting to consider, in passing, the meridian of reference of the modern Jewish calendar. This calendar exists in two forms, and the more exact form, attributed traditionally to Rabbi Adda, uses the same lunation period found in Ptolemy's *Almagest*, al-Battānī's tables, and in a number of others. In this way, by examining the lunation times, one finds that the meridian of reference is Baghdad not, as one might have expected, Jerusalem.

4.1. *The Toledan Tables*

The Toledan Tables were composed around A.D. 1080 in Toledo, by al-Zarqāla[42] and various colleagues. Although it was very widely influential in its Latin versions, it is curiously unknown in Arabic sources. However, one finds two works which might be interpreted as revisions of the Toledan tables. These are the *Almanach* of al-Zarqāla,[43] and the *Zīj* of Ibn Kammād, the latter known only in a Latin version.[44]

There was evidently an Arabic original in which the tables of mean motions were drawn up for the meridian through Toledo, and it was the Latin version of this, together with a long text giving instructions in the use of the tables, which quite dominated the Latin astronomical literature from the late twelfth century until the beginning of the fourteenth.[45] The first Latin acquaintance with them, however, depended on versions recalculated for a variety of more useful

42 Ibrahīm b. Yaḥyā al-Naqqāsh, Abu Isḥāq, (Ibn) al-Zarqāla. The true form of the name Zarqāla is discussed at length by Millás-Vallicrosa (n. 8 above), pp. 15–17, who dismisses the popular form Zarqālī. He regarded it as compounded from an Arabic root *z-r-q* and a Romance suffix 'ello/el', as illustrated by the names Gabirol and Faquirol. This is reasonable, but must one follow Millás-Vallicrosa who then proposes Zarqāla = 'blue-eyed', from the sense *zariqa*, 'to be blue'? A. von Kremer, 'Beiträge zur arabischen Lexicographie', *Sitzungsberichte der Kaiserlichen Akademie der Wissenschaften in Wien*, philosophisch-historisch Klasse, 103 (1883), pp. 181–270, and 105 (1884), pp. 429–504, gives *zarrāq* as 'diviner', 'astrologer' (p. 249). Note also the fourteenth-century Damascene astronomer Muḥammed b. ʿAlī b. Ibrahīm b. Zuraiq al-Khairī (H. Suter, *Die Mathematiker und Astronomen der Araber und ihre Werke*, Abhandlungen zur Geschichte der mathematischen Wissenschaften, 10 (Leipzig, 1900), p. 173).

43 The *Almanach* of Zarqāla was edited by Millás-Vallicrosa (n. 8 above) from MS München, arab. 853, and from the Spanish version, Paris, Bibliothèque de l'Arsenal, 8322. The latter manuscript contains also the Spanish version of al-Battānī's *Zīj*. Incomplete Latin versions are noted by Millás-Vallicrosa (n. 8 above), pp. 151–2, and another is found in MS, London, British Library, Royal, 7.F.VIII, fols 180–91. 'Almanach' certainly does not appear to have an Arabic etymology, but, in spite of ingenious efforts, no clear explanation has emerged. It is found in Maghribī Arabic, but is not common; indeed it is much commoner in Latin. Arabic examples are noted by Millás-Vallicrosa (n. 8 above), pp. 348–51, but as he notes, ibid. p. 73, the name *al-manākh* is not used in the Arabic text of Zarqāla's *Almanach*. If the word is Romance in origin, one might consider, for example, a derivation from Spanish *manejar*, 'to handle', 'to operate', so that a *manākh* would be 'handy' or 'procedural' tables. The word *al-manākh* occurs in modern Arabic with the meaning 'climate', but it is accepted that this is a degenerate form of *al-munākh* (from the root *n-w-kh*) which originally meant a place which a traveller on a camel might find nice enough for a rest, hence 'a nice climate', or simply 'climate'.

44 Ibn Kammād, an early thirteenth-century astronomer of the Maghrib and Andalusia, composed a *zīj*, extant in a Latin version in MS Madrid, Biblioteca nacional, 10023, whose parameters may be regarded as an improvement on those of the Toledan tables, but these improvements may only represent later work by Zarqāla himself. See G. J. Toomer, 'The Solar Theory of az-Zarqal: A History of Errors', *Centaurus*, 14 (1969), pp. 306–36.

45 A general survey of the contents of the tables, and the manuscript sources, is given by G. J. Toomer, 'A Survey of the Toledan Tables', *Osiris*, 15 (1968), pp. 1–174.

meridians, such as Marseilles and Hereford, chosen by individual translators. Only subsequently did the principal version emerge, due apparently to Gerard of Cremona.

A recent investigation[46] shows that the mean motions of these tables were the result of a complex interweaving of three components:

1. The *Zīj* of al-Battānī, from which were computed the synodic longitudes of the Moon and the planets; it was also the source of the equations of Sun, Moon and planets.
2. Observations of the Sun in Toledo (or Cordova), ca. 1080.
3. The theory of accession and recession in a new and precise form, evidently due to al-Zarqāla.

The new solar observations were interpreted with the aid of the new version of accession and recession, in order to arrive at a new set of parameters for both elements, and in particular a mean sidereal Sun. It is then just a matter of adding the various synodic differences to this sidereal Sun, in order to obtain the corresponding sidereal longitudes of the Moon, and other results. Thus all the tabulated longitudes are sidereal.

These new tables were very precisely in agreement with observations of the Sun and Moon at the time of composition, but since no new planetary observations were incorporated, their predicted positions were inaccurate. Since the lunation period is the same as that of the *Zīj* of al-Battānī, one can readily compute the difference between the longitudes of the respective meridians of reference, which is 45. On the scale which was commonly used, with Baghdad at 80, we have al-Raqqa at 73;30 and Toledo at 28;30.

Both the Toledan tables and the theory of accession and recession have been investigated from time to time, but many mistaken notions persist, especially of the latter topic, because of the very real scientific and historical obscurities in all the material, both Arabic and Latin.[47] In scientific terms, the theory of accession and recession provides a model of the precessional difference between tropical and sidereal mean longitudes of the Sun and of the other motions. This is a problem area which extends back to Babylonian astronomy, and to which the first definitive solution had been provided by Hipparchus and Ptolemy. However, there were other solutions, including the so-called 'trepidation', in which the precessional difference was taken to vary periodically instead of steadily increasing. The term *trepidatio*, however, occurs, to my knowledge, not before the early fifteenth century, when it referred to a version incorporated in the Alfonsine tables.[48] In the interests of historical precision that term is best

[46] My own, which remains to be published.

[47] For a recent survey see R. Mercier, 'Studies in the Medieval Conception of Precession', *Archives internationales d'histoire des sciences*, 26 (1976), pp. 197–220, 27 (1977), pp. 33–71, but many of the conclusions given there have been superseded by more recent work of mine, yet to be published.

[48] For example, the *Tractatus de accessu et recessu octave spere*, written in Montpellier in 1450 by Nicolas de Comitibus, consulted in MS Firenze, Biblioteca Medicea Laurenziana, Ashburnham, 208(134), pp. 407–17: '...que tendit ab occidente in orientem tercio motu accessus et recessus quo trepidat nunc in orientem nunc in meridiem nunc in occidentem nunc in septentrionem' (p. 408[a]); 'Ut autem motus huius trepidationis intelligi queat' (p. 411[b]).

avoided in discussions of the Toledan tables, where instead *accessio et recessio* (Arabic, *al-iqbāl wa' l-idbār*) is the usual term.

The theory of accession and recession is defined by the account in the tract *De motu octave spere*[49] which came into circulation during the twelfth century. From the last quarter of the century it is almost invariably attributed to Thābit b. Qurra, the ninth-century Ḥarrānian astronomer and mathematician, but one may show that this attribution is out of the question. The reasons in the most summary form are as follows:

1. No Arabic source, in my experience, attributes the model to him.[50]
2. An earlier Latin version is quoted extensively by Johannes Hispalensis, who attributes it to al-Zarqāla.[51]
3. When the tropical mean Sun is computed by means of this model combined with the sidereal mean Sun of the Toledan tables, the best agreement with observation occurs ca. A.D. 1080.[52]

Although nothing is extant to support an attribution of such a model to Thābit, his grandson Ibrāhīm b. Sinān b. Thābit proposed a model involving a second ecliptic controlled by an epicycle, which was intended to explain both the decreasing obliquity and the increasing longitude of the Sun's apogee.[53] Part of this model is incorporated by al-Zarqāla in a proposal of his to explain the variation of the obliquity of the ecliptic. This is known from a tract of his extant only in Hebrew.[54] Ibrāhīm is not noted expressly by al-Zarqāla, but his work is clearly recognisable in a passage in the *Tadhkira* (cap. ii, sect. 4) of Naṣīr al-Dīn al-Ṭūsī, as at least one of the commentators realized.[55]

The introductory text (or so-called canons) of the tables is interesting. It contains for example an influential section on trigonometry, treating at least the direct- and reverse-sine (modern versine). This particular treatment is a direct descendant of Brahmagupta's *Khaṇḍakhādyaka*. In both the canons and the Sanskrit text the values of the sine function are expressed by giving the differences of the function at intervals of 15° in a circle of radius 150, that is $150(\sin a_n - \sin a_{n-1})$, $a_n = 15n$, for n = 1 to 6. A further link with the Sanskrit text is provided by the curious term *gardaga* (or *kardaga*), which in this context

[49] F. J. Carmody, *The Astronomical Works of Thabit b. Qurra* (Berkeley, 1960), pp. 84–113.

[50] Although Ibn Yūnis, 'Le Livre de la grande table hakémite. Extraits du MS de Leyde', ed. A.-P. Caussin de Perceval, *Notices et extraits des MSS*, VII (Paris, 1804), pp. 16–240 (59), refers to a tract by Thābit on accession and recession, the later quotation from that tract, also given by Caussin, p. 115, shows that this is quite distinct from the *De motu octave spere* and serves only to illustrate Thābit's concern for this problem area, something we already knew from his tract *De motu solis*: Carmody (n. 49 above, p. 102).

[51] J. M. Millás-Vallicrosa, 'Una obra astronómica desconocida de Johannes Avendaut Hispanus', *Osiris*, 1 (1936), pp. 451–75. In this tract Johannes refers to the Toledan tables, from which he quotes a number of parameters. This is an important notice.

[52] To be published; see n. 47 above.

[53] The tract is extant in Arabic in MS Patna, Bankipore, Khuda Baksh Oriental Public Library, 2468. The continuity of the tract is disturbed by faulty binding; it now occupies fols 118–24v, 323, 1, and all of this apart from fol. 1 has been printed, the details being given by J. P. Hogendijk, 'Rearranging the Arabic Mathematical Astronomical Manuscript Bankipore 2468', *Journal for the History of Arabic Science*, 6 (1982), pp. 133–59.

[54] Millás-Vallicrosa (n. 8 above), pp. 239–343.

[55] See the commentary by al-Ḥafarī, in MS London, India Office Library, Loth 747, fol. 69.

means essentially one sixth of the quadrant. The term derives, surely, from the Sanskrit *gṛhārdhajyā* (*gṛha-ardhajyā*), '(half-)chord of a half-sign', or sin 15°. Although *rāśī* is normally used for 'zodiacal sign', the less common *gṛha* is used in this passage of the Sanskrit work (book III, verse 6a).[56] The *Khaṇḍakhādyaka*, at least in some derivative form, was ordinarily referred to in Arabic as *al-Arkand*, just as the *Brāhmasphuṭasiddhānta* became the *Sindhind*.

4.2. *Raymond of Marseilles and Roger of Hereford*

When the Toledan tables first appear in Latin, they are seen in the version prepared by Raymond of Marseilles around 1140.[57] Here the mean longitudes are recalculated accurately for the Latin calendar, and for the meridian through Marseilles, taken as $16^1/_2$° east of Toledo or 45° on the scale of longitude normally employed. The calendar is the usual Latin form, with the year measured from midnight between the end of December and the beginning of January. The radices are given for every twenty-eighth year, from A.D. 0 to 1904. The grouping of the years by twenty-eight seems to follow an earlier practice used for the Syrian calendar and which is found in the *Zīj* of al-Khwārizmī, and the *Almanach* of al-Zarqāla. The tables of Marseilles betray such a source, because the bissextile years are placed third, seventh, etc., which is erroneous for this version of the Latin calendar; the bissextile years should be fourth, eighth, etc.

In his long preface, Raymond tells the reader of the need for new and accurate tables, as a basis for an almanach which he wishes to compose. He castigates two unnamed astrologers for their incompetent computation of the first stationary point of the retrogradation of Mars in 1140, following its 'combustion', that is, its conjunction with the Sun on 27 October 1139. In these remarks, he demonstrates his own competence in the use of the complex tables of planetary equations. This is no mean feat, particularly when one considers the general state of Latin education at the time. One should recall Walcher's effort to come to terms with the fact that a circle is divided into 360 degrees. His time for the true conjunction is several days in error, but when he gives the longitude of Mars as 212;21 on 27 October 1139, at seventeen hours after midnight, he is in error by only one minute of arc, compared with an exact calculation according to the Ptolemaic theory. His longitude of the first station is 100;46, compared with the exact result 100;50.

In this version of the Toledan tables, there is no reference to accession and recession, and no indication that Raymond understood the distinction between tropical and sidereal longitudes.

[56] *The Khaṇḍakhādyaka (an Astronomical Treatise) of Brahmagupta, with the Commentary of Bhaṭṭotpala*, ed. and trans. by Bina Chatterjee, 2 vols (Delhi, 1970), II, p. 94.

[57] Haskins (n. 7 above), p. 96 ff. The tables and preface are found in two manuscripts: Paris, Bibliothèque nationale, lat. 14704, fols 110–35ᵛ, and Cambridge, Fitzwilliam Museum, McClean 165, fols 44–7. The preface alone is found in Oxford, Corpus Christi College 243, fol. 53.

In two of the manuscripts which are used elsewhere in this survey[58] one finds a version of the Toledan tables under the name of Roger of Hereford. This English scholar who flourished ca. 1175 is also the author of an original and influential treatise on the Latin calendar. His tables are computed for a meridian situated 4;30 west of Toledo, or 21;0 west of Marseilles. In his prefatory remarks[59] he makes it clear that he has derived his radices by a suitable addition to those of the earlier tables for Marseilles. His own begin in '1092', that is, midnight at the end of December 1092, and then proceed in steps of twenty-eight years. He notes the total solar eclipse of 13 September 1178, which he very probably observed, and for which he noted the times according to the meridians of Arin, Toledo, Marseilles and Hereford. It is not clear whether he determined the difference 21;0 between the meridians of Marseilles and Hereford by a comparison of the observed times of this eclipse.

5. *Tables for Pisa and London*

Tables for London and Pisa are referred to in a number of places in the literature of the twelfth and thirteenth centuries, but hitherto no one has brought together the various materials in order to determine the texts and parameters. Altogether, I find that some twenty manuscripts have a bearing in one way or another on this theme. The close relation between the tables for London and Pisa is not evident at first, but emerges after some investigation.

Perhaps it would be as well at the outset to dispose of a popular misconception. Duhem noted a reference to the London tables which he found in a copy[60] of a tract written ca. 1300 by Andalo di Negro, which he linked with no logic whatever to a table of a very common type found on fol. 1ᵛ of the same manuscript, and which was used to facilitate the conversion of dates from Latin to Hijra Eras, beginning with the year A.D. 1232. Duhem asserted that the London tables must have been composed for the year A.D. 1232, and so originated a mistake which has proved to be as popular as it is wrong.

Tables expressly for the meridian of London, which we denote L for brevity, are found in two English manuscripts,[61] the former of which includes also tables of equations, and a number of short chapters which give instructions for the use of the tables. The radices are calculated for Noon, 1 March 1150, which is denoted '1149', that is, 1149 complete years, and for subsequent years in steps of twenty. The mean longitudes are all tropical, and the solar motion in twenty years (7305 days) is 20^r 0;9. This implies a year of very nearly $365^1/_4 - ^1/_{131}$

[58] Madrid, Biblioteca nacional, 10016, fols 73–85 and London, British Library, Arundel 377, fols 86ᵛ–87. The second manuscript includes only a short text and a selection from the tables of mean motion.

[59] Arundel 377, fol. 86ᵛ: 'quod et factum est in presenti tabula super motus Massilie: tum propter annos domini et bissextos, tum propter menses romanos. Maluimus enim hec quam annos Arabum et eorum menses propter difficultatem sequi, eo quod inusitata sint apud nostrates'.

[60] MS Paris, Bibliothèque nationale, lat. 7272, fols 60–67ᵛ. In cap. v (fol. 67ᵛᵃ he says that 'civitas super quam fundate sunt iste tabule est Londonie', an odd remark, since this text does not otherwise appear to be the usual set of instructions. The passages from this part of the work have been published by L. Thorndike, *The Sphere of Sacrobosco and its Commentators* (Chicago, 1949), pp. 463–75. See P. Duhem, *Le Système du monde* (Paris, 1914), III, p. 237.

[61] London, British Library, Arundel 377, fols 7–35 and Oxford, Bodleian Library, Digby 114, fols 25–32ᵛ.

days. At epoch the mean Sun is 345;26,38. The same chapters of instructions are found in two Bodleian manuscripts,[62] where in each case they are accompanied by yet another text[63] in which it is said that Robert of Chester made the tables for London, and which adds various details which agree with the features of tables L. This text, however, is written around 1170, and therefore by someone later than Robert himself. The date 1170 is fixed by the author when he cites the solar eclipse of 21 May 1137, for he notes the interval from that date to 1170 complete years.

The short chapters which were referred to above are clearly related to the Latin *Zīj* of al-Khwārizmī, and in such a way as to confirm Robert as their author. For example cap. xiii on retrogradation is repeated almost word for word. In others the opening sentence or other details are reproduced. Moreover the Arabic term *kusūf* (eclipse) is transcribed *elkuszuf*, which is the spelling peculiar to Robert's version of the *Zīj*, in cap. xxxiii. Evidently Robert has tried to take what he can from the earlier work, even though the underlying Ptolemaic theoretical context is so unlike the Indian.

In a tract on the astrolabe[64] Robert makes an interesting remark in the penultimate chapter whose title is 'De equatione domorum .xii. iuxta Albucasim Almacheritam (Abū'l-Qāsim al-Majrīṭī)', beginning:

> De ratione coequationis .xii. domorum in libro canonum que super Toletum et civitatem Londoniarum edidimus prout tractatus exposcebat ratio tractavimus. Unde in hoc libro qualiter ipse inveniantur per astrolabium docere curabimus. (Canon. Misc. 61, fol. 22v)

The reference is clearly to cap. xxxvi in the *Zīj*, version *N*, which has the same title, 'De equatione .xii. domorum', and which is of course by al-Majrīṭī. On the other hand, there is no such chapter, or corresponding table of *Equatio casarum*, in the extant material for the tables of London or Pisa. If Robert's 'book of canons for Toledo' are the tables referred to as the 'first part' in the text noted above (n. 63), there is a difficulty again over the date, for Robert would not have composed tables beginning with a year as late as 1169, although this may be due to ignorance on the part of the anonymous author. There are in fact references to Toledo to be found here and there in our material, enough to suggest that a version for Toledo may well have existed. For example, in the principal copy of

62 Savile 21, fols 91–95v and Ashmole 361, fols 13ra–19vb.

63 Ashmole 361, fols 19vb–28; Savile 21, fols 86–91. The opening words: 'Incipit pars altera huius operis que videlicet ad meridiem urbis Londoniarum iuxta Albaten Haracensis summam per Robertum Cestrensem contexitur', as well as a difficult passage further on (Ashmole, fol. 24rb; Savile, fol. 88v): 'ea namque eius pars que ad meridiem civitatis Toleti constituitur a .mclxix. anno domini incipit et ab eodem termino annus domini per .xxo. (sic Ash.; Sav. '28') colligens...altera vero eius pars cuius videlicet ratio ad meridiem urbis Londoniarum contexitur ab anno domini .mcl. sumpsit exordium et ab hoc termino annos domini per vigenum coacervans...', have been discussed by M. Steinschneider, 'Die europäischen Übersetzungen aus dem arabischen bis Mitte des 17. Jahrhunderts,' *Sitzungsberichte des Kaiserlichen Akademie der Wissenschaften in Wien*, philosophisch-historische Klasse, 149 (1904) and 151 (1905), reprinted together, Graz, 1956, pp. 68–9, and Haskins (n. 7 above), p. 123. The author, who could not have been Robert himself, is unaware that the London tables were based not on the *Zīj* of al-Battānī, but on that of al-Ṣūfī. He clearly indicates that the 'first part' (whose author is not actually named) was a set of tables on the meridian of Toledo, using the Latin calendar starting in January, the years beginning 1169 (complete ?) and grouped by 28, and somehow based on Ibn Ezra.

64 MS Oxford, Bodleian Library, Canon. Misc. 61, fols 12–22v. Haskins (n. 7 above), p. 123, in his comments about the opening words does not take the point about the title of the chapter or the reference to the *Zīj*.

the London tables (MS Arundel 377), the correction to the mean motion of Mars is given corresponding to the difference 10;30 between the meridians of Toledo and London. There are several sets of tables with years beginning in January, and grouped by 28, but they are adaptations of the Toledan tables for various meridians: Marseilles, Hereford, Ferrara and Novara.

I turn to another set of tables (which will be denoted P), which are found in three manuscripts.[65] These differ from the tables L in only one respect, namely that they are drawn up for an unnamed meridian situated 14° to the east of that for tables L, viz., London. This difference between the longitudes of the meridians is fixed as usual by a straightforward analysis of the mean motions. The radices begin at Noon, 1 March 1170, that is, the date of the second row of the tables L. One of the copies of these tables fills both sides of the fourth folio of the MS Paris, Bibliothèque nationale, lat. 16208, inserted among the leaves carrying the Latin version of the *Zīj* of al-Khwārizmī for the meridian of Toledo, which is described in section 1.7 above.

Now there are three texts known which were composed for these tables. The first has been printed, in an edition by Millás-Vallicrosa,[66] and is certainly by Abraham Ibn Ezra. This is a long discursive effort composed in 1154, in which the author relates his composition of tables for Pisa, including many interesting references to other astronomical tables and texts. There is, however, neither a set of rules for the use of the tables, a geometrical *theorica*, nor a full listing of parameters. He has drawn on al-Zarqāla's tract on the solar year.[67] We may list briefly the few precise details of the tables for Pisa which are given in the work:

1. They are derived from the *Zīj* of al-Ṣūfī, as 'verified' by al-Zarqāla in A.H. 482/A.D. 1089.
2. The epoch adopted is Noon, 1 March, at Pisa, situated at 33° relative to Baghdad at 80°. The first year listed is A.D. 1089.
3. The Sun's motion in 7305 days is 20^r 0;9.
4. The years are given in steps of twenty, with leap years at the second, sixth, etc.

These details would be consistent with an identification with either of the tables L or P.

The second text is another long and discursive work, similar in general style to the first, but anonymous. It is found uniquely in the MS Cambridge, Fitzwilliam Museum, McClean 165, fols 67–76^v. Here also the author writes for one particular set of tables, which were composed in 1143, following the *Zīj* of al-Ṣūfī as verified by al-Zarqāla. Many other details can be extracted from the

[65] Paris, Bibliothèque nationale, lat. 16208, fol. 4–4^v; ibid., lat. 16207, fols 19–36; Oxford, Bodleian Library, Seld. sup., 90, fols 30–4.

[66] *El libro de los fundamentos de las Tablas astronómicas de R. Abraham Ibn Ezra* (Madrid and Barcelona, 1947).

[67] In his tract on accession and recession, al-Zarqāla refers to his 'Suma relativo al Sol': see the Spanish translation of the fourteenth-century Hebrew version (Paris, Bibliothèque nationale, héb. 1036) made by Millás-Vallicrosa (n. 8 above, p. 279 etc.). This work on the Sun, which is not extant, is presumably the same as that noted in the present Latin text as *De anno Solis*.

text, enough indeed to show convincingly that it is in fact the tables P, not L, which are relevant. The author says, for example, that in 1029 (complete) the mean Sun is 344;30,20, which is consistent with P, not L. The author names al-Zarqāla's *De anno solis* as one of his sources. The text is not dated, but it is stated that 1149 (complete) is before 'the present year'. It is also anonymous, but since the author refers impersonally to the tables, and also to 'annos domini Ihesu Christi' (fol. 68), he cannot be Ibn Ezra, and indeed must be a Christian. It is reasonable to infer that:

1. Both texts are for the same tables, viz., those through the meridian of Pisa, identified as the tables P.
2. The tables for Pisa were composed in 1143 by Ibn Ezra, on the basis of those of al-Ṣūfī.
3. The first text was composed in 1154, but the second may have been composed before or after that date.

The third text, one expressly attributed to Ibn Ezra, is found in one of the manuscripts[68] in which the London tables are found. This is a relatively brief set of instructions for the use of the Pisan tables, whose meridian is given the longitude 33°, while the radices were evidently given for years beginning in 1149 complete. The conversion to the meridian through Winchester and Angers, at 24°, is indicated, and it is striking that the difference 9° is so inconsistent with the London–Pisa difference of 14° assumed by the compiler of the London tables.

At this stage in the development of Latin astronomy, geographical longitudes reflect a mixture of influences from Ptolemy's *Geography*, and from contemporary researches. Ptolemy's coordinates for Pisa and London are respectively $33^1/_2$ and 20, essentially the values used in our tables P and L. On the other hand the larger figure 24 used for Angers and Winchester must reflect a genuine determination of the difference Pisa-Angers, for the true difference 10;56 is not very different from the assumed 9°. Something is known of Ibn Ezra's extensive travels, and Fleischer, as quoted by Millás-Vallicrosa,[69] determined that he was in Tuscany, and in Lucca in particular, in the period 1141–5, when the tables for Pisa were composed by him.

The *Zīj* of al-Ṣūfī is not extant in Arabic, but it is noted by Ibn Yūnis. In his *Ḥakimī Zīj*, cap. v,[70] Ibn Yūnis says that al-Ṣūfī fixed the movement of the Sun in a Persian year (365^d) at 359;45,40,2. This implies a movement in 7305 days equal to 20^r 0;9,2,19. Among Arabic sources known to me this is about as close as one can come to the motion used in the Latin tables for Pisa and London. Ibn Yūnis himself used a mean Sun which entails a movement in twenty years equal to 20^r 0;9;2,53. The tables of al-Ṣūfī were not singled out especially by Ibn Yūnis, and it is curious that they should have been chosen by Ibn Ezra. Al-Ṣūfī

68 Arundel 377, fols 56^v–63.

69 Millás-Vallicrosa (n. 66 above), p. 17.

70 Caussin (n. 50 above), pp. 150, 154.

was well known as the author of a star catalogue, noted for its elaborate figures of the constellations, and this is the only work of his known in Latin.[71]

6. *The Zīj of al-Battānī*

This important handbook, written around A.D. 880 by Abū ʿAbdallāh Muḥammad b. Jābir b. Sinān al-Battānī, has a set of tables drawn up for the meridian of al-Raqqa, near Ḥarrān. While it has been established that the whole work, both tables and text, came to be known to Latin scholars in the twelfth century, there are nevertheless problems in understanding the history of its translation and diffusion. Plato of Tivoli translated the text ca. 1136, but none of the extant manuscripts[72] of his version contain the tables. The importance of the work to Spanish Islamic astronomers can hardly be doubted, and yet only one Arabic manuscript is extant, Escorial, Real Biblioteca, arabe 908, which Nallino judged to belong to the end of the eleventh century, or beginning of the twelfth.[73] There is also a Spanish version, Paris, Bibliothèque de l'Arsenal, 8322, originating from the court of Alfonso el Sabio.[74]

One can prove that the tables were known to Plato and used by him, for he ended his translation of Bar Ḥiyya's *Liber embadorum* with a list of the positions of Sun, Moon, and planets for 6 August 1145, all computed accurately from this *Zīj*.

	Plato of Tivoli	*Zīj* of al-Battānī	Toledan tables
Sun	140;15	140;9	139;36,5
Moon	342;20	342;27	343;23,42
Saturn	38;57	38;58	38;19,30
Jupiter	26;52	27;37	26;32,37
Mars	207;15	207;7	207;20,48
Venus	182;29	182;26	181;59,8
Mercury	134;45	134;36	134;3,24
Node	90;1	90;1	89;20,42

In this table the true (not mean) positions are computed from the tables of al-Battānī and from the Toledan tables, both of which were available to Plato. It is clear beyond question that he used those of al-Battānī.

It is much more difficult to form a settled opinion about the translation of

[71] P. Kunitzsch, 'Ṣūfī Latinus', *Zeitschrift des Deutschen Morgenländischen Gesellschaft*, 115 (1965), pp. 65–74.

[72] Nallino (n. 3 above), I, pp. l–lvii, referred to the following copies: Erfurt, Bibliotheca Amploniana, Q.350; Milano, Ambrosiana, H 109 sup.; Paris, Bibliothèque nationale, lat. 7266; ibid., lat. 16657; Venezia, Biblioteca nazionale Marciana, lat.VIII.14 (Valentinelli XI.60); Città del Vaticano, Biblioteca apostolica, lat. 3098. It is also found in Oxford, Bodleian Library, Canon. Misc. 61, ibid., Digby 40 and Digby 51 (incomplete). An edition of Plato's translation would clearly be very useful.

[73] Nallino (n. 3 above), I, p. lxi. This was of course the basis of Nallino's superb edition.

[74] The Spanish version has been edited by G. Bossong, *Los Canones de Albateni* (Tübingen, 1978), who analyzes the language of the translation in *Probleme der Übersetzung wissenschaftlicher Werke aus dem Arabischen in das Altspanische zur Zeit Alfons des Weisen* (Tübingen, 1979).

al-Battānī attributed to Robert of Chester. The attribution occurs in two remarks. The first is written by Hermann of Carinthia in the preface to his translation of Ptolemy's *Planisphere*.[75] Hermann alludes in the briefest way to his own translation of 'al-Khwārizmī'[76] and to Robert's translation of 'al-Battānī', without in either case naming the works in question. He seems to be saying that they would be more careful than Ptolemy in noting that a certain arc on the sphere measures the distance between the equator and the tropics (23;51). It is difficult to see the real point of Hermann's interjection of these two Arabic names, and Nallino took the view that this was merely a display of learning. The extant works of al-Battānī and al-Khwārizmī are nowhere concerned with the theory of the astrolabe; but then he may simply have had in mind that they used smaller values of the obliquity. The other remark which attributes a version to Robert was quoted above in section 5. There again, as I explained, one hesitates to accept the information at face value. Burnett, however, has not quarrelled with the attribution, and his extensive and convincing arguments in favour of Hermann's use of al-Battānī's text leave one in no doubt that he, at any rate, knew it.[77]

An interesting sign of the use of al-Battānī's *Zīj* in Andalusia is provided by Maimonides's tract on the visibility of the New Moon,[78] the phenomenon which marks the start of the Jewish month. Neugebauer, in his commentary to the edition cited, drew attention to the derivation of many of the parameters from al-Battānī. These are sometimes rounded off unsatisfactorily. For example, the rates of motion of the Sun and Moon are drawn from al-Battānī's table of the motions in days, which provides only a rough approximation to the exact rates implied by the tables of radices. It is interesting, however, to find that the radices assumed by Maimonides are calculated with greater exactness from al-Battānī. They are calculated for 6 p.m., 22 March 1178, the point which marks the start of 3rd Nisan 4938 in the Jewish calendar, but for a meridian 12;30 west of

[75] It is quoted by Nallino (n. 3 above), I, p. xlix, and by Burnett (n. 30 above), p. 9, n. 40, who offers the emendation *opera* for *ope* in the passage: 'Metitur itaque (ait Ptolemeus) descriptio nostra utrumque arcum NG, et GH partibus XXIII punctis fere LI ex eis quae CCCLX totum ABGD circulum metiuntur, quae pars est distantia utriusque tropici a circulo aequinoctiali. Quem locum a Ptolemaeo minus diligenter perspectum cum Albatene miratur et Alchoarismus: quorum hunc quidem ope<ra> nostra Latium habet, illius vero commodissima translatione Roberti mei industria, Latinae orationis thesaurum accumulat.'

[76] Hermann refers elsewhere to his translation, according to a remark which was found by M. Steinschneider, *Die hebräischen Übersetzungen des Mittelalters und die Juden als Dolmetscher* (Berlin, 1893), p. 568, n. 224, in Hermann's version of the *Introductorium maius* of Albumasar; see Haskins (n. 7 above), p. 44 and Lemay, *Abu Maʿshar and Latin Aristotelianism in the Twelfth Century* (Beirut, 1962), p. 37, n. 1: 'In tercio et secundo celeres, quorum plus fi alcurdaget azerea (in the margin: 'in seccionibus festinis') secundum fi alcurdaget albatia (in the margin: 'tardis') tractatur que in translatione nostra zigerz Alchuarismi sufficienter exposuimus'. This is found in MS Napoli, Biblioteca nazionale, VIII.C.50, fol. 43. The terms *azerea* (= *al-sarīʿa*, 'rapid') and *albatia* (= *al-baṭīʾa*, 'slow') occur in MSS *O* and *C* of the *Zīj* of al-Khwārizmī as *zare* (*O*), *care* (*C*) and *beteia* (*O*), *batiae* (*C*); Suter (n. 3 above), table 24 (Solar equation). On the term *curdaget* see section 4.1 above. Haskins mistranscribed *festinis* as *formis*.

[77] Burnett (n. 30 above), pp. 30 and 333–7.

[78] *The Code of Maimonides. Book Three, Treatise Eight. Sanctification of the New Moon*, trans. S. Gandz, with comm. by O. Neugebauer (New Haven, 1967).

al-Raqqa.[79] The intended meridian was Jerusalem for which Maimonides gave the longitude 66°, that is 7;15 west of al-Raqqa.

7. *Ptolemy's Almagest at Béziers*

A tract on the astrolabe was written by Rudolph of Bruges, pupil of Hermann of Carinthia, and dedicated to 'John David'. It is known in a number of manuscripts, as noted by Haskins.[80] Rudolph records an observation he has made of the Sun's altitude at Béziers in 1144:[81]

> In the year 1144 at Béziers in Gotia on the Mediterranean Sea, I wanted to make a plate for the latitude of that place. On 24 April the Sun was in the fourth degree of Taurus, when its altitude was 62 degrees, which when divided by 6 gives 10 divisions and a sub-division amounting to two degrees. At the point on the meridian line corresponding to this degree of the Sun put 'p'.

The true longitude of the Sun on that day according to modern calculation is 40.6, when the declination is 15.1. Consequently Rudolph's altitude measurement 62 entails that his latitude was 90 – (62 – 15.1) = 43.1, which is agreeably accurate for Béziers.

His solar longitude – 34th degree – is of course calculated for that day from some unknown tables. The following are the results of mean and true longitudes according to four possible sources:

Source	Mean	True
Ptolemy	33;38	34;51
al-Khwārizmī	29;45	31;23
al-Battānī	39;29	40;54
Toledan tables	30;59	32;24

Somewhat surprisingly, it appears that Ptolemy's tables, either from the *Almagest*, or the *Handy Tables*, were used. Although the *Almagest* had not yet been translated, it is mentioned in Adelard's translation of the *Zīj* (preface, cap. xxiii, and also in a note in MS *C*, in a later hand, attached to cap. xxi), and certainly Hermann and Robert of Chester were actively studying it, as we know

[79] The comparison of the mean longitudes for this date is as follows:

	Maimonides	al-Battānī 12;30 west
Sun	7;3,32	7;3,35
Moon	31;14,43	31;14,41
Lunar argument	84;28,42	84;28,38
Lunar node	–180;57,28	–180;57,14

[80] Haskins (n. 7 above), p. 56. P. Kunitzsch, 'On the Authenticity of the Treatise on the Composition and Use of the Astrolabe Ascribed to Messahalla', *Archives internationales d'histoire des sciences*, 31 (1981), pp. 42–62, has examined this tract in relation to the popular Latin text on the astrolabe ascribed to Messahalla.

[81] London, British Library, Cotton, Vesp. A.II, fol. 36vb; Oxford, Bodleian Library, Digby 51, fol. 27va. In the latter the date is given wrongly as 'mclxiiii'.

from the Preface to Robert's translation of al-Kindī's *Judicia*,[82] and from the numerous references to it in Hermann's *De essentiis*.[83]

A version of the *Handy Tables* of Ptolemy had been available for some time in the *Praeceptum Canonis Ptolomei*.[84] This includes tables for Sun and Moon, and an introductory text translated from a Greek original which was composed ca. A.D. 540, and which gives a rather lame and imprecise account of the use of the tables. The tables repeat exactly the numerical entries of the *Handy Tables* for the Sun and Moon, both means and equations. However, as part of the translation, the Greek month names (Thoth, etc.) are replaced by so-called Latin equivalents beginning with September. These replacements are of course incorrect, since the Greek tables were computed for the Egyptian year of 365 days. Further the Era of Philip for which the *Handy Tables* are computed is not defined precisely in the text, and Latin scholars would know its relation to the Era of the Incarnation only by means of those chronological tables which came in translations of Arabic *zīj*es. It is clear therefore that this version of the *Handy Tables* could never have been the source of any Ptolemaic calculation, such as that which appears to have been made by Rudolph of Bruges. The date of the Latin translation of the *Praeceptum* is quite uncertain, but is not later than the eleventh century. Since Apamea is used in an example, it may well be that the Greek original was composed in Syria. The author also refers to Theon at one point, and one might regard the whole as a gloss on a text of Theon, rather than a gloss directly on Ptolemy.[85]

[82] Haskins (n. 7 above), p. 121.

[83] Burnett (n. 30 above), pp. 28–9.

[84] A number of copies are known, of which the following were consulted: Chartres, Bibliothèque municipale, 498, fols 174–89^{v} and 238–40^{v}; Oxford, Corpus Christi College, 283, fols 66–81^{v}; London, British Library, Harley 2506, fols 55va–69ra. The first of these contains terms written in uncial Greek script, and in a Greek orthography consistent with a sixth-century provincial origin. I am grateful to Anne Tihon for correspondence about this copy, and reference should be made to *Le 'Grand Commentaire' de Théon d'Alexandrie aux Tables faciles de Ptolémée*, ed. J. Mogenet and A. Tihon, Studi e testi, 315 (Vatican City, 1985). Note that the first two manuscripts contain copies of the *Zīj* of al-Khwārizmī, as listed in section 1.1.

[85] I am grateful to the British Academy for the generous support which made possible numerous visits to libraries here and abroad.

APPENDIX

*Arabic Words and Latin Equivalents**

	Arabic	*L*, *Cc*	*O*	*C*	*N*
1.	ʿalāma, *plural* ʿalāmāt		elaalemet[t] nota[(t)]	elaelamet[t] notula[(t)] dies	notula[(t)] dies[t]
2.	auj	elaug auc[t] sublimatio	elaug sublimatio[(t)] sullimatio[(t)]	elaug sublimatio[(t)] sullimatio[(t)]	sublimatio[(t)] sullimatio[t]
3.	buht		elbuht[(t)] stupor motus[t]	elbuht[(t)] stupor cursus[t] motus[t]	elbuht motus diurni et horarii atque momentanei[t]
4.	burūj	signa[(t)]	elburug[(t)] signa[(t)]	elburug[(t)] signa[(t)]	signa[(t)]
5.	daqāʾiq	dakaice (*L*) dekaice (*Cc*) minuta[t] puncta[t]	dakaice partes[(t)] minuta[t]	dakaice[(t)] dakaike[(t)] puncta[t]	minuta[(t)]
6.	faḍl, *plural* fuḍūl		elfudhul[t] superatio[(t)]	elfudhul[(t)] superatio[(t)]	superatio[t] residuum
7.	falak		felek[(t)] rota[(t)] circulus	felek[(t)] rota[(t)] circulus	rota[(t)] circulus[(t)]
8.	falak al-mustaqīm		elfelek elmustakim[(t)] circulus equinoctialis circulus rectus	felek elmustakim[(t)] circulus equinoctialis rota directa[(t)]	circulus equinoctialis
9.	ḥāṣil	elhahehil (*L*) elbahebil (*Cc*) obtentum <centrum> ultimum	hacil obtentum	hacil obtentum <centrum> ultimum	
10.	ḥiṣṣa	elheza[(t)] argumentum	elheza[(t)] argumentum[(t)]	elheca[(t)] hezat[t] argumentum[(t)]	argumentum[(t)]
11.	ijtimāʿ		elistima eligstimaat[t] coniunctio[(t)]	elistima elistiamaat[t] coniunctio[(t)]	coniunctio[(t)]
12.	istiqbāl		elistikbel elisticbelet[t] oppositio[(t)]	elistikbel elisticbelet[t] oppositio[(t)]	oppositio[(t)]

* Note that 't' refers to the tables, while '(t)' refers to text and tables jointly.

Arabic	*L*, *Cc*	*O*	*C*	*N*
13. jadwal	pagina tabula^{t}	gedval$^{(t)}$ pagina$^{(t)}$ tabula$^{(t)}$	gedval^{t} pagina tabula$^{(t)}$	tabula$^{[t]}$
14. jaib		elgeib$^{(t)}$	elgeib$^{(t)}$	sinus$^{(t)}$
15. janūbī	genubi	genubi	genubi australis meridionalis	australis meridionalis
16. jauzahr	draco$^{(t)}$	egauzehar$^{(t)}$ elgavzehar^{t} draco$^{(t)}$	eliauzehar eliavzehar^{t} draco^{t}	draco$^{(t)}$
17. kabīsa		elkebice$^{(t)}$	elkebice	elkebice
18. kusūf *plural* kusūfāt	eclipsis (*Cc*) eclypsis (*L*)	elkucufet^{t} eclypsis eclipsis$^{(t)}$ defectus$^{(t)}$	elcuzuf elkusufet^{t} elkucufet^{t} eclipsis$^{(t)}$ defectus$^{(t)}$	elkuszuf eclipsis$^{(t)}$ defectus$^{(t)}$
19. mabsūṭa	singuli^{t}	plani^{t}	mabzota^{t} plani$^{(t)}$	plani$^{(t)}$ residui^{t}
20. mail	megil obliquatio inclinatio^{t}	megil obliquatio$^{(t)}$	elmeil$^{(t)}$ obliquatio$^{(t)}$	obliquatio$^{(t)}$
21. majmūʿa	adiuncti^{t}	elmagmoa^{t} collecti$^{(t)}$	maimoa^{t} maiora^{t} collecti$^{(t)}$	collecti$^{(t)}$
22. mankūs		elmankuz diminutus	elmankut elmakuz diminutus	diminutus
23. maṭraḥ al-shuʿāʿ		matarech ascovva^{t} proiectio radii^{t} radiatio	matrak alshoa^{t} matareh eshoa^{t} proiectio radii^{t} radiatio	proiectio radii^{t} radiatio
24. mauḍūʿ		locus^{t}	maudho^{t} locus^{t}	locus^{t}
25. muʿaddal	examinata	examinata$^{(t)}$	muaddala^{t} definita$^{(t)}$	definita^{t}
26. mukth		mora^{t}	elmuht elmukdh^{t} mora^{t}	elmuht mora$^{(t)}$
27. sāʿa, *plural* sāʿāt	hora$^{(t)}$	zahat^{t} hora$^{(t)}$	caat^{t} caht^{t} hora$^{(t)}$	hora$^{(t)}$

Arabic	*L, Cc*	*O*	*C*	*N*
28. suqūṭ		elzukut$^{(t)}$ etc. elzukuz^{t} casus$^{(t)}$	elciicuth$^{(t)}$ elciikuf elciicut^{t} elzukut^{t} casus$^{(t)}$	elszukuf casus$^{(t)}$
29. suṭūr al-ʿadad	regule numeri^{t}	semite semite numeri/orum^{t} vie numeri^{t} pagina numerorum^{t}	cotoraladet catareladad^{t} catureladad^{t} semita/e numeri/orum^{t} vie numeri/orum^{t} sulci numerorum^{t}	linee numeri$^{(t)}$
30. taʿdīl	tadil examen directio^{t}	tadil$^{(t)}$ examen examinatio$^{(t)}$	tadil$^{(t)}$ examinatio$^{(t)}$	coequatio$^{(t)}$
31. tadwīr		eltedewir^{t}	elthedewir^{t}	
32. thāniya	secunde minutie$^{(t)}$	elthenie secunde$^{(t)}$ minutie	zenie elthenie$^{(t)}$ minutie secunde$^{(t)}$ minute^{t}	secunde$^{(t)}$
33. wasaṭ	elwacat elwazat medialitas medius motus^{t}	elwazat medialitas$^{(t)}$ medium	elwacat medius cursus^{t}	medius cursus$^{(t)}$
34. zīj	ezic	ezic ezich$^{(t)}$ opus$^{(t)}$	ezich	ezeig chanonum

VIII

The lost Zīj of al-Ṣūfī in the twelfth-century tables for London and Pisa[*]

1. INTRODUCTION

Many Zījes, or astronomical handbooks, are known from Islamic sources. The earliest is that of al-Khwārezmī (mid ninth century), and by the time of 'Abdu'l-Raḥmān ibn 'Umar, al-Ṣūfī (died A.D.986) a large number are known. The Zīj al-Ḥakimī, of ibn Yūnis (ca.A.D.1000) is one of the finest illustrations of Islamic astronomy. It is based not only on some exceptionally accurate and careful observations, but as an exposition it is particularly full and critical. The work is useful to the historian because of the numerous reports of the details of earlier work. In Ch. 5 Ibn Yūnis gives details of solar parameters according to eleven different Islamic astronomers who have carried out their own independent observations. These begin with Yaḥyā ibn Abī Manṣūr, the principal author of the Zīj al-Mumtaḥan, compiled in Baghdad in the later ninth century, and end of course with the work of Ibn Yūnis himself. In a reference to al-Ṣūfī, he reports his finding that in a Persian year (365 days), the Sun moves through the angle 11s 29;45,40,2. This implies a mean solar year equal to

$$365 \text{x} 360/359;45,40,2 = 365.2423582 \text{ days.}$$

This is an accurate value, in line with the rapid improvements in accuracy at that time as a result of efforts to determine the length of the year. No other parameters are quoted from the work of al-Ṣūfī, and it is clear that this particular work did not stand out above others in the view of Ibn Yūnis. From a number of the other writers, he quotes not only the motion of the Sun, but also its equation and apogee, as well as the obliquity of the ecliptic.

---

[*] This study was originally published in the *Lectures of the Conference of al-Ṣūfī and Ibn Nafīs, 5-8 Oct. 1987, University of Jordan*, Beirut: Dār al-Fikr, 1991, pp. 38-72, but with numerous misprints and formatting problems. In this corrected version the opportunity has been taken to add some points of updating.

2. THE TABLES FOR LONDON AND PISA

2.1 *The Zīj of al-Khwārezmī*

Scientific Latin astronomy begins in the early twelfth century with the first translations from Arabic sources in Spain. Once again the Zīj of al-Khwārezmī stands at the head of the tradition, evidently studied intensely in Islamic Spain long after it had ceased to be of interest in the Middle East.

It was translated in full by Adelard of Bath, some time in the decade 1125-35, following an earlier shorter version by Petrus Alfonsi, a Jew converted to Christianity, and living in England. A final revision was made by Robert of Chester ca.1140, who found Latin words for virtually all of the Arabic technical terms, many of which had been merely transcribed. We are fortunate in having a number of twelfth-century MSS which preserve for us much of this early material (Mercier 1987a).

The Zīj of al-Khwārezmī was perhaps the first, but by no means the only Zīj which came to be studied by Latin scholars in the twelfth century. Robert of Chester became involved with another set of tables which, like the Zīj of al-Khwārezmī, was brought from Arabic sources by another Jewish scholar Abraham ibn Ezra, who was much more eminent than Petrus Alfonsi. We know these tables in two versions, which differ only in that they are referred to two different prime meridians, Pisa and London. Our information is drawn from some 20 Latin MSS in the libraries of London, Oxford, Cambridge, Paris and Madrid. The MSS are listed at the end of the paper, where each is given a symbol, such as L1 or P1 (for London or Paris), and there are also listed the various tables and texts which are found therein, with cross-references to the MSS.

We have the tables themselves, in both versions, the usual sets of instructions arranged in short chapters, and certain related texts of great importance for understanding the tables. The tables are attributed to al-Ṣūfī in two of these.

2.2 *Description of the tables*

The Tables for Pisa (text T.I, see section 9) are known from three MSS, two in Paris, and one in Oxford. The copy in P1, MS latin 16208 of the Bibliothèque Nationale, appears to be the oldest, but the others are not simply direct copies of it. The fact that they are for the meridian through Pisa is not really explicit in these copies, but is known from two texts which may be shown to have been written with these tables in mind. The tables of mean motions for collected years are gathered together in Table I. The years for which the motions are given are '1169 to '1289 , in steps of 20. Each of these years refers to the number of completed years in a version of the Latin calendar, reckoning the year from Mar.1, so that '1169 means A.D.1170, March 1, Noon. The mean longitudes are linear functions of the time, and are known when we have the value of the motion at a certain date, that is the 'radix' of the motion, and also the rate of motion. In this case the radices are taken from the first row of each table, '1169 , and the rate is expressed as the motion during 20 years (7305 days).

The tables for London (text T.I) are known from two MSS, in London and Oxford, respectively. The copy in L1, Arundel 377 of the British Library, is much more complete; it contains not only the tables of equations, but is also the only copy of any of these tables which has also the 'canons' or short chapters giving the instructions for usage. The motions are again given in steps of 20 years, but beginning in '1149 , that is A.D.1150 March 1. The longitude of London is given as 19 degrees in the table for the mean Sun, according to the principal copy, the manuscript L1. The tables of collected years are in Table II. In this case there is an explicit reference to London as the prime meridian.

Apart from the tables of motion for collected years, there are found the usual tables for individual years (up to 20), for months, days, and hours.

These various copies are in need of emendation, arising from errors which are partly arithmetical, and partly scribal. It is possible to make a nearly complete emendation, depending on three lines of criticism: internal arithmetical consistency, data available from the various related texts, and comparison with the scientific calculation of the positions, at that time, of the Sun, Moon, and planets. The last approach is made through an examination of the 'deviation' curves, the deviation being the difference between the medieval mean longitude, and the modern one, and interpreted in the light of Islamic and other medieval material, which serves as a guide to the level of accuracy generally attained by astronomers then.

The two versions of the tables show inconsistencies in many details. For example one sometimes finds that errors crept into the motions computed for individual years, which go up to 20, and that this incorrect value is carried over to the construction of the table for collected years. This happened with the lunar node in the London tables, in which the calculator found 1r 46;17,17 for 20 years, instead of the correct rate 1r 46;19,17 as used in the Pisan version. The deviation curve shows that the latter is correct. Even though the difference is slight, amounting to only 6 seconds of arc p.a., this is large compared with the sort of error associated with this parameter. In the Zīj al-Hakimi, for example, the deviation from the true value is about 1 second p.a.

The Pisan tables were certainly developed before those for London, but the copies available to us cannot be quite identical with the original. The fact that the motions are given for the years from 1170 is a strong indication that they were compiled after the London tables. Uncertain as we are about the detailed history of our copies, it is tempting to see them as influenced in part by the London version. Both have a value for the lunar motion which is clearly wrong: 267r 133;33,34 is proposed as the simplest correction to 267r 133;43,34, the only emendation to be proposed which is not already found somewhere in the texts. The implied excess of 30 seconds p.a. in the lunar motion is altogether too large to have been accepted at that time. With the emendation however, the implied synodic month is then 29.53059312 days, extremely close to the value 29.53059414, the astonishingly accurate value inherited from Babylonian practice, and often found in Islamic material.

After these various emendations have been established one can determine the actual mean parameters, and these are collected in Tables III and IV, below, in section 8. In Table III there are given also the values of the radices for '1170 calculated from the radices of the Pisan tables after a shift of 14 westward. The agreement of these calculated radices with those for London is satisfactorily exact, particularly for the Sun and Moon, where the discrepancy is no greater than 0.6 arc second. This difference of 14 degrees in the meridian of reference is the only difference, really, between the tables for London and Pisa.

The equations are defined according to the usual Ptolemaic theory, transmitted to the Islamic world essentially by means of Ptolemy's Handy Tables, the prototype of every Zīj. Islamic astronomers were interested above all in the Sun and Moon, and generally left unaltered the Ptolemaic parameters of the planetary equations. In our present tables of the equations, the only non-Ptolemaic value is the equation of the Sun, 1;59,10. This is however the value given to it by al-Battānī, whose tables were certainly known to Latin astronomers at this time, and we cannot be at all sure that the present tables of equations were not borrowed from al-Battānī. Other values of the solar equation reported by Ibn Yūnis were

Yaḥya b. Abī Manṣūr	1;59
Sind b. 'Alī	1;59,51
Abū'l-Qāsim Aḥmad b. Mūsā b.Shākir	2;0,8
Ibn al-Amājūr, al-Turkī	2;0,20
Banū Amājūr, and Banū Mūsā	2;0,50
Ibn Yūnis (Zīj al-Ḥākimī)	2;0,30

2.3 *The attribution to al-Ṣūfī*

One can certainly question whether, apart from the change of reference meridian, all the present parameters of the tables for Pisa and London are just the same as in his original Zīj. Since we have only one parameter from the Arabic source, that provides the only numerical comparison which we can make with the tables for Pisa and London. There, the Sun's motion is given as 20r 0;9 in 7305 days (20 years). From the motion given by Ibn Yūnis from al-Ṣūfī's Zīj, we would have 20r 0;9,2,18. Among Ibn Yūnis's other sources there are some which also agree well, as the following list shows; Caussin (1804), pp. 56, 150-8.

	in 365 days	*in 20 years*
Banū Mūsā	11s 29;45,39,58,2	20r 0;9,0,59
Abū'l-Qāsim Aḥmad b. Mūsā b.Shākir	11s 29;45,40	20r 0;9,1,39
Muḥ. b. Aḥmad b. Yūsuf al-Samarqandī	11s 29;45,39,48	20r 0;8,57,38
(MS reading uncertain)	11s 29;45,39,58	20r 0;9,0,59
Banū Amājūr	11s 29;45,39,45	20r 0;8,56,38
al-Ṣūfī	11s 29;45,40,2	20r 0;9,2,18
Ibn Yūnis		
(noted in discussion)	11s 29;45,39,54	20r 0;8,59,39
(Zīj al-Ḥākimī)	11s 29;45,40,3,44	20r 0;9,2,53

Clearly some of these motions are even closer to the motion 20r 0;9,0 given in the Latin tables. Although Ibn Ezra had some knowledge of Ibn Yūnis, as we see in Text II, the Zīj al-Ḥākimī was unavailable in the West, and nor were the astronomical works of the other authors just named.

The bibliographical studies in the Fihrist of Ibn al-Nadīm, and by Ibn al-Qifṭī make no reference to any Zīj by al-Ṣūfī, as is clear from the discussion by Sezgin (1978, p.212).

3. Mathematical analysis

3.1 *Deviation curves*

The mean longitudes of any Zīj may be profitably compared with the modern mean longitudes. The term 'deviation' is here used to designate the difference: medieval minus modern. There are nine deviations representing Sun, Moon, Lunar apogee, Lunar node, Mercury, Venus, Mars, Jupiter and Saturn. An interesting example is provided by the Zīj of Ibn Yūnis (Fig.1). The deviations for the Sun and Moon, are extremely close to zero around the time of the observations of Ibn Yūnis; indeed that for the Sun is very close to zero for a great many centuries. If this solar model were used now the error would hardly

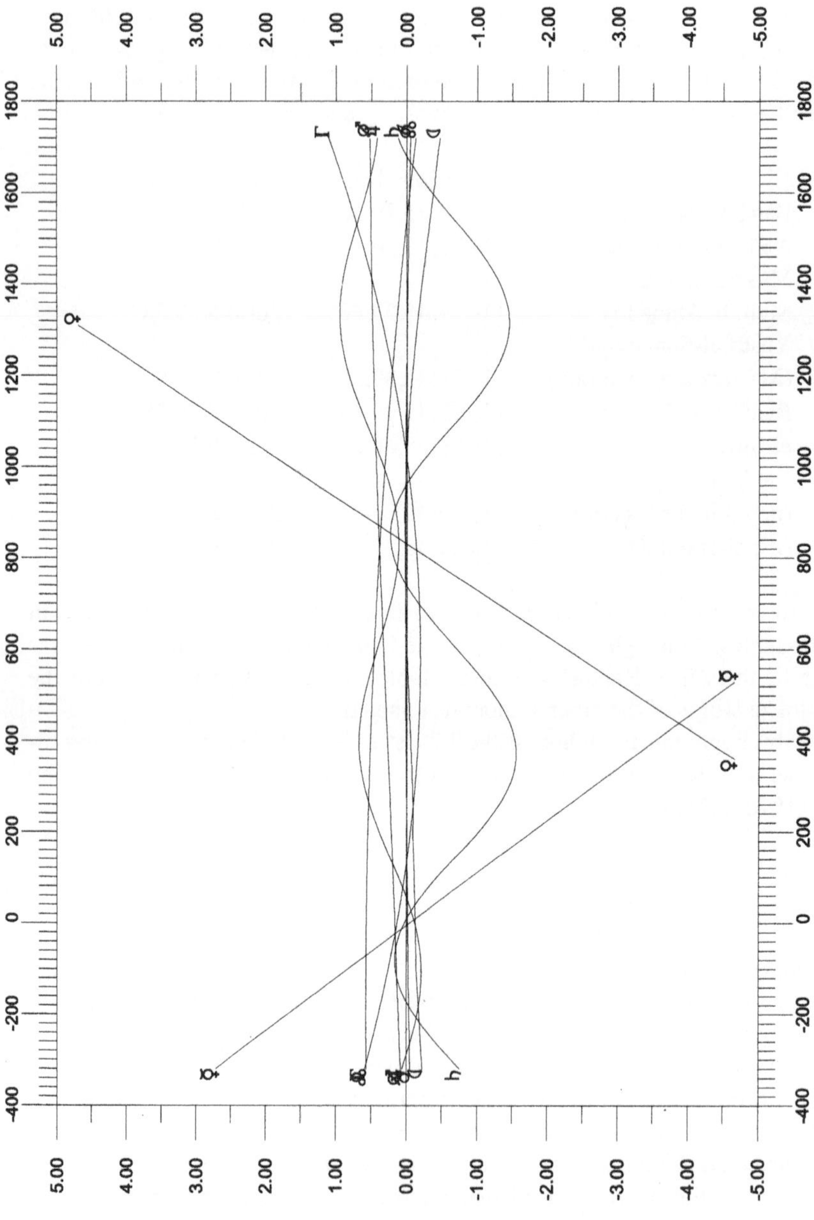

IBN YUNIS (mean)
Meridian: 31.250
Deviations

Fig. 1

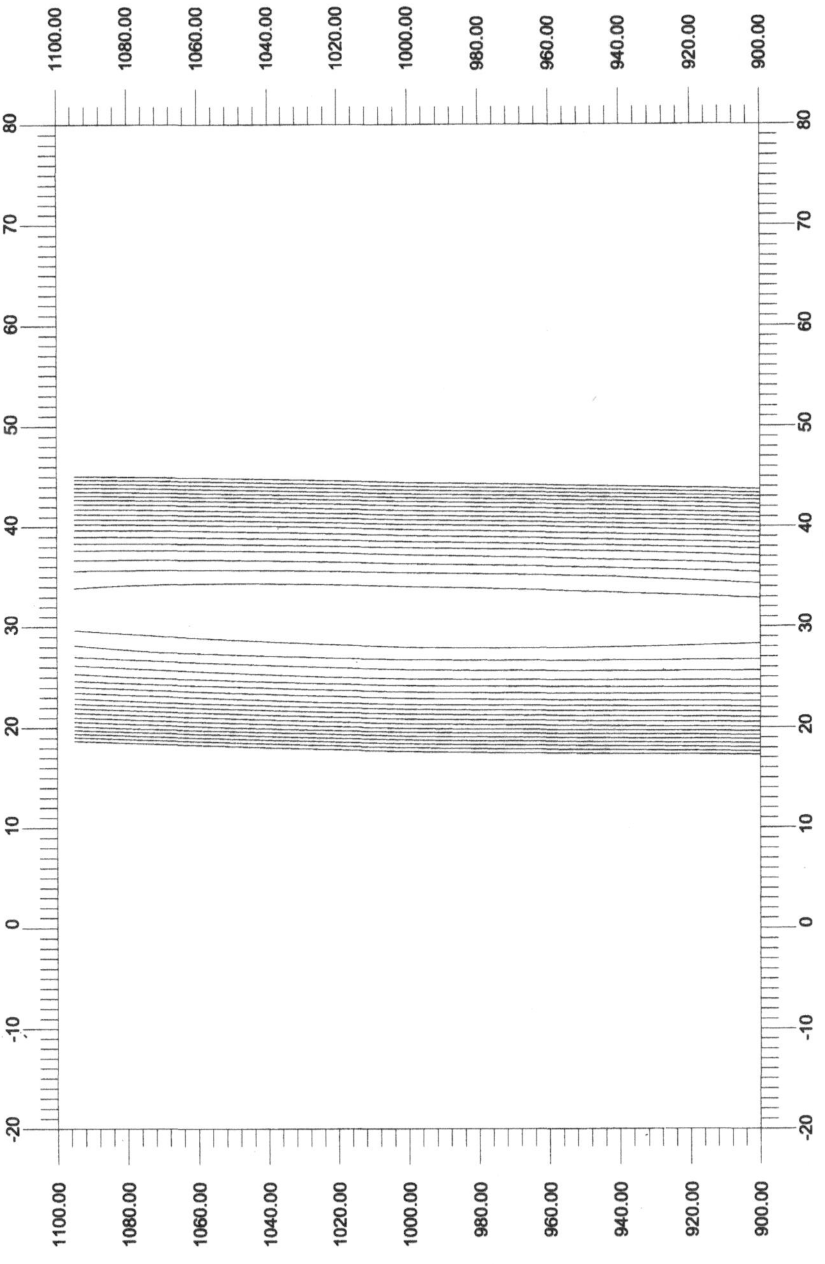

Fig. 2

IBN YUNIS (mean) statistic 1111000000
Optimum: Year 1006.1, Meridian 31.1

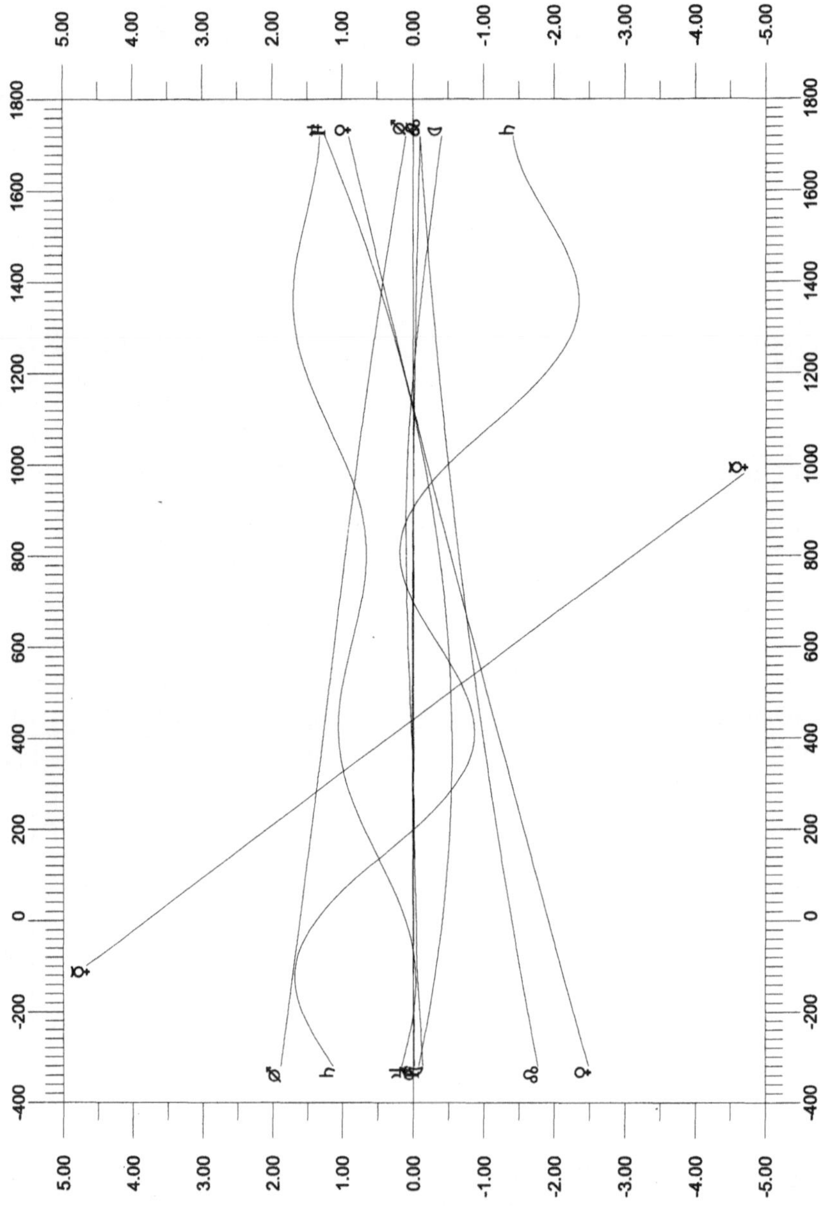

PISAN TABLES (mean)
Meridian: -2.500
Deviations

Fig. 3

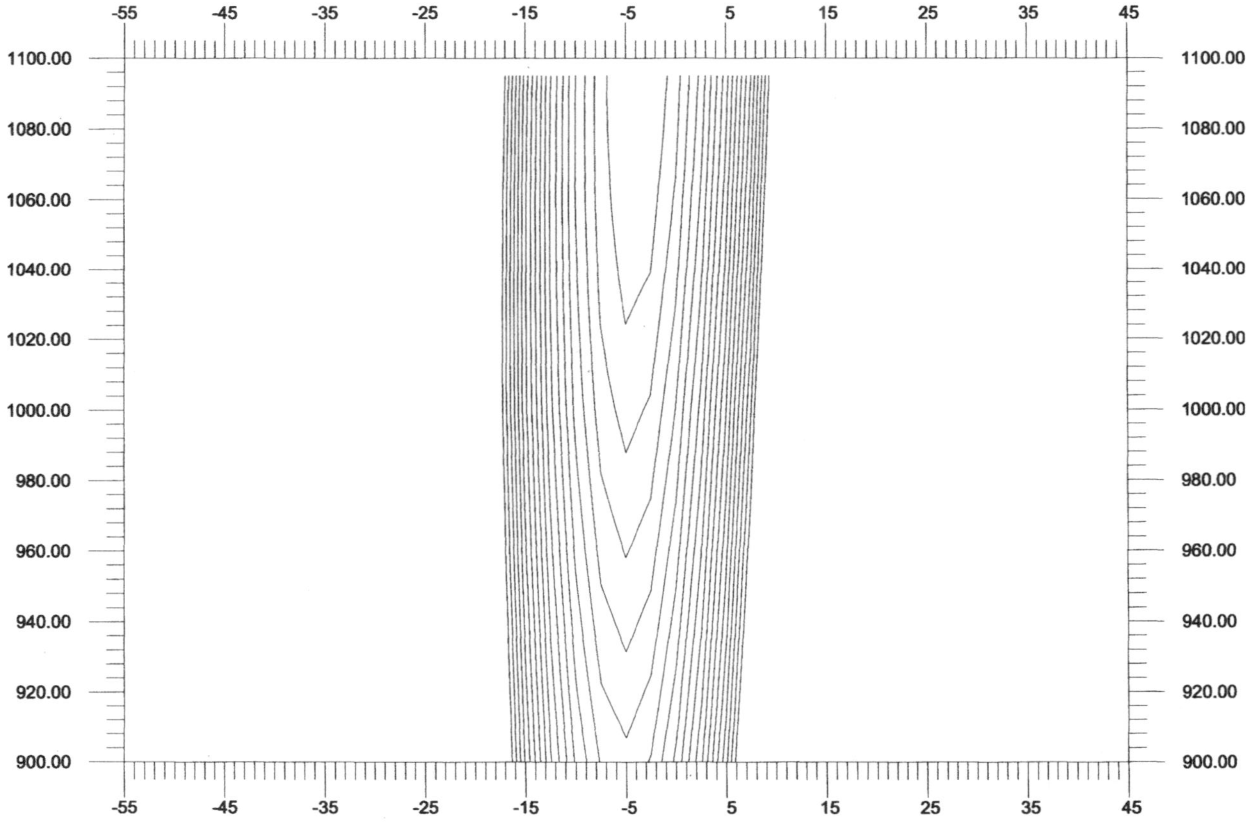

PISAN TABLES (mean) statistic 1111000000
Optimum: Year 1113.5, Meridian -3.6

Fig. 4

exceed five minutes of arc. This is a clear testimony to the extremely high standards of observations maintained at the time. Errors in the other deviations are rather greater, partly due to the inherent difficulties, partly to a relative lack of interest in the planets. Accurate parameters for Mercury always proved to be beyond the grasp of medieval astronomers.

The date at which the observations were made by Ibn Yūnis, as well as the longitude of the observatory, can be estimated by a procedure of optimisation.[1] The level contours are shown in Fig. 2, based on the deviations of the Sun and Moon, with the statistic (1111000000). From this we have the joint determination of the optimum year and meridian, 1006 and 31;5, which is in accordance with the known position of his observatory in Cairo, for which the modern longitude is 31;15.

Turning to the Pisan tables, for which the deviations are shown in Fig. 3, we can see that the Sun and Moon are again very accurate, but that there is decidedly less accuracy in the planetary parameters than there was with Ibn Yūnis. There is no really clear convergence towards an optimum date. If we were to consider the Sun, Moon and Lunar apogee, for which the contour levels are shown in Fig. 4, then the date and meridian would be A.D.1113 and -3;36. Note that when there are only three deviation curves there is a unique determination of date and meridian. Whatever the choice of statistics, however, the optimum meridian is very roughly -3. In any case Pisa does not emerge as the actual place of observation, a point discussed below, section 6. The convergence of the curves for Sun, Moon, Lunar apogee and Venus towards A.D.1100 should be regarded as accidental.

3.2 *Determination of meridian of the original observations*

Since the Pisan tables are best at around the Greenwich longitude -2, while the Greenwich longitude of Pisa is 10;24, it is clear that the observations were not made there originally. With the aid of the following table we can determine the actual site of the observations.

	Medieval	*Modern*	*Medieval (Ptolemy) scaled to Pisan Tables*
Baghdad	80	45	44
Pisa	33;30	10;24	-2;30
London	20	0	-16;30
Toledo	10	-4	-26;30

[1] This is explained in articles I and V in the present volume.

In the third column the longitudes are determined in relation to that fixed by the optimum meridian of the Pisan tables on the Greenwich scale, together with the differences taken from the medieval scale. Since these differences are much greater than their modern counterparts, the resulting scale will agree with the modern one only at the true place of the observations. We can conclude that that place was Baghdad, or somewhere in the vicinity.

In more formal terms, let

λ = Greenwich longitude of the place X
Λ = Medieval longitude of the place X
λ_0 = optimum meridian for the Tables given for place X_0
$$\delta(X) = \Lambda(X) - \Lambda(X_0) - (\lambda(X) - \lambda_0)$$

The place X_1 where $\delta(X_1) = 0$ is the place where the observations were established originally. The tables at X_0 were calculated by making use of the longitude difference $\Lambda(X_0) - \Lambda(X_1)$.

That the underlying observations were made in the Middle East may be shown also in another way. The parameters for the lunar elongation, which are quoted in Table III, imply that the mean luni-solar conjunction occurred 11.876 days previous to the Era 21848460. If this precise date is compared with the moment of the same conjunction as implied by the parameters of the Zīj of al-Battānī, whose meridian is at the longitude 73;15, we obtain from the resulting time difference that the meridian of the Pisan Tables is 33;28,32. If compared instead with the Zīj al-Mumtaḥan, we obtain 33;32,8. In either case we obtain essentially the longitude 33;30 of Pisa according to Ptolemy, and so the conclusion is clear, that this assumed longitude of Pisa, according to Ptolemy, was employed in order to reduce luni-solar parameters of one of the ninth-century Zījes established at Baghdad or al-Raqqa, instead of employing any fresh observations made at Pisa. Ibn Ezra is not quite correct in saying that the longitude of Pisa was taken as 33.

3.3 *Determination of the Era of the Zīj of al-Ṣūfī*

The apogees given for the Era of the London tables are evidently calculated from values given for an earlier date, for each of the planetary longitudes ends in ;33,18, except for the Sun and Venus for which it is 86;20,18. We would expect therefore that they were calculated from integral values situated at a date placed earlier than 1150 March 1 by an amount

$$y(n) = 66(n + 0;33,18) = 66n + 36.63 \text{ years},$$

where n is an integer.

Since we would expect the result to lie within al-Ṣūfī's lifetime (died A.D.986), and to correspond to an integral number of years in either the Hijra or Yezdajird era, the solution is easily found, namely, n = 2,

$$y(2) = 168.63 = 168;37,48$$

The Julian Date (J.D.) of the Era 1150 Mar 1 is 2141155, so that the J.D. of the earlier date is 2141155 - 168.63 x 365.25 = 2079563, which is 370 A.H. complete plus 10 days. The small discrepancy of 10 days can be ignored. There is no solution in terms of A.Y. The Latin date corresponding to 2079563 is A.D.981 July 4, about 5 years before the death of al-Ṣūfī. We can conclude that the Era of al-Ṣūfī's Zīj was 370 A.H. complete, and that the longitudes of the apogees were smaller by 2;33,18 than those of the London tables. We have no separate information about the apogees used in the Pisan tables.

In Table V there are given the positions of the apogees calculated from those of al-Battānī's Zīj. He gives the date merely as 1191 Dhū'l-Qarnain, which is the Seleucid Era (Western form), counted from A.D.-311 September 30. I have chosen an increment 4;5,18 which brings the resulting apogees as close as possible within the uncertainties allowed by the date given by al-Battānī, and also by the uncertainty in his figure for the apogee, which may be 82;14 or 82;15, but is here assumed to be 82;15. Even so, it is clear that the resulting positions of the apogee do not agree completely with those of the London Tables. One can be sure that al-Battānī's Zīj was not the direct source, although some of his results may have been used by al-Ṣūfī. A further argument against the possibility that the source is the Zīj of al-Battānī, is provided by the fact that the planetary nodes given in the London tables are all different from those of al-Battānī, who follows the Handy Tables of Ptolemy.

4. TEXTS ASSOCIATED WITH THE TABLES

The texts in question are of various types. Separate 'canons' (text IV), are written only for the London tables. They consist of a series of short chapters which serve as instructions for the use of the tables.

Appearing along with these canons for the London Tables, in two of the copies, is a short tract (text VI), in which are discussed special problems of calendrical conversion, etc.

For the Pisan Tables there are three texts (I, II, III), two of which are by Abraham ibn Ezra.[2] While written with these tables in mind, many other matters are discussed, and much light is shed, by the way, on problems of interest at the time. Canons for the Pisan tables are included in these texts.

Robert of Chester wrote a tract on the astrolabe (Text V) in which in the last chapter he makes a reference to his work on tables for Toledo and London.

The London tables are mentioned in passing also in the thirteenth-century tracts on calendrical and astronomical theory (texts VII and VIII). Finally there is a tract on the sphere by Andalo di Negro, here listed as Text X, which also mentions the London tables, without actually adding anything to our information, but it is of interest because it was the occasion for Duhem's incorrect dating of these Tables at A.D.1232.

5. The texts for the Pisan Tables

There are three texts designed as introductions for these tables, and these are denoted I, II, and III in this paper. Texts I and III are by Abraham ibn Ezra, while II is an anonymous work by a Christian, but clearly dependent on the work of the former.

5.1 *Text I*

This has been edited by Millas Vallicrosa, and contains a widely discursive introduction to the Pisan tables, which Ibn Ezra claims as his own compilation. After lengthy discussions about the length of the year, and its correct manner of definition, a problem which was always of major interest to Islamic astronomers, there are chapters detailing the procedures of calculation of the true planets, etc. There is never a clear line drawn between calculation and geometrical interpretation, in marked contrast with the style of the usual Zīj, which followed the tradition of Ptolemy's Handy Tables.

A certain amount of precise information about the tables can be fixed.

1. The text is written in 1154.[3]
2. The longitude of Pisa is 33.
3. The mean Sun follows the determination of al-Ṣūfī, as verified by al-Zarqāla for the date A.D.1090 March 1 (A.H.482 complete).

---

[2] The text III′, which was not noted in the first publication of this article, is clearly based on III, but includes extraneous material, such as a paragraph on accession and recession, and cannot be taken as seriously representing the work of Ibn Ezra on the tables for Pisa and London.

[3] Millas Vallicrosa (1947), p.78.

4. The tables in collected years are arranged in steps of 20 years (7305 days), with the leap years in the 2nd, 6th, etc.
5. The Sun's motion in 20 years is 20r 0; 9, and the year is given by 365 1/4-1/131.
6. The longitude of Angers is determined by a solar eclipse observed at Bordeaux (which is on the same meridian, the modern longitudes being 0;32 West and 0;34 West, respectively). There are four total or near-total solar eclipses, 1133 August 2, 1140 Mar 20, 1147 October 26 and 1153 January 26; the third or fourth are most likely because of the dates. In Text VII however the partial eclipse of 1137 May 21 is referred to.
7. A list of contents of the tables is given. Apart from the usual basic tables for the Sun, Moon and planets, this included various other tables which one is accustomed to find in a Zīj.[4]
8. There is a slight indication that the lunar and planetary parameters also were drawn from al-Ṣūfī, or at least that they were available from his work. The annual motion of Jupiter according to al-Ṣūfī is said to 'have no seconds' in comparison with the value 30;21,40 given by the Magistri Probationum. In the Pisan tables, in fact, it is 30;21,49,15.

The solar motion given in paragraph 5 above is quoted from the edition (Millas Vallicrosa, 1947), p.97, lines 2-3, where however one must choose the alternative reading '9 minutes' from MS O7 instead of '2 minutes'. In fact references to the solar motion and solar year in this text are dreadfully confused, and must be compared with details available from the other material which I have consulted.

In this work Ibn Ezra cites a great many names, including, among Islamic astronomers, al-Battānī*, al-Khwārezmī*, Ibn Abī Manṣūr*, al-Marwarūdhī, Māshā'allāh, Maslama, Ibn al-Muthannā*, al-Nairīzī, Qusṭa ibn Lūqqā, Ibn al-Ṣaffār*, Banū Shākir*, Ibn Sīna*, al-Ṣūfī*, Thābit ibn Qurra*, Ibn Yūnis* and Zarqāla*. Those marked with * are cited frequently. It is possible, perhaps, that he knew the Zīj of Ibn Yūnis, which contains so much information, and from which he might have picked up less common names such as al-Marwarūdhī, as well as a certain amount of his numerical data, for Ibn Yūnis cites many parameters from his predecessors.

Ibn Ezra says that al-Ṣūfī observed the Sun over a period of 40 years.[5]

---

[4] Solar declination, ascensions, etc., planetary synodic appearances, fixed stars in polar-equatorial coordinates, parallax, lunar visibility, conjunction and opposition, and eclipses.

[5] Millas Vallicrosa (1947), p. 86.

It is clear that he was quite familiar with al-Zarqāla's work on the Solar Year, as well as his tract on accession and recession; the latter is extant in a Hebrew version, translated by Millas Vallicrosa. Ibn Ezra rejects with some scorn the model of accession and recession. He also rejects al-Zarqāla's conclusion that the anomalistic year differed from the sidereal year.

5.2 *Text II*

This anonymous work is found only in the MS C1. In a way, it is quite similar to Text I, in that it opens discursively with a discussion of the definition and determination of the solar year, and leads eventually to instructions in the use of the Pisan tables. One might believe on a first reading that it is also written by Abraham ibn Ezra, being an earlier or later version of Text I. On close examination however, one sees many differences in detail, signs of a poor grasp of technicalities, and substantial errors of fact. Even though the author refers to 'our tables composed in 1143 (f.70v), he also refers to 'our Lord Jesus Christ' (67r), which would not be expected of the Jewish writer Ibn Ezra. It is not clear therefore, whether we can accept without reservation the original year of composition 1143, because this is not directly the word of Ibn Ezra. There are also significant variations between the two texts in the spelling of proper names, as the following comparisons show.

Text I	**Text II**
elcaurezmus	algoarismus
albumasar, albumazar	abymasar
thebit, tebith, tebit	thebit, thebith, bencore
abencine	ebecine
albatheni, albateni	albatheni, albatenus
azarchel, azerchel, acerchel	azarchel, ezirchel, ezarchel
abracaz, abrachix, abracax	abracaz
azophi, azofi	assophius

Even when one allows for the variations within Text I it is clear that the author of Text II followed a rather different style of transcriptions.

It is certain that the work is written with the Pisan tables in mind, even though Pisa is not named, nor is its longitude given. The rate of motion is stated to be 20r 0;9 in 20 years (68v), the tables of collected years being arranged in steps of 20 years (68v). The tables are stated to have been composed according to those of al-Ṣūfī (68r, 70r), who had carried out observations of the Sun over a period of 44 years (68v), his result having been confirmed by Zarqāla (68v), whose tables followed these. The initial year of the tables of mean motion is A.D.1029 (70r, 71r), when the mean Sun is 344;30,20 (71r), which implies the mean longitude 345;33,20 for the year 1170, in complete agreement with the motion given in the copy of the Pisan tables known to us, which begins at the year 1170. In the copy O3 of the London tables it is interesting to find that tables for collected years begin in 1029 complete.

The author's account is less reliable when he tells us about the observations of Zarqāla. For he says that 20 years separate Zarqāla's observation of the Sun in A.D.1120 from his own in 1140. The former found the Spring Equinox to be at 2 hours after Noon on 14 March, the latter 2 hours before Noon on 14 March. We know with certainty that Zarqāla's work was done ca.1080, and neither the modern calculation of the Spring Equinox nor that given by the Pisan tables, occurred at those times. It is interesting that the Spring Equinox of A.D.1081 occurred at approximately 1;30 hours after Noon on March 14, although it does seem unlikely that this fact could have found its way obscurely into the present text. Our author in fact presents here a somewhat fictitious calculation, involving moreover several arithmetical errors, and even then does not obtain the mean Sun for 1140 as he seems to intend.

The reported date 1143 of composition of the Pisan tables is consistent with what we know of Ibn Ezra's travels. Fleischer, as quoted by Millas Vallicrosa (1949, p.17) determined that he was in the region of Tuscany in Italy, and in Lucca in particular, in the period 1141-5.

5.3 *Text III*

This relatively brief work is found in MS L1, which is also important as containing the principal copy of the London tables, and is attributed expressly to Abraham ibn Ezra. This Text is an introduction to the use of the Pisan tables, Pisa being named and given the longitude 33. The initial year of the tables is stated to be 1149 complete, the era used in the London tables. He remarks that the longitude of Angers is 24, and that Winchester is on the same meridian. We recall that in Text I, he had fixed the longitude of Angers from an observation of an eclipse in Bordeaux. It is not clear why he should have been interested in the longitude of Angers.

This is an interesting step in the process by which the geographical longitudes in Western Europe were being corrected at this time. Ptolemy's longitude of Pisa is retained, but the positions of Bordeaux (18), Angers (18;50) and Winchester (19;20) are now changed to 24. This reduces the average distance between this meridian and Pisa from 14;16 to 9, the correct average distance being 11;12. The longitude of the London tables was stated as 19 degrees (see section 2.2) according to Robert of Chester, that is 33-14 = 19; he was evidently unaware of Ibn Ezra's improvement.

6. Texts for the London Tables

6.1 *Text IV*

This text is a series of chapters giving instructions for the use of the tables. It survives in three copies, one of which is found with the principal copy of the London Tables, while the other two accompany Text VI. These are written in the style normally found in the Arabic Zījes, and which go back to Ptolemy's Handy Tables. Here, moreover, there is a clear dependence on some of the chapters of instructions in the Latin version of the Zīj of al-Khwārezmī: two chapters are copied almost without alteration, Ch. 13 on planetary retrogradation and stations, and Ch. 17 on planetary latitudes. Elsewhere an opening phrase is copied, as in Ch. 31 on luni-solar conjunctions, which begins 'Querenti coniunctionum seu oppositionum'.

It is not surprising that Robert of Chester, who is clearly responsible for the London Tables, should make use, wherever he could, of his version of the older Zīj. The older one used Indian methods rather than Ptolemaic planetary models, so that the instructions for the true longitudes would have to be rewritten for the London Tables. For the planetary latitudes, however, there was a tendency to continue to use the Indian model, because Ptolemy's tables were very difficult to use and understand. This is the case here, so that Ch. 17 is copied without alteration, and this practice continued even when the Toledan tables came into common use in the later twelfth century. We will see in the discussion of Text V that in his tract on the astrolabe, Robert of Chester also appears to have incorporated Ch. 36 into the part of his work which dealt with tables for Toledo, on the calculation of the houses.

6.2 *Texts V and VI*

Taking Text VI first, two copies are known of this. In O1 it precedes Text IV, and in O2 it is follows it. The author is concerned to give a rather didactic account of the chronological calculations which are a normal preliminary to the use of astronomical tables. He discusses the conversion between Arabic and Latin calendars, and the problems following on the various choices of the beginning of the year in the Latin calendar. The application of these methods to the tables of the planets is then described, both for a year after 1150, as well as for one before; in the latter case he uses the solar eclipse of 1137 May 21 as an example. A final chapter mentions the variation of the planetary apogees, and he gives the table of the solar apogee. He also notes that lunar and solar eclipses can be used in order to fix the longitude of a place, and includes a short table of geographical coordinates, which is of some historical interest, in giving both Ptolemaic and revised longitudes for a number of places in Spain and Western Europe. A selection from this table in given in Table VII, section 8.

The work opens 'The second part of this work begins, namely that established for the meridian of the City of London, by Robert of Chester, on the basis of the treatise by al-Battānī of al-Raqqa'. This 'second' part is clearly the London tables, and the author here specifies the epoch 1150 March 1 Noon, the arrangement of the collected years in sets of 20, and the tract ends with the table of the solar apogee, as given elsewhere for the London tables.

The author's opening remark shows his belief that the London tables are based on al-Battānī, so that he is unaware of Ibn Ezra's explanation of their derivation from the Zīj of al-Ṣūfī.

An interesting calculation is made regarding the solar eclipse of 1137 May 21, for the author calculates the interval from that moment to the beginning of 1170, assuming by the way that the collected years are in groups of 20. He must have in mind the London tables which collects the years in that way, but in which the epoch is 1150. The reference to '1170 then suggests that the author is actually writing around that date. If that is the case, he cannot be Robert of Chester himself, but someone who comes after him, and who writes here an introduction of his own to the canons of the London tables, which are placed next to this preface.

If that much is clear, there is nevertheless confusion regarding the tract referred to in these pages as the 'first' part of this work. This first part is stated to be for a set of tables based on the work of Ibn Ezra and for the meridian of Toledo (O2, 20ra, 24rb) using the Latin calendar beginning in January (O2, 20ra) in 1169 (presumably complete) with years grouped by 28 (O2, 24rb). Nothing is known among surviving texts of a set of tables answering to these requirements. There are a number of Zījes extant in Latin, all versions of the Toledan tables, which are drawn up for the Latin calendar reckoned from January, and with the years grouped by 28, but for none of these is the meridian Toledo, the meridians in fact being Marseille, Hereford, Ferrara, and Novara. A version of the Zīj of al-Khwārezmī is known from MS P1, in which the Latin calendar is used, reckoned from January, and probably for the meridian of Toledo, but here the years are grouped by 20 (Mercier, 1987a).

In his tract on the astrolabe, Text V, Robert makes an interesting remark in the penultimate chapter, whose title is 'De equatione domorim .xii. iuxta Albucasim Almacheritam (Abu'l-Qāsim Maslama al-Majrīṭī)', beginning:

> De ratione coequatione .xii. domorum in libro canonum que super Toletum et civitatem Londoniarum edidimus prout tractatus exposcebat ratio tractavimus. Unde in hoc libro qualiter ipse inveniantur per astrolabium docere curabimus.

This is a clear reference firstly to cap.xxxvi entitled 'De equatione .xii. domorum' in the Zīj of al-Khwārezmī, as compiled by Robert of Chester from al-Majriti's version of the Zīj. It is also a clear reference to Robert's complete work for both Toledo and London, thus verifying the remarks made in Text VI. There is, however, no chapter on this topic in the extant material of the tables for London, so that it must have been included in the missing portion of Robert's work.

There are references to Toledo to be found here and there in the material we are studying. In the principal copy of the London tables, L1, the correction to the mean motion of Mars is given, corresponding to the change in meridian from London to Toledo. This correction is 0;0,53 (the MS has 53 'minutes' in error for 'seconds'), implying a change of meridian longitude equal to 10;7. In the copy of the Pisan tables in P1, f.4v, a note to the table of the solar apogee gives the correction 0;1,43 to the mean solar longitude due to the same shift of meridian from London to Toledo. This implies a change of meridian equal to 10;27. These shifts correspond essentially to the Ptolemaic difference 10;30 between London and Toledo. More importantly they reveal a connection between these tables for London and certain others for Toledo.

6.3 *Text VII*

This is a very long tract found uniquely in the manuscript C2. To some extent it resembles Text I in style, in that it goes into considerable depth and detail in its account of the underlying geometry, without keeping separate the basic procedures needed in calculating from the tables. Judging by the style of the tract, and the age of the MS, the work would appear to be of the early thirteenth century, and is unfortunately anonymous. The London tables are clearly the object of the account, as the author makes very clear, specifying the place, and the epoch 1150 March 1. He thinks also that they are based on the work of al-Battānī, and mentions no other possible source, nor the work of Robert of Chester and Ibn Ezra.

The opening phrase 'Circulus solis dicitur esse ecentricus..' reminds one forcibly of the popular tract attributed, without any good reason, to Gerard of Cremona, and usually referred to as the Theorica Planetarum.

6.4 *Text VIII*

This tract, which is found in at least three manuscripts, is a type of 'theorica planetarum', very much resembling Texts I, III and VII, in that it provides rules for the use of the tables, but these rules are embedded in an account of the geometric models. The work is of the late twelfth or early thirteenth centuries, and the author writes mainly with the Toledan tables of al-Zarqāla in mind, and notes also the Tables for Toulouse (Tolosa) derived from them. He mentions the London tables at one point, only to note that exceptionally in that case, time is measured from Noon of the first day of the Era (Noon March 1), rather than from the preceding noon, such as Noon December 31. He also gives the longitude of London as 32;30, compared with Toledo at 28;30, very different from the intervals of about 10 noted above, section 6.2. The author is unknown, and one may with some confidence reject the attribution to Roger of Hereford, which is made in some copies.

6.5 *Text IX*

This is an early fourteenth-century tract by Andalo di Negro, found in the manuscript P3, and printed by Thorndike. It would seem to be incomplete, but in any case includes five chapters, of which the first four deal with the most elementary concepts of spherical astronomy. It was for this reason printed by Thorndike along with commentaries on the Sphere of Sacrobosco. The fifth chapter on the other hand defines some elementary notions relating to tabular astronomy, such as radix, mean motion, etc. A number of tables are named, viz., those for Hyspanis (? Seville), London, Genoa, Toledo, Paris, Marseilles, Pisa, Palermo and Constantinople. It is curious that he says 'these tables' on two occasions when he mentions the London Tables, for which the year is reckoned to begin at Noon on 1st March. The reason for this notice might be clearer if we had the work in a complete form.

This was the occasion for a mistake by Duhem, who dated the London Tables at 1232. Duhem found in the same MS, on its first leaf, a common type of table for calendrical conversion, which happened in this case to begin in the the year A.D.1232. With no logic whatever he inferred that the London Tables were designed for that year, and unfortunately this mistake has had a wide currency.

7. MISCELLANEOUS TEXTS CONCERNING AL-ṢŪFĪ

7.1 *al-Ṣūfī on the Astrolabe*

In the, k. al-Amal bi'l-Asturlab (Ṣūfī [1962]) one of his works on the astrolabe, al-Ṣūfī nowhere makes any reference to a Zīj, either to his own or any other, and indeed not even to his own Catalogue of Stars. This is in spite of its length (386 chapters) and its general completeness as a technical account of the instrument.

7.2 *The Mudkhal al-Aḥkām*

This tract was noticed in a manuscript of the India Office Library, Loth 733, fol.1-17, where it is attributed to al-Ṣūfī. During the course of this tract on astrological applications of astronomy, such as the ascendent, and the revolution of the years of the world (fol.11v-12), he discusses the use of three Zījes, viz., al-Shāh, al-Khwārezmī and al-Mumtaḥan, but nowhere mentions his own work.

7.3 *al-Zarqāla*

In the references to al-Ṣūfī in Texts I and II, his work is said to have been verified by Zarqāla: secundum Azofi composui, et Azarchelis probatio fuit in anno 482 ab helligera, (Millas Vallicrosa, 1949, p.87). Nowhere however in any of the tracts known to have been written by him, or reasonably attributed to him, does one find any reference to al-Ṣūfī. Millas Vallicrosa (1943) has published his almanac, and his tract on the motion of the fixed stars (extant in a Hebrew version). In the latter where many earlier figures are named, it is disappointing not to find something corresponding to the Latin reference. I also find nothing in any of the versions of the canons to the Toledan tables, for which he is said to have been responsible.

7.4 *The Almanac of Abraham Zacut*

In the Preface to this Almanac written ca.1480 (edited by Cantera Burgos(1931)), Zacut discusses the fixed stars and their motion, and makes clear references to the star catalogue of al-Ṣūfī which was translated into Spanish under the aegis of King Alfonso el Sabio ca.1255. In Ch. 9 of this Preface Zacut has some most interesting remarks about precession, and the equation of the eighth sphere, eventually rejecting versions of accession and recession (Zarqāla) and trepidation (Alfonsine Tables), in favour of the simple linear precession as al-Ṣūfī prescribed. Zacut refers to no other work of al-Ṣūfī, and his remarks do not suggest that he knew any other.

8. Selection of Tables and Parameters

TABLE I

Tables for Pisa
Mean motions extracted from MS Paris BN latin 16208, fol.4-4v.

YEAR	SUN	SUN	MOON	MOON
		(apogee)		(arg.)
1169	11,15;33,20	2,26;20,18	4,10;19,53	5,2;31,53
1189	11,15;42,20	2,26;39,18	8,24;3,27	6,12;17,41
1209	11,15;51,20	2,26;57,18	1,7;47,1	7,22;3,31
1229	11,16;0,20	2,27;15,18	5,22;24,35	9,1;51,21
1249	11,16;9,20	2,27;33,18	10,6;4,20	10,11;37,11
1269	11,16;18,20	2,27;51,18	2,17;56,54	11,21;23,1
1289	11,16;27,20	2,28;9,18	7,1;40,28	1,1;8,51
YEAR	NODE	SATURN	JUPITER	MARS
1169	0,25;42,32	11,6;10,14	1,10;18,57	9,8;0,11
1189	1,22;31,47	7,10;50,16	9,17;35,22	4,26;19,21
1209	2,19;21,4	3,15;30,18	5,24;41,47	0,14:38,31
1229	3,15;22,23	11,20;10,20	2,1;58,12	8,2;57,41
1249	4,12;11,40	7,24;50,22	10,9;14,27	3,21;16,51
1269	5,9;42,55	3,14;8,24	6,16;31,2	11,9;36,1
1289	6,5;30,12	0,4;0,28	2,23;47,21	6,21;55,11
YEAR	VENUS	MERCURY		
	(arg.)	(arg.)		
1169	8,3;13,43	4,2;17,50		
1189	2,6;57,2	4,16;46,40		
1209	8,10;40,20	5,1;15,30		
1229	2,14;23,38	5,15;44,20		
1249	2,18;7,56	6,0;13,10		
1269	8,21;50,14	6,14;42,0		
1289	8,25;33,32	6,29;10,50		

TABLE II

Tables for London
Mean motions extracted from MS London BL Arundel 377,fol.9v-31v

YEAR	SUN	SUN (apogee)	MOON	MOON (arg.)
1149	11,15;25,38	2,26;20,18	11,27;7,4	3,23;9,30
1169	11,15;34,38	2,26;39,18	4,10;50,38	5,2;51,0
1189	11,15;41,1	2,26;57,18	8,23;9,4	6,12;32,30
1209	11,15;52,38	2,27;15,18	1,8;17,46	7,22;14,0
1229	11,16;1,38	2,27;33,18	5,22;0,20	9,1;55,30
1249	11,16;10,38	2,27;51,18	10,5;44,24	10,11;37,0
1269	11,16;19,38	2,28;9,18	2,19;27,28	11,21;18,30
1289	11,16;28,38	2,28;27,18	7,3;11,32	1,1;0,0
YEAR	NODE	SATURN	JUPITER	MARS
1149	11,27;53,23	3,1;30, 5	5,3;4,21	1,19;49,12
1169	0,24;40,40	11,6;10,5	1,10;20,21	9,8;1,22
1189	1,21;27,57	7,10;50,5	9,17;36,21	4,26;20,32
1209	2,18;1,14	3,15;30,5	5,24;52,21	0,14:39,42
1229	3,15;2,31	11,20;10,5	2,2;8,21	8,2;58,52
1249	4,11;49,48	7,24;50,5	10,9;14,24	3,21;18,2
1269	5,8;37,5	3,29;30,5	6,16;40,21	11,9;34,12
1289	6,5;24,22	0,4;10,5	2,23;52,21	6,27,56,22
YEAR	VENUS (arg.)	MERCURY (arg.)		
1149	1,22;31,51	3,17;56,13		
1169	8,3;15,9	4,2;25,3		
1189	2,6;58,17	4,16;53,53		
1209	8,10;41,45	5,1;22,43		
1229	2,18;25,3	5,15;51,33		
1249	2,22;8,21	6,0;20,23		
1269	8,25;51,39	6,14;49,13		
1289	8,29;34,57	6,29;18,3		

TABLE III

Summary of mean motions in the Tables for Pisa; emendations are placed below the values taken from the MSS

Planet	Radix A.D.1170 (2148460)	Motion in 20 years	Radix for London (1170)
SUN	345;33,20	20r 0;9,0	345;35,37.9
SUN(apogee)	86;39,18	0r 0;18,0	
MOON	130;19,53	267r 133;43,34	130;50,37.7
	do.	<267r 133;33,34>	
MOON (arg.)	152;31,53	265r 39;46,30	153; 2,22.1
	<152;20,31>	<265r 39;41,30>	<152;51, 0.1>
LUNAR NODE	-25;42,32	-1r -26;49,17	-25;42,39.4
MERCURY (arg.)	122;17,50	63r 14;28,50	122;25, 4.9
VENUS (arg.)	243;13,43	12r 183;43,18	243;15, 9.3
MARS	278; 0,11	10r 228;19,10	278; 1,24.4
JUPITER	40;19,57	1r 247;16,25	40;20,8.6
SATURN	336;10,14	0r 244;40,2	336;10,18.6
The following are calculated from the emended values			
MOON(apogee)	157;59,22	2r 93;52,4	157;59,37.6
VENUS(mean)	228;47, 3	32r 183;52,18	228;50,47.3
MERCURY(mean)	107;51,10	83r 14;37,50	108; 0,42.9

TABLE IV

Summary of mean motions in the Tables for London; emendations are placed below the values taken from the MSS

Planet	Radix A.D.1150 (2141155)	Radix A.D.1170 (2148460)	Motion in 20 years
SUN	345;25,38	345;34,38	20r 0; 9, 0
	<345;26,38>	<345;35,38>	20r 0; 9, 0
SUN(apogee)	86;20,18	86;39,18	0r 0;18, 0
MOON	357;7,4	130;50,38	267r 133;43,34
	<357;17, 4>	do.	<267r 133;33,34>
MOON (arg.)	113;9,30	152;51,0	265r 39;41,30
LUNAR NODE	-357;53,23	-24;40,40	-1r -26;47,17
	do.	<-25;42,40>	<-1r -26;49,17>
MERCURY (arg.)	107;56,13	122;25,3	63r 14;28,50
VENUS (arg.)	52;31,51	243;15, 9	12r 183;43,18
	<59;31,51>	do.	do.
MARS	49;49,12	278;1,22	10r 228;19,10
JUPITER	153;4,21	40;20,21	1r 247;16, 0
SATURN	91;30,5	336;10,5	0r 244;40, 0
The following are calculated from the emended values.			
MOON (apogee)	64;7,34	157;59,38	2r 93;52, 4
VENUS (mean)	44;58,29	228;50,47	32r 183;52,18
MERCURY (mean)	93;22,51	108, 0,4	183r 14;37,50

TABLE V

Parameters related to the equations of the Sun, Moon, planets

	equation of centre	equation of argument	apogee London	node London (1150)	al-Battānī apogees for 1150[6]
Sun	1;59,10	86;20,18	86;20,18		
Moon	13;8	5;1			
Saturn	6;31	6;13	249;33,18	103;12	248;33,18
Jupiter	5;15	11;3	176;33,18	82;1	168;33,18
Mars	11;24	41;9	131;33,18	21;54	131; 3,18
Venus	1;59	45;59	86;20,18[7]	59;27	86;19,18
Mercury	3;2	22;2	205;33,18	21;10	205;33,18

TABLE VI

Geographical Coordinates

modern	Ptolemy (Geography)		long.	lat.	Modern long.
Seville	Ἵσπαλις	(II,4,14)	7;15	37;50	-5;59
Cordoba	Κορδύβη	(II,4,11)	9;20	38; 5	-4;46
Toledo	Τώλητον	(II,6,57)	10	41	-4; 2
Bordeaux	Βουρδίγαλα	(II,7,8)	18	45;30	0;34
Angers	᾿Ιουλιόμαγος	(II,8,8)	18;50	49;20	0;32
Winchester	Οὐέντα	(II,3,28)	19;20	52;20	1;19
London	Λονδίνιον	(II,3,27)	20	54	0;0
Pisa	Πίσσαι	(III,1,48)	33;30	42;45	10;24

---

[6] See section 3.3.

[7] MS L1: ;2,18.

TABLE VII

Geographical Coordinates in Text VI

Modern	Latin	Longitude	Latitude
Baghdad	Baldac	81	33
al-Raqqa	Aracta	73	38
Pisa	Pisis	33	45
Cordoba	Corduba	28 [9]	35
Seville	Hispalis	27 [10]	35
Toledo	Tholetum	28 [11]	40
Saragossa	Cesaraugusta	29 [12]	42
Bougie	Bugea	32 [15]	30
Morocco	Marroca	17 [0]	30
	Osca (1)	27 [10]	40
	(Sus al-Aqsa)		
Valencia	Valencia	29 [12]	39
Denia	Denia	28 [11]	39
England	Anglia	18	52

(1) The mouth of the river Sous in West Morocco, Φθούθ ποταμοῦ ἐκβολαί in Ptolemy's Geography IV,1,3.

This is a selection from the list in MS O2, fol.25. Attached to some of the longitude figures are also other values, written as superscripts in the manuscript, but shown here in [ ]. While such superscripted values are the older, essentially Ptolemaic, longitudes, the main figures appear to represent revisions appropriate to Spain and North Africa, and these generally differ from the main Arabic lists, known from Kennedy (1987). The other figures given without such superscripts are essentially unrevised Ptolemaic longitudes.

9. List of Manuscripts and Texts

The tables are T.I and T.II, and the texts are I to IX. The relevant folio numbers are given after the reference to the table or text, such as T.I 7-35.

Manuscripts

London BL
- L1 Arundel 377
  - T.I 7-35
  - III 56v-63 (Millas Vallicrosa (1947) p.61)
  - IV 7-35
- L2 Cotton Vesp.A.II
  - I 27-31vb

Oxford Bodley
- O1 Savile 21
  - VIII 42r-61v
  - VI 86-91
  - IV 91v-95v
- O2 Ashmole 361
  - IV 13ra-19vb
  - VI 19vb-28vb
- O3 Digby 114
  - T.I 25-32v
- O4 Seld. sup. 90
  - T.II 30-4
- O5 Bodley 300
  - VIII 1-19va
- O6 Canon. misc. 61
  - V 12-22v
- O7 Digby 40
  - I 51-88v

Cambridge
- C1 Fitz. Mus. McClean 165
  - II 67-80v
- C2 UL Kk 1.1
  - VII 192-211v
- C3 UL Add 6866
  - T.I (solar apogee) 150v

Paris Bibliothèque Nationale
- P1 latin 16208
  - T.II 4-4v
- P2 latin 16207
  - T.II 19-36

P3 latin 7272
IX 60-67 (ed. Thorndike (1949) p.462-475)
P4 latin 15171
VIII 136-157

Madrid Biblioteca Nacional
M1 10009
T.I & T.II (Saturn only) 1
M2 10053
I 67ra-76va (Millas Vallicrosa (1947) p.73-167)

Berlin Staatsbibliothek
B1 Latin 307
III' 27

TABLES AND TEXTS

TABLES

T.I (London)
L1, O3, C3, M1, O1 & O2 (solar apogee)
T.II (Pisa)
O4, C3, P1, P2, M1

TEXTS

I Cognitum est corpus solare habere
L2, O7, P5, M2
II Ptolemaeus et multi sapientium
C1
III Volens in tabulis operari primo scias
L1
III' Cum volueris operari in tabulis
B1
IV London canons
L1, O1, O2
V Si gradum solis in singulis diebus anni
O6

VI Promissa vero electe
O1, O2
VII Circulus solis dicitur esse ecentricus non quia centrum non habet
C2
VIII Diversi astrologi secundum diversos annos
O1, O5, P4
IX Quoniam ad ymaginandum circulos spere
P3

References

Billard (1971). *L'Astronomie Indienne. Investigation des textes Sanskrits et des données numériques*, R. Billard. Ecole Française d'Extrême Orient, Paris, 1971.

Caussin (1804). 'Le Livre de la grande table hakémite. Extraits du MS de Leyde', ed., A.-P. Caussin de Perceval, Notices et extraits des MSS, VII Paris, 1804.

Cantera Burgos (1931). 'El judicio Salmantino Abraham Zacut. Notas para la historia de la Astronomia en la España medieval', F. Cantero Burgos. *Revista de la Academia de Ciencia de Madrid* 28 (1931), Madrid.

Duhem (1914). *Le Système du Monde*, P. Duhem. Paris, 1914; Vol.III, p.237.

Haskins (1927). *Studies in the History of Medieval Science*, Charles Homer Haskins, Harvard, 1927, repr. 1955.

Kennedy (1987). *Geographical Coordinates of Localities from Islamic Sources*, E.S. Kennedy and M.H. Kennedy. Veröffentlichungen des Instituts für Geschichte der Arabisch-Islamischen Wissenschaften, herausg. F. Sezgin. Reihe A: Texte und Studien, Bd. 2. Frankfurt am Main, 1987.

Mercier (1987a). 'Astronomical Tables in The Twelfth Century', Raymond Mercier, in *Adelard of Bath: An English Scientist and Arabist of the Twelfth century*, edited by Charles Burnett, Warburg Institute Texts and Studies, London, 1987 [= Study VII in the present volume].

Mercier (1987a). 'Meridians of Reference of Indian Astronomical Canons', Raymond Mercier. *History of Oriental Astronomy*, Proceedings of IAU Colloquium, New Delhi 1985, Cambridge, 1987 [= Study V in the present volume].

Millas Vallicrosa (1947). *El libro de los fundamentos de las Tablas astronomicas de R. Abraham Ibn Ezra*, edicion critica, con introduccion y notas por Jose Ma. Millas Vallicrosa, Madrid-Barcelona, 1947.

Millas Vallicrosa (1943). *Estudios sobre Azarqueil*, Jose Ma. Millas Vallicrosa, Madrid-Granada, 1943-1950.

Sezgin, F. *Geschichte der arabischen Schriftums*, Band VI Astronomie. Leiden, Brill, 1978.

Ṣūfī (1962). *Kitāb al-'Amal bi'l-Asṭurlāb*, by Abu'l-Raḥman b. 'Umar al-Ṣūfī, edited from the unique manuscript 2493 of the Bibliotheque Nationale, Paris, under the supervision of Muḥ. Abu'l-Mu'īd Khān of the Da'iratu'l-Ma'ārif al-Othmānia, Hyderabad, Deccan, 1962.

Thorndike (1949). *The Sphere of Sacrobosco and its Commentators*, Lynn Thorndike, Chicago, 1949.

IX

THE GREEK 'PERSIAN SYNTAXIS' AND THE ZIJ-I ĪLKHĀNĪ

1. *Introduction*

A set of Greek astronomical tables, known as the Persian Handy Tables, or the Persian Syntaxis, has been known for many years to historians of astronomy, at least since a brief analysis was published by Boulliau in 1645 (1). There have been a number of notices since then, but no adequate study of the scientific content of the tables, nor a correct identification of the Persian source. The most serious notice of the exposition of some 45 chapters, if not of the tables themselves, and of closely related Greek material, was given by Usener (2, vol. 3, p. 323–371). Since then, Pingree (3) has approached the problem of identifying the source, but his conclusions are demonstrably wrong, as we shall see, and his work marks a somewhat retrograde step.

The lists of fixed stars which accompany the main tables have been editd precisely by Kunitzsch (4), who related them to Arabic and Persian sources.

George Chrysococces, *ca* 1346, wrote an exposition, in some 45 chapters, of the Syntaxis, and in the introduction to it (printed by Boulliau (1), Usener (2, vol. 3, p. 356–357) and Lampros (5)), he recounts how Gregory Chioniades, like so many adventurous scholars in Western Europe in the previous century, went in search of Islamic Science. He set out for Persia, passing en route by Trebizond. After some initial remarks, Chrysococces continues:

> In a short while he was taught by the Persians, having both consorted with the King, and met with consideration from him. Then he desired to study astronomical matters, but found that they were not taught. For it was the rule with the Persians, that all subjects were available to those who wished to come to study, except astronomy, which was for Persians only. He searched for the cause, which was that a certain ancient opinion prevailed among them, concerning the mathematical sciences, namely, that their King will be overthrown by the Romans, after consulting the practice of astronomy, whose foundation would be first taken from the Persians. He was at a loss as to how he might come to share this wonderful thing. In spite of being wearied, and having much served the Persian King, he had scarcely achieved his objective; when, by Royal command, the teachers were gathered. Soon Chioniades shone in Persia, and was thought to be worthy of the King's honour. Having gathered many treasures, and organised many subordinates, he again reached Trebizond, with his many books

> on the subject of astronomy. He translated these by his own lights, making a noteworthy effort. There are in fact other books of the Persian Syntaxis which he translated, those having certain examples with the years systematically at the beginning. However, he handed on the Syntaxis alone, the best and most accurate of all, as our teacher said, who appeared to be speaking the truth. He translated separately the commentary, which was taken from the Persians by word of mouth alone. In this way the Syntaxis, called the Handy, was produced (2, p. 357).

Chrysococces does not say where precisely Chioniades went, but we know that it was to Tabrīz, in Ādharbāyjān, if not also to other places in that region. For there is a collection of sixteen letters written by Chioniades, which have been edited by Papadopoulos (6), and from these it is apparent that he was appointed Bishop of Tabrīz at the time of Alexius II Great Comnenus (1297–1330). Papadopoulos (6, p. 177) argues that the letter which provides this information was written *ca* 1303. Tabrīz was not only one of the three Īlkhānid capitals, along with Marāghah and Ardabīl (7, p. 159), and probably the principal one, but it was besides a focus of Latin Christian missions, and also for Venetian and Genoese trade (8, vol. 2, p. 107 seq.). Bar Hebraeus, a Nestorian Bishop, went there in 1282 to supervise the completion of a new church (9, p. xxv). Rashīd al-Dīn relates that it was a centre of intellectual life (10, p. cxxvi). Marco Polo passed this way on both his outward and return journeys. The details of the itinerary from Trebizond to Tabrīz are discussed by Heyd (8, vol. 2, p. 119) and by Yule and Cordier (11, vol. 3, p. 162).

If only Chrysococces' teachers had been more forthcoming about the nature of the tables for which he wrote his exposition, later historians would not have been set the problem of discovering their Persian origins. Nevertheless, if one tried to suggest the most likely Persian author, it would certainly be Naṣīr al-Dīn al-Ṭūsī, whose Zīj-i Īlkhānī was composed at Marāghah *ca* 1270. No other work was more likely to be cited to someone visiting northwestern Iran at the end of the thirteenth century in search of the most up to date astronomical tables. Usener was led to make the same observation (2, vol. 3, p. 352).

The main purpose of this paper is to demonstrate that all the tables of the Sun, Moon and planets of the Syntaxis are drawn virtually without alteration from the Zīj-i Īlkhānī. The truth of this is at once evident to anyone who will make the comparison, and it is not a little surprising that some 700 years should have passed before this was done.

Chrysococces shows no sign of knowing the Persian source, and even went so far as to say that the Zīj was drawn up for Dvīn, the old capital of Armenia, a city which in fact had been swept away around 1236 by the Mongol invasion (12). It is virtually impossible to believe that Chioniades, however, was unaware of the observatory at Marāghah. Consider the career of his contemporary, the

Nestorian Bishop Bar Hebraeus, who similarly went in search of Islamic science. He not only spent some time in Tabrīz, as noted above, but made several extended visits to Marāghah, where he wrote some of his scientific books, such as one on Euclid. He referred to Naṣīr al-Dīn and his observatory, and also wrote a work of his own on astronomical tables (9, pp. xxiii, xxxiii–vi, 451).

On close examination of the Greek literature, one does indeed find references both to the Zīj-i Īlkhānī and to Naṣīr al-Dīn. Neugebauer (13; 14, p. 28) has noted in the MS Vat. gr. 1058 (f. 268v, 270r) computations of solar eclipses for the years 1295–1296, according to the συντάξεως τῆς ἐλκανη; he did not consider the possible relation to the tables of Chrysococces. In his *Astronomy*, Theodore Melitoniotes, *ca* 1360, refers to χοντζα νασυρες, but he was evidently unaware that the Syntaxis, which he gives in full, is due to him (2, vol. 3, p. 338).

This ignorance has persisted to our own time. Boulliau made no mention of al-Ṭūsī or Marāghah in his discussion of the Syntaxis. Tannery remarked, " As far as one might judge by these extracts ⟨of Boulliau⟩, these tables, whose real author would be difficult to name, appear in any case to be earlier that the Ilkhanian tables due to Nasir al-Din " (15, vol. 4, p. 289). Kunitzsch, who studied the short lists of fixed stars which one finds variously attached to the MSS of the Syntaxis, noticed that one of these was translated from the Zīj-i Īlkhānī. He did not concern himself with the present point, that all the tables of the Sun, Moon, and planets are also taken from it.

Pingree (3) has noted the Greek translations of the Zīj al-Sanjarīa nd of the Zīj al-'Alā'ī, where there is no problem of identification, because the titles are given clearly in Greek transcriptions, and indeed one can make a direct comparison with the extant Arabic text of the former work. However, Pingree went further to say that he had thereby laid the foundation for an understanding of the source of Chrysococces' tables:

> These tables — those belonging to the Zīj as-Sanjarī as well as those belonging to the Zīj al-'Alā'ī — are the source of many of the tables upon which Chrysococces wrote his exegesis; this fact proves that the texts in V, v, and L are translations by Gregory Chioniades (3, p. 142; for the sigla, *ibid.*, p. 151).

However, not only did he fail to examine the Zīj-i Ilkhānī, as the most likely source, but his method of comparison between the tables of the Syntaxis and his other works was uncritical and altogether superficial. He was, also, so far insensitive to the scientific content of the tables of mean motions that he always fails to give their initial values at the epoch date (the so-called *radices*). It is not unusual to find that the mean longitudes in two different Zījes differ not in the rate of change, but only through a constant difference. In this way a simple adjustment was made by an observer in order to fit a later observation. In any case, the rates of motion in the Syntaxis actually differ substantially from those of the two other tables in question.

Pingree's proposals have unfortunately been treated as established fact in a number of reviews, such as those of Kennedy (16, p. 678) and Mogenet (17).

To conclude this introduction, we recall Yabuuchi's conclusion (18) that a Chinese work, the Ch'i Cheng T'ui Pu (Motion of the seven star-planets), which we have in a revised version of A.D. 1482, was derived eventually from the Zīj-i Īlkhānī. Yabuuchi says further that the tables Hui-hui Li-fa (回 回 曆 法, 2309, 2309, 3930, 1762*; ' Muslim astronomical table ') of the *Ming Shih* (19, p. 745–880) are in total agreement with the Ch'i Cheng T'ui Pu. I have examined these tables of the *Ming Shih*, and found that while the tables of equations are indeed a straightforward Chinese translation from some Zij, they are altogether different from the tables of equations of the Zīj-i Īlkhānī. The difference lies not merely in the fact that the equations are symmetrically positive and negative, in contrast to the purely additive arrangement adopted by al-Ṭūsī, but arises from differences in all the underlying parameters. I give here all the maxima of the first and second equations according to the Hui-hui Li-fa.

Star	first equation	second equation
Sun	2; 0,47	———
Moon	12;31	4; 5
Saturn	6;19	5;40
Jupiter	5; 4	10;23
Mars	11;25	36;45
Venus	2; 1	44;58
Mercury	2;25	19;56

These are all different from the corresponding values in the Zij-i Īlkhānī. Indeed there is no agreement with any Islamic source known to me. Yabuuchi had no access to the Persian MSS of the Zīj, and argued, very unsatisfactorily, from the fact that the Chinese source gives 89;21 as the Solar apogee for the year A.Y.660, or A.H.660, as if the mere mention of a date in the thirteenth century pointed necessarily to Naṣīr al-Dīn al-Ṭūsī. There is nevertheless a strong likelihood that the Zij-i Īlkhānī was known to the Chinese. The instruments of the observatory of Marāghah are described in some detail in the *Yüan Shih* (20).

* The numbers refer to Matthews' *Chinese-English Dictionary*

2. *Descriptive remarks on the Zīj-i Īlkhānī and the Syntaxis*

Both the Zij Īlkānī and the Syntaxis are widely available. The manuscripts which were consulted are listed below.

Zīj-i Īlkhānī

Z.1 London BL Or. 5572
Z.2 London BL Or. 7464
Z.3 London BL Add. 7698
Z.4 Cambridge Browne O.2 (7)
Z.5 Oxford Bodl. Hunt 143

Syntaxis

S.1 Escorial Σ.I.11
S.2 London BL Burney 91–2
S.3 Oxford Bodl. Barocci 58
S.4 Oxford Bodl. Barocci 63
S.5 Oxford Bodl. Barocci 100
S.6 Oxford Bodl. Barocci 166
S.7 Oxford Bodl. Canon gr 81
S.8 Oxford Bodl. Holkham 109
S.9 Paris BN gr 2107
S.10 Paris BN gr. 2401
S.11 Paris BN gr 2461

The Catalogues of the various libraries give descriptions of these manuscripts, although rarely in any detail where astronomical tables are concerned. Some of the Greek manuscripts are described also in the volumes of the *Catalogus codicum astrologorum graecorum*, but that series tends strongly to select astrological material to the exclusion of astronomical material. An interesting exception to the rule is provided by Revilla who gives all the titles of the tables in S.1 (21, pp. 245–252).

While the Zīj is a comparatively stable text, the Syntaxis is not, particularly when one includes the expository material occurring in the various manuscripts. Nevertheless there is a stable core, and it is just this core which is the translation from the Zij-i Īlkhānī. This core consists of the whole of the set of tables of mean motions and equations, with two exceptions. These exceptions are the tables of dated mean longitudes, which are calculated for years later than those of the Zīj, and secondly the tables for the latitude of the inner planets.

Kennedy has given a resumé of the Zij-i Īlkhānī in his well known *Survey* (22, p. 161), but it is necessary to give a fully detailed list of the contents for our present purpose. Not all copies of the Zīj include either the geographical or the stellar lists. Of the five manuscripts listed above, only Z.2, Z.4 and Z.5 include both. In Z.3 tables for Kaid are given; these apparently do not belong to the authentic text. In Z.2 there is additional material of interest for the history of the Mongols in the introduction; this has been translated by Boyle (24). There are always tables of oblique ascensions, length of day and solar altitude at noon for the latitude (37; 20) of Marāghah, together with tables of oblique ascension for latitudes from 1 to 54.

The Syntaxis is found sometimes with other expositions than that of Chrysococces. There is one that is attributed to Isaac Argyros, for example, which is found in some of the manuscripts listed. These expositions contain worked examples for various dates; in the exposition of Chrysococces, these are for A.D.1346–7. Apart from the central collection of tables, other material is found, but it is interesting that there is never any astrological section, even though it might have been taken from Book IV of the Zīj. The geographical table of the Zīj seems to be quite unknown in the Greek versions, but in the Syntaxis we have instead various extracts from Ptolemy's Geography, and frequently also a list of 88 places extracted from the list of about 250 places in the Zīj al-Sanjarī. This is in turn taken from an even longer list of 603 places in al-Bīrūnī's Qanūn al-Mas'udī. I intend to publish the Greek list elsewhere.

There are three different lists of the fixed stars which one finds attached unpredictably to the various copies of the Syntaxis. Kunitzsch (4) has carefully edited and analysed them. The large number of manuscripts which he examined included only S.9 and S.10, but nothing new emerges from the other nine manuscripts in this respect. The three types are

I. 25 stars selected from the list of 43 in the Zīj al-Sanjarī.

II. 60 stars copied without alteration from the Zīj-i Īlkhānī.

III. 30 stars conforming to the ancient astrological tradition of λαμπροι ἀστέρες.

3. *Indexes to the Zīj-i Īlkhānī and to the Syntaxis*

All the titles of the Books, chapters and sections, and also of all the tables are listed below. In the case of the Syntaxis, the Escorial manuscript S.1 is followed, and the appropriate folio numbers are given. The Books, etc., are numbered as in the texts. The tables of the Zīj are given, on the other hand, an obvious serial notation, although in Book 2, where there is no chapter division, the tables are grouped according to subject matter.

In the Syntaxis the tables are referred to the previous numbers of the tables of the Zīj. If this number is put in parentheses, there is some small difference between the two. For example, the tables T.2. n. 1 give the dated mean longitudes calculated according to the Zīj, but for later dates.

Zīj-i Īlkānī

Preface

Discourse on the star observatory, and on Zijes and almanacs

Discourse on the form of this observatory, newly built, and on other observatories

Book 1	On eras
Introduction	On the initial points of eras used at present
Chapter 1	Exposition of the Qitai era, on knowing its years and months
Section 1	On the explanation of the divisions of the night–day of people related to the Qitai
T.1.1.1.1	The twelve earthly branches in Turkish and Chinese, with Persian equivalents
T.1.1.1.2	On knowing the beginnings of the *shih* and *k'e* in terms of the *feng* of the night–day in sexagesimal form
Section 2	On the consideration of cycles of days among the Qitai
T.1.1.2.1	The sexagenary cycle in Chinese
Section 3	On knowing the years of the Qitai and their divisions
T.1.1.3.1	The beginning of every division of the divisions (*chieh–ch'i*) of the year, calculated in days and *feng*
Section 4	On the consideration of the cycles of years among the Qitai and their eras
Section 5	On knowing the beginnings of the divisions (*chieh–ch'i*) of the years of the Sun which occur in each year
T.1.1.5.1	The beginning of *Li–ch'un* in several years (A.Y.633–657)
T.1.1.5.2	The beginnings of the divisions of the year *taqïqu* corresponding to A.Y.642
Section 6	On knowing the beginning of *Ārām*
Section 7	On knowing the argument of the Sun and the Moon
T.1.1.7.1	The two cycles taken from the beginning of the cycle *shang yüan*, constructed as an example (A.Y.633–657)
Section 8	On finding the equation of the Sun
T.1.1.8.1	Equation of the Sun, namely what they call *tānkāz*
Section 9	On finding the equation of the Moon
T.1.1.9.1	Equation of the Moon, namely what they call *tānkāz*
T.1.1.9.2	Multiples of *shuo shih*, the length of the month of lunation
T.1.1.9.3	Multiples of *chuan chung ch'a*, the increase of the lunar argument during the synodic month
T.1.1.9.4	Multiples of *chuan chung*, the anomalistic month
T.1.1.9.5	Length of sum of half–night and the day in each month

T.1.1.9.6	Multiples of the *feng* in the days of the month
Section 10	On knowing the first day of each month of the year
T.1.1.10.1	The first day of each month in year *taqīqu* corresponding to A.Y.642
T.1.1.10.2	Intercalary years (Turkish and Chinese) A.Y.633–657
Section 11	On the fourth cycle of the Qitai people
Section 12	On knowing the Qitai year from the Arabic year
T.1.1.12.1	Tabulation of Qitai years during A.H.641–661
Chapter 2	On the eras used by other peoples
Section 1	On the Rūmī era
T.1.2.1.1	First day of the month in the Rūmī era
Section 2	On the Arabic era
T.1.2.2.1	First day of the month in the Arabic era
Section 3	On the Fārsī era
T.1.2.3.1	First day of the month in the Fārsī era
Section 4	On calculating each one of the three from the other
T.1.2.4.1	The number of days in each era in multiples of 120 (60) 1860 years
T.1.2.4.2	The number of days in each era in multiples of 1 (1) 60 years
T.1.2.4.3	The names of the months of each era
Section 5	On the new era called Malikī, after Jalāl al-Daulah Malik-shāh
T.1.2.5.1	First day of the month of Malikī years
T.1.2.5.2	Numbers of Malikī intercalary years
T.1.2.5.3	Initial points of incomplete Malikī years from the beginning of that era in the longitude 95, *al-jazā'irī*, according to the calculation of the new observatory
Section 6	On the Jewish era
T.1.2.6.1	The initial point of the Jewish year
T.1.2.6.2	On knowing the initial point of the Jewish years and the properties of each one, 1500 (11) 1839 Era of Philip
Section 7	On the special days in each era
T.1.2.7.1	On the beginning of the Christian feasts
Book 2	On knowing the institution of the Royal Observatory and its position in latitude and longitude
Section 1	On calculating the mean Sun and the mean longitude of other planets at any time
Section 2	On the true position of the seven stars and the node in longitude
Section 3	On the latitude of the Moon and the five planets

Section 4	On the *qismah* of the true position of the stars	
Section 5	On the *niṭāqāṭ* of the stars and the stations of the planets	
Section 6	On knowing the semi-diurnal period	
Section 7	On knowing conjunctions and oppositions	
Section 8	On knowing the lunar eclipse	
Section 9	On knowing the solar eclipse	
Section 10	On knowing the time of arrival of the Sun at a given point in the sphere of the signs, and the horoscope at that time	
Section 11	On knowing the appearance of the new Moon, and the appearance and disappearance of the planets	
Section 12	On the *mizājāt* of the Moon, and its *istiqbālāt* and the *ittiṣālāt* of the stars, and their *taḥwīlāt*	
Section 13	On fixing the houses	
Section 14	On the positions of several fixed stars and their motion	
Section 15	On the longitudes and latitudes of several famous cities	
T.2.1.1	Sun	Centre and apogee A.Y.600–700
T.2.1.2		Centre and apogee in months and days A.Y.
T.2.1.3		Centre in epagomenal days A.Y.
T.2.1.4		Centre in centuries A.Y.
T.2.1.5		Centre in minutes of one hour
T.2.1.6		Apogee for single years, decades and centuries A.Y.
T.2.1.7		Change of meridian from 82
T.2.1.8		Equation
T.2.1.9		Equation of days and nights
T.2.1.10		Parts of centre in the circle
T.2.2.1	Moon	Centre argument and mean A.Y.600–700
T.2.2.2		Centre argument and mean in months and days A.Y.
T.2.2.3		Centre argument and mean in epagomenal days A.Y.
T.2.2.4		Centre argument and mean in centuries A.Y.
T.2.2.5		Centre argument and mean in minutes of one hour
T.2.2.6		Change of meridian from 82
T.2.2.7		First equation
T.2.2.8		Second equation, variation and proportional minutes
T.2.2.9		Third equation
T.2.2.10		Equation of days and nights to be subtracted from Sun
T.2.3.1	Node	Mean A.Y.600–700
T.2.3.2		Mean in months A.Y.
T.2.3.3		Mean in epagomenal days A.Y.
T.2.3.4		Mean in centuries A.Y.
T.2.3.5		Mean in minutes of one hour

T.2.3.6		Change of meridian from 82
T.2.3.7		Latitude of Moon in terms of argument of the Moon
T.2.4.1	Eclipse	Lunar eclipse: digits, time of incidence, duration
T.2.4.2–6		Lunar parallax in climates three to seven
T.2.4.7		Solar eclipse
T.2.5.1–6	Saturn	Mean centre and argument
T.2.5.7		First equation
T.2.5.8		Second equation
T.2.5.9		Proportional minutes and variation
T.2.5.10		Proportional minutes of latitude
T.2.5.11		Northern and southern maxima of latitude
T.2.6.1–6	Jupiter	Mean centre and argument
T.2.6.7–9		Equations
T.2.6.10–11		Latitude
T.2.7.1–6	Mars	Mean centre and argument
T.2.7.7–9		Equations
T.2.7.10–11		Latitude
T.2.8.1–6	Venus	Mean centre and argument
T.2.8.7–9		Equations
T.2.8.10		Proportional minutes of latitude
T.2.8.11–13		First, second and third latitude
T.2.9.1–6	Mercury	Mean centre and argument
T.2.9.7–9		Equations
T.2.9.10–13		Latitude
T.2.10.1	Niṭāqāt of Sun, Moon and planets	
T.2.10.2	Stations of the planets	
T.2.10.3	Appearance and disappearance of the planets	
T.2.11.1	Ascensions for latitude 37;20	
T.2.11.2	Right ascension (intervals of 1°)	
T.2.11.3	Length of day for latitude 37;20	
T.2.11.4	Altitude of Sun at noon for latitude 37;20	
T.2.12.1	Movement of the Sun in minutes of *buht*	
T.2.12.2	*Ijtimā'āt* and *istiqbālāt*	
T.2.12.3	*Mumāzajāt*	
T.2.12.3	*Ittiṣālāt*	
T.2.13.1	Names of well known fixed stars with their longitudes, latitudes, magnitudes and temperamenta	
T.2.13.2	Continuation of the last according to different observers	
T.2.13.3	Movement of the fixed stars in single years, decades, centuries	

T.2.14.1	Names of places, with their longitudes from *Jazā'ir-i khālidāt*, and latitudes from the equator
T.2.14.2	Extract from the Almagest of Ptolemy of regions in order of increasing length of day, 12 (0;15) 21 hours
Book 3	On knowing the times and the ascendant at any time
Section 1	On knowing the sine and sagitta
Section 2	On knowing the shadow (tangent)
Section 3	On knowing the declination of the sphere of signs from the equator
Section 4	On knowing the distance of every point from the equator
Section 5	On knowing the maximum altitude of the stars
Section 6	On knowing the equation of days and the arc of days
Section 7	On the ascension at the terrestrial equator, that is the ascension of the right sphere
Section 8	On the ascension at a place
Section 9	On the ascension at transit
Section 10	On the rising and setting of stars
Section 11	On knowing the ascension of the ascendant from the culmination of the star
Section 12	On understanding the reverse process, namely to find the culmination
Section 13	On knowing the ascendant from the hour
Section 14	On correcting for the time of mid-day
T.3.1	Sine
T.3.2	Tangent 0 (0;1) 45
T.3.3	Tangent 45 (0;20) 89;50
T.3.4	Second tangent
T.3.5	First declination
T.3.6	Second declination
T.3.7	Right ascension (steps of 1;0)
T.3.8	Oblique ascension (steps of 1;0) for latitudes 0 (1) 54; obliquity 23;35.
T.3.9	On knowing the Sun's arc from the altitude of the Sun for the latitude 37;20
T.3.10	Equation of days and night
Book 4	On the remaining astronomical calculations
Chapter 1	On what concerns the ascendants of the nativities
Section 1	On *namūdārāt*
Section 2	On the 'horizon' of stars according to their positions in the ascendant

Sintaxis

Exposition by the Physician George Chrysococces of the Persian Handy Tables, together with the Tables

11.	On the hours from noon	7r
12.	On the calculation of the (true) longitude of the Sun	7v
13.	On the distance between cities, by which one is separated from another, and how it is calculated	8v
14.	On the calculation of the (true) longitude of the Moon	9v
15.	On the equation of time of the Sun and Moon	11v
16.	On proportion	11v
17.	On the hours from noon, and how they are calculated	12r
18.	On finding the days of simple syzygies of Sun and Moon	12r
19.	On the table of conjunctions and full moons	13v
20.	On the method (of finding) the conjunction and opposition of the Sun and Moon	14v
21.	On the calculation of the longitude of the nodes, namely ☊ and ☋, which the Persians call head and tail	19v
22.	On the latitude of the Moon	20r
23.	On eclipses, and the beginning of the lunation	21r
24.	On the eclipse of the Sun, called *kousouf* in Persian	23r
25.	On the calculation of the (true) longitude of the five planets	25v
26.	On the transit of the stars ♄ ♃ ♂ ☉ ♀ ☿ ☾ and ☊ determined in day and hour	28r
27.	On the calculation of the latitude of the outer planets ♄ ♃ ♂	30v
28.	On the calculation of the latitude of the inner planets ♀ and ☿	32r
29.	On drawing up the ephemerides (the almanac)	34r
30.	On the amount by which each star is moved in a day	35r
31.	On determing the stars for the list of the ephemerides	36r
32.	On the alignements among the stars, their names, and their proximity and distance from the Sun	37r
33.	On the right and left portions of alignements, in which the stars are left and right	38r
34.	On appearance and disappearance, first and second stations, as well as combustions and *akronukta*	39r
35.	On how, consequently, one should find the hours at which the stars are aligned with the Sun (by degrees)	40v
36.	On how one should find the time when the six stars ♄ ♃ ♂ ☉ ♀ and ☿ are aligned	42v
37.	On finding the hour when the planets which do not move in a straight line are aligned (by degrees)	45v
38.	On finding the horoscope, namely the degrees of the ascendent	46r
39.	On the degree of mid–heaven	47r
40.	On the horoscope, noon, and midnight	47v

TABLES

Mercury radices	(T.2.9.1)	106r
mean motion and equations	T.2.9.2–9	107r
proportional minutes of latitude	T.2.9.10	111r–112v
equation of latitude	(T.2.9.11)	112v
first latitude	T.2.9.12	111r–111v
second latitude	(T.2.9.13)	112r–112v
Length of day (for longitude), latitude 36		72v, 113r
Length of day (for longitude), latitude 36 (obliquity 23;51,20)		113v
Length of day (for longitude), latitude 42		114r
Length of day (for longitude), latitude 43		114v
Ascension of Sun at mid-heaven		115r
Ascension of right sphere		115v
Ascension of oblique sphere, latitudes 16,17,24,25,30, 31,35,36,37,40,41,42,43,45,46,48,49	T.3.8	116r–125r

4. *Detailed notes to the list of contents*

The following notes are certainly not intended to provide a complete commentary to all the technical terms which occur in the titles listed above, but only to fill in a few of the gaps within material which is assumed to be largely familiar. Since Ulugh Beg's Zıj is so closely dependent on this work, one can usefully refer to Sédillot's translation and commentary (24) for many details. Kennedy's resumé should also be consulted, particularly where tables are concerned, for Sédillot says nothing about them (apart from a few tables in Book 1).

4.1 *The Sino–Turkish calendar according to the Zīj-i Īlkhānī*

This account of the Sino–Turkish calendar is apparently the source of a number of other Persian and Arabic accounts, including that of Ulugh Beg (Zīj–i Jadīd–i Sulṭānı). Modern studies begin with that of Ideler (25), and include a recent article by Kennedy (26), who was concerned with the equations of Sun and Moon. Bazin (27) has composed a long study of medieval Turkish calendars, which touches on certain aspects of this one.

A number of Chinese and Turkish terms, in Persian transcription, occur in this account. Some of these, such as the names in the cycle of 12 animals, are clear and well known. Others could only be fixed after consulting the technical literature on the Chinese calendar, as explained at the end of the list. The numbers of the Chinese characters refer to Mathews' Dictionary.

The Chinese are referred to by the name Qitai (قتا), which is properly the name of a nation situated to the north-east of China. The name however had come to have a much wider application, including at least northern China, or the whole country. For a full account, see Pelliot's *Notes on Marco Polo* (28; vol. 1, s.v. ' Catai ').

جاغ *shih* (5780) 時: ' unit of time equal to two hours '. The 12 *shih* are put into one to one relation with the 12 earthly branches and with the cycle of 12 animals, given in T.1.1.1.1.

كه *k'ê* (3322) 刻: ' unit of time equal to one eighth of *shih* '.

فنك *feng*; *fen* (1851) 分: ' unit of time equal to 1/10,000 of the day '. The same term is used in Chinese for other divisions of time, including the usual ' minute '. al–Ṭūsī always expresses the number of *fen* sexagesimally.

chieh–ch'i (795, 554) 節 氣. These are the 24 divisions of the year, marked by the passage of the Sun through the intervals of 15° of longitude. al–Ṭūsī does not give the transcription, but only refers to them as ' divisions ' (*aqsām*).

ليچون *li–ch'un* (3921, 1493) 立 春: ' beginning of Spring ', the first of the *chieh–ch'i*, when the Sun's longitude is 315°. This occurs on Jan. 29, not Feb. 5 as in the modern Chinese calendar.

داقوق *taqïqu* is the name of the tenth animal, the cock, in the Turkish list of twelve animals; see von Gabain's Glossary (29). The corresponding year is A.Y.642, A.D.1273. It is not clear why this year is chosen for detailed illustration.

ارام *Arām*: ' first month of the Turkish year '. The word is not, however, Turkish, but apparently a loan word from Iranian *rām* (both Middle and New Persian), meaning ' peace ', ' joy ', as referring to the New Year festivities. In Turkish the initial ' r ' is not tolerated. This suggestion was made by Bazin (27, p. 433–434).

شانكون *shang yüan* (5669, 7707) 上 元: ' first beginning ', viz., of the first of three sixty year cycles, beginning in A.D.1264, A.Y.633. The next two are

چونكون *chung yüan* (1504, 7707) 中 元: ' second beginning ',

خاون *hsia yüan* (2520, 7707) 下 元: ' third beginning '.

Gaubil (30; vol. 3, p. 16, 80, etc.) gives *shang yuen* as the name of the Grand Conjunction (like the Indian Kaliyuga), for which various dates were proposed. The choice of that date does not seem, however, to disturb a generally assumed congruence of the sixty year cycles, so that A.D.1264, for example, would always begin a new cycle. The Persian text also uses ون to transcribe *wan* (7030) 萬, 10,000. The alternative readings *yüan*/*wan* seem to be required according to the context.

تانكاژ *tānkāž*. This must be the Turkish, not the Chinese, equivalent of ' equation '. Only the manuscript Z.5 gives this reading, the others giving يانكاژ *yānkāž*. However *tānkāž* must be correct; von Gabain gives *tänkäš* ' to compare ' (29, Glossary). The roots *teŋ*, *teŋes* ('equal', ' to be equal ') are discussed by Clauson (31).

شوجه *shuo shih* (5940, 5821) 朔實: 'synodic month', literally 'full lunation'. The value used is 29.5306 days.

جنجون *chuan chung* (1431, 1500) 轉終: 'anomalistic month', literally 'whole revolution'. The value used is 27.5556 days, evidently an approximation to 248/9. This value occurs earlier in a group of Babylonian texts and is also given by Geminos and by Varāhamihira (32, p. 480, 602, etc.; 26).

جنجونشا *chuan chung ch'a* (1431, 1500, 105) 轉終差: 'difference in the anomalistic month'. The value used is 17.7750 days, obtained as nine times the excess of the synodic over the anomalistic months. The factor nine is part of the special method of computing the lunar equation, where the usual 'argument' as such is not used.

The last three terms in Chinese are determined by the usage in the calendrical chapters, especially ch. 54, of the *Yüan Shih*, the Dynastic History of the Mongol period (33; vol. 4, p. 1191, 1213–4). There one is given an account of the new calendar of Kuo Shou Ching, the *Shou Shih Li*, where the parameters differ somewhat from those of the Sino–Turkish calendar. Gaubil, who quotes at length from this part of the *Yüan Shih* (30, vol. 3, p. 69, 72, 73) gives the lunar terminology in French transcription,

shuo shih: *cho-tche*, 'nombre de la nouvelle lune'.
chuan chung: *tchouen–tchong*, 'anomalie entière'.
chuan ch'a: *tchouen–tcha*, 'différence de la révolution'.

There are some indications in the Chinese sources of this luni–solar calendar. The length of the year is 365.2436 days, the same value which is found in a series of Chinese calendars established in the Sung, Southern Sung, Chin, and Yüan dynasties. From the list of calendars given by Yabuuchi (34, p. 340–345; and 35) we note that calendars employing this length of year are distributed as follows

Sung	1065–1127	five calendars
Southern Sung	1168–1190	two calendars
Chin	1182–1234	Reformed Ta Ming calendar
Yüan	1215–1280	

As to the term 'Ta Ming', it is used of a fair number of calendars, as Yabuuchi explains in some detail. The earliest reference goes back to the fifth century calendar of Tsu Ch'ung Chih, who used a length of year equal to 365.2428 days. The 'Reformed Ta Ming' (Chung–hsiu Ta Ming, 重修大明, 1509, 2794, 5943, 4534) is not apparently related to that, nor related in any clear way to others denoted 'Ta Ming' which just preceeded it, and for which numerical details are not available. Important evidence is given in the Chan–jan chü–

shih wên-chi of Yeh-lü Ch'u-ts'ai (36, p. 117–118), a Chinese advisor who accompanied Genghis Khan on his Western expedition in A.D.1220. Yeh-lü evidently undertook to establish an observatory (in Samarkand, according to Yabuuchi) and a new calendar, having found certain faults in the current calendar, to which he refers simply as 'Ta Ming'; that was presumably the Reformed Ta Ming calendar. The latter continued in use, nevertheless, with the Mongols until another reform *ca* 1280, carried out by Kuo Shou Ching and others. Therefore, because of the agreement in the length of the year with the Reformed Ta Ming, and because the latter was in regular use under the Mongols, it may be concluded that this is the source of the calendar described by al-Ṭūsī.

4.2 *The equations of motion*

The equations of Sun, Moon, and planets represent the usual Ptolemaic theory, but they have all been rearranged in a such a way that the quantity tabulated has only to be added, never subtracted. The benefits are obvious enough for a user who has no use for negative quantities. This technique was already old by the time of al-Ṭūsī, and was used, for example, by al-Bīrūnī for the Sun and Moon. An analysis of ibn A'lam's similar procedure has been given by Kennedy (37). The rearrangement requires that certain constants be subtracted from all the mean longitudes, as explained further in section 5.

In his *Tadhkira fī 'ilm al-Hay'a*, al-Ṭūsī developed a new lunar theory in which he adhered to the principle that rotation is always uniform with respect to the centre of rotation, rather than about an equant. This model has been carefully analysed by Hartner (38), and I have calculated the resulting equations. Since there are small differences from the equations calculated from Ptolemy's model one can be sure that the lunar equations given are not calculated from the new model.

4.3 *Lunar parallax, tables T.2.4.2–6*

Parallax tables in the Syntaxis are placed adjacent to tables of eclipse geometry, just as in the Zīj-i Īlkhānī, and it is surprising to discover that they are not exactly the same. For those of the Syntaxis are identical instead to those given in the Zīj of al-Battānī (39, vol. 2, p. 95), and differ because they are computed with a different value of the obliquity: 23;30 according to al-Ṭūsī, and 23;35 according to al-Battānī. There are tables in the Zīj-i Īlkhānī in which the older value of the obliquity is used, namely the tables of oblique ascension placed at the end. This is plainly the source of a selection of such tables given in the Greek version. However, the remaining tables of the Zīj in which the new obliquity is used, are not accepted in the Greek version, where the equation of time, declination, length of day and right ascension are calculated from the old obliquity.

It is strange and puzzling to discover this resolute use of the old and superseded value of the obliquity, amounting apparently to a deliberate rejection of al-Ṭūsī's value. As just noted, al-Battānī employed the value 23;35, and he is included in the list of Persian and Arabic astronomers given by Meliteniotes (2, vol. 3, p. 338). However, he cannot be the source for all these tables of oblique ascension, etc.; he gives that quantity only for the latitude (36°) of al-Raqqa.

4.4 *Table T.3.9*

This gives the Sun's hour angle as a function of its altitude. More precisely, if $\alpha(h, H)$ is the hour angle measured from the meridian, when the actual altitude is h and the meridian altitude H, then the tabulated angle is $\alpha(0, H) - \alpha(h, H)$. It is a simple matter to derive

$$\cos \alpha\ (h,\ H) = \frac{\sin h + \sin \phi \cos (H + \phi)}{\cos \phi \sin (H + \phi)}$$

where ϕ is the geographical latitude. In this table the latitude is naturally taken as 37;20. Goldstein (40) has described another table of this type where the latitude was 36.

4.5 *The latitude of the inner planets*

In the Zij, the calculations of the latitude of the inner planets follow those of the Almagest, except that there is a substantial change in the 'deviation' of Venus. In Ptolemy's approach, the latitude is the sum of three components, the ' inclination ', the ' deviation ' and the ' slant ', which we denote β_1, β_2 and β_3; see Pedersen (41, p. 368 seq.). The tabulation in the Zij differs from that of Ptolemy in various ways, some of the calculations which he leaves to the reader being already carried out and tabulated here. Thus β_1, the ' first latitude ' is given expressly instead of being derived by the reader from the proportional minutes. Further β_3, the ' third latitude ' for Mercury, is expressly divided into two branches, as Ptolemy prescribes, respectively 0.9 and 1.1 times the function tabulated in the Almagest. These branches are used according to whether the corrected centre lies in the ranges 90 to 270, or —90 to 90. There is, however, one real change from the Almagest, in that β_2 for Venus, in the half of the epicycle nearer the earth, is about 1.4 times the values given in the Almagest, reaching —8;40 instead of —6;22.

On turning to the later writers, ibn al-Shāṭir, al-Kāshī and Ulugh Beg, we see that they also differ from the Almagest in the same way as al-Ṭūsī. al-Kāshī and Ulugh Beg reformulate β_3 for Mercury, by dividing the proportional minutes into two branches, instead of so dividing the function of the corrected argument. The tables of ibn al-Shāṭir and Ulugh Beg have been consulted in the MSS Oxford Bodl Hunt 304 and Greaves 5, respectively; those of al-Kāshī in the MS India Office Library Ethé 2232.

On turning to the Syntaxis, we find numerous differences from the Zīj-i Īlkhānī. These begin with the terminology, for now β_2 and β_3 are the 'first' and 'second' latitudes, while β_1 is the 'equation of latitude'. Only β_2 is identical in both cases to Ptolemy's functions, and so different from al-Ṭūsī in the case of Venus. β_1 is different altogether, and much smaller in magnitude, although still positive for Venus and negative for Mercury. β_3 is again divided into two branches, but not as Ptolemy prescribes. For Venus, these branches are respectively 0;10 greater and smaller than the function given by Ptolemy; for Mercury, they are approximately 0;30 greater and smaller.

It is evident that from the mid-thirteenth century there was a move towards a more experimental approach to the latitudes of the inner planets, and this continued at least until the time of Ulugh Beg. Probably one should regard the special features in the Syntaxis as something obtained from Islamic proposals, otherwise unknown to us, rather than as an innovation due to Byzantine students.

Neugebauer (14, p. 27) has described briefly the latitude tables of the Zij al-ᶜAlā'ī, as found in Vat. gr. 1058, but it is not quite clear to me whether this could be the source of the tables in the Syntaxis.

5. *Parameters of mean longitudes*

The table in this section gives all the parameters of the mean motions for the Zīj-i Īlkhānī and the Syntaxis. The epoch date of the Zīj, denoted here t_T, is taken from the first entry for A.Y.600 complete (A.D.1232 Jan.18, J.D.2171063). The rates of motion are taken from the tables which give the motion for multiples of the century. The Syntaxis gives the latter tables unchanged, and in fact the only changes found there in regard to the mean motions arise from the fact that the dated entries are for later years. Those dates vary from one manuscript to another, but are earlier in the Escorial MS than in the others consulted, and the radices quoted here in the table are taken from it. The entries are given for A.Y.710–720 and 791–838 (all current), with some small variations; the lunar node, however, is given for A.Y.715, 755, 791–838. The epoch date t_C used in the table is therefore chosen to be A.Y.709 complete (A.D.1340 Dec. 21, J.D.2210848), except for the lunar node, where it is A.D.714 complete (A.D.1345 Dec. 20, J.D.2212673).

The usual symbols for the Sun, Moon and planets represent the tropical mean longitudes. The apogees, also measured from the equinox of date, are denoted by ϖ with the appropriate suffix. The Zīj follows a convention whereby certain constants are added or subtracted from the centre or argument with the aim of making all the equations positive. These constants are all given in the left hand column, added or subtracted as required. From the three entries for each planet one may compute the mean Sun; here there is not perfect consistency, for the entries for Jupiter and Mars give 304;1,58 and 303;41,7 instead of the value 304;1,6 obtained otherwise.

It is quite evident from a comparison between the last two columns, that the longitudes of the Zīj-i Īlkhānī serve for the computation of the radices in the Syntaxis, without any alteration of the reference meridian.

Mean longitude	Zīj-i Īlkhānī			Syntaxis
	epoch at t_T	motion in 100×365 days	position at t_O	epoch at t_O
☉ − ϖ☉	215;36,15	99^{r}334;41, 3	188; 0,35.7	188; 0,36
ϖ☉ − 2; 0,30	86;24,21	100/70	87;57,46.7	87;57,47
☾ − 7; 40	226;46,22	1335^{r}338;23,18	289;36,57.8	289;36,58
☾ − ϖ☾ − 13; 8	245; 1,27	1324^{r}231;52,28	195;22, 2.3	195;22, 0
2 (☾ − ☉)	220;50,31	2472^{r} 4;33, 3	38;36, 8.5	38;36, 9
☊	218; 5,34	-5^{r}132;51,44	174;37,47.4	174;37,47*
♄ − ϖ♄ − 14; 0	112;19,38	3^{r}141;20, 3	3;34,53.3	3;34,54
☉ − ♄ + 6; 0	295;48,19	96^{r}193;21,40	16;58, 8.0	16;57,24
ϖ♄	248;53, 9	100/70	250;26,34.7	250;26,35
♃ − ϖ♃ − 18; 0	282;43,14	8^{r}152;52, 4	348;32,47.2	348;30
☉ − ♃ + 6; 0	191;38, 1	91^{r}181;49, 0	98;12,45.9	98;15
ϖ♃	177;40,43	100/70	179;14, 8.7	179;15
♂ − ϖ♂ − 12; 0	96; 0,54	53^{r} 47;12,22	64;40,10.8	64;40
☉ − ♂ − 12; 0	73;10, 5	46^{r}287;28,41	76;55, 8.9	76;55
ϖ♂	134;30, 8	100/70	136; 3,33.7	136; 4
☉ − ϖ♀ − 2; 0	224;20,56	99^{r}334;41, 3	196;45,16.7	196;45
♀ − ☉ + 2; 0	140; 8,22	62^{r}182;57,44	188;22, 5.8	188;22
ϖ♀	77;40,10	100/70	79;13,35.7	79;14
☉ − ϖ☿ − 4; 0	88;55,56	99^{r}334;41, 3	61;20,16.7	61;20
☿ − ☉ + 4; 0	164;50, 0	314^{r}357; 4,24	287;35,35.8	287;38
ϖ☿	211; 5,10	100/70	212;38,35.7	212;38

* t_O is here replaced by A. Y. 714 complete.

The Persian and Greek terms employed in the tabulation of the mean longitudes are compared in the next table; P represents any planet.

		Zīj-i Īlkhānī	Syntaxis	translation
☉ $- \varpi_{☉}$		markaz	μέση	centre
$\varpi_{☉}$		auj	ὕψωμα	apogee
☾		wasaṭ	μέση	mean
☾ $- \varpi_{☾}$		khāṣṣah	ἴδιον	argument
2 (☾ − ☉)		markaz	κέντρον	centre
☊		taqwīm-i ra'as	κίνησις τοῦ ἀναβιβάζοντος καὶ καταβιβάζοντος	motion of node
planets				
outer	inner			
P $- \varpi_P$	☉ $- \varpi_P$	markaz	κέντρον	centre
☉ − P	P − ☉	khāṣṣah	ἴδιον	argument
ϖ_P	ϖ_P	auj, bu'd al--ab'ad	ὕψωμα	apogee

6. *The meridian of reference of the Syntaxis*

The Zij-i Īlkhānī is plainly referred to the meridian with the longitude 82. This is indicated in every section of the tables, for there is always provided a small table giving the increments to any mean longitude which are needed when the meridian is shifted from 82. These small tables are also given in the Syntaxis, but with the difference that the meridian is now assigned the longitude 72. This change reflects only a change in the conventional position of the origin of geographical longitude, for the Zīj follows Ptolemy, whereas the Syntaxis uses the Arabic convention, which measured the longitude from the eastern limit of the Ocean. For many places the difference of longitudes is simply 10. Chrysococces refers to his meridian of reference in a number of places. At the beginning of Ch.13, for example, he writes; " This Zīj, namely the Syntaxis, is produced for the longitude 72, in the city of Khazaria named τυβήνη ". The name τυβήνη transcribes Dvīn, formerly the capital of Armenia. This place, however, had been destroyed by the Mongols around 1236 (12, s.v. Dvīn), and it may be inferred that Chrysococces chose this place simply because it is the only one in the Greek list of places with the appropriate longitude. This list includes 88 names, and is an extract from that of the Zıj al-Sanjarī (ca. A.D. 1135), of al-Khāzinī, which includes approximately 250 names. The latter is, in turn, an extract from the great list of the Qānūn al-Mas'ūdī (ca. A.D. 1030), of al-Bīrūnī (42, vol. 2, p. 547-579). I have been able to consult only an unsatisfactory version of the Zīj al-Sanjarī, Brit. Lib. MS Or.6669. Another copy of the Zīj is found in the MS Vat. ar. 761, but this does not include the geographical table. The Brit. Lib. manuscript is throughout in a poor state, with a number of pages torn or missing, as Kunitzsch (4, p. 390) has noted in his discussion of the star lists. The geographical table is on f. 104v–106r, where it is clear that a page is missing between 104 and 105, with the loss of about 95 names. In the Greek MSS of the Syntaxis, one finds also a list of names extracted from Ptolemy's Geography.

I give here the places in Armenia and Ādharbāyjān according to both the Zīj al-Sanjarī and the Greek extract.

Zīj al-Sanjarī			Greek list
Armenia	long.	lat.	*ἀρμενία*
Barda'	63; 0	43; 0	μπαρταάν
Baylaqān	64; 0	39;50	
Akhlāt	64;50	39;40	
Darband-i Khazrān	66; 0	41; 0	χασρὰν
Shirvān	67;30	40;50	
Ādharbāyjān			*ἀρδεμπεκάν*
Dvīn	72; 0	38; 0	τυβήνη
Nashwi-Nakhchuvān	72;20	38;30	
Ardabīl	73; 0	37;50	ἀρτουέλ
Marand	73; 0	37;55	
Tabrīz	73;10	37;40	
Marāghah	73;10	36;25	
Khuwī	73;20	38;20	

The coincidence of the coordinates shows clearly that τυβήνη transcribes Dvīn. One should note also that β in usually pronounced /v/.

It is strange that Tabrīz and Marāghah are not included. Perhaps the list had been translated before 1236 when Dvīn was still of importance to Christians, perhaps of more interest than Tabrīz or Marāghah.

Pingree's proposal that τυβήνη referred to Tabrīz has gained some currency, but it is plainly wrong. This proposal is not only absurd on phonetic grounds, but it is clearly disproved by an inspection of the above-mentioned extract from the lists of places. He first made this proposal (3, p. 144; and 43, p. 325) when he published a correction to a note by Neugebauer (13).

The latter, when analysing a calculation of a lunar eclipse of 1295, given in a Greek text, for a place named as ταυρές, with the coordinates (82; 38), proposed to identify this place with the Mesopotamian city named δάρας. The identity, however, of ταυρές and Tabrīz is clear to any scholar of Marco Polo, and besides, the coordinates are the same as those given in the Zīj-i Īlkhānī for Tabrīz. Pingree, not content to correct Neugebauer's identification, went on to introduce his own red herring, by identifying τυβήνη also with Tabrīz. This was asserted on the ground that (1) τυβήνη is taken as the meridian by Chrysococces, while (2) Chioniades translated the tables at the court of the Persian king, which (3) was certainly Tabrīz. Of these points, only (1) is clear. For the place of the translation is not really known, nor is it of more than somewhat circumstantial interest, and in any case the Īlkhānids used not only Tabrīz, but also Marāghah as a capital. Pingree further remarks (43, p. 325) that Chrysococces says that the court of the Persian king was held at τυβήνη. I am not aware that Chrysococces said any such thing about the location of the Persian court; he refers to τυβήνη only in connection with the meridian of the tables. Of course, none of this confusion would have arisen if the Zīj-i Īlkhānī had been correctly recognized as the source of the Syntaxis, thus fixing beyond doubt the meridian of reference as Marāghah.

It remains as a curious fact that the geographical table of the Zīj-i Īlkhānī went unnoticed by the Greek translator. In that list, Dvīn no longer appears. Under Armenia, one finds Āmid, Mayyāfāriqīn, Malāzgird, Akhlāt, Arjij and Ānī.

7. *Conclusion*

The fourteenth-century Persian Syntaxis, with the Exposition of Chrysococces is analyzed with a view to establishing the Persian source. The tables of the Syntaxis are shown to be a Greek transcription of the tables of the Zīj-i Īlkhānī of Naṣīr al-Dīn al-Ṭūsī. Complete indexes are provided for both works, including all the chapters and all the tables, and in such a way as to reveal the exact correspondence between the Persian and the Greek tables. A summary of all the mean parameters is given, according to both Persian and Greek texts. This shows that no change of meridian was made in the Greek version.

The Zīj opens with an account of the Sino-Turkish calendar of the thirteenth century, and the Chinese characters are supplied for the various Chinese technical terms which appear in Persian transcription.

Some notes are provided for various other problems, including the geographical meridian of reference from the Greek point of view.

Acknowledgement. – I am grateful to the British Academy for grants which made possible numerous visits to Libraries, both here and in Spain. I also wish to express my gratitude to Prof. Kunitzsch who saved me from a number of linguistic blunders.

BIBLIOGRAPHY

1. I. Bullialdus: *Astronomia philolaica.* Paris, 1645.
2. H. Usener: " Ad historiam astronomiae symbola ", *Kleine Schriften.* 4 vols, 1912–1914.
3. D. Pingree: " Gregory Chioniades and Palaeologan Astronomy ", *Dumbarton Oaks Papers,* no. 18 (Harvard, 1964), 135–160.
4. P. Kunitzsch: " Das Fixsternverzeichnis in der ' Persischen Syntaxis ' des Georgios Chrysokokkes ", *Byz. Zeit., 57* (Leipzig, 1964), 382–411.
5. S. P. Lampros: " Ta 'up' arithmon 91 kai 92 kataloipa ", *Neos 'Ellēnomnēmōn, 15* (Athens, 1921), 332–336.
6. I. B. Papadopoulos: " Gregoriou Chioniadou tou astronomou epistolai ", *'Epistēmonikē 'Epetēris tēs Philosophikēs Scholēs tou Panepistēmiou Thessalonikēs, 1* (1927), 151–206.
7. G. Le Strange: *The Lands of the Eastern Caliphate.* London, 1905, etc.
8. W. Heyd: *Histoire du commerce du Levant au Moyen-Âge.* 2 vols. Leipzig, 1936.
9. *The Chronography of Gregory Abu'l Faraj. Bar Hebraeus.* Translated from the Syriac by Ernest A. W. Budge. Oxford, 1932.
10. E. Quatremère: *Histoire des Mongols de la Perse par Reschid-ed-din.* Paris, 1836.
11. Sir Henry Yule and Henri Cordier: *Cathay and the Way Thither.* 4 vols. Repr. Taipei, 1966.
12. *Encyclopaedia of Islam.* 2 (new edition: Leiden and London, 1965), 678–681.
13. O. Neugebauer: " Taures or Daras ", *Bull. de l'Acad. Roy. de Belgique* (Classe de Lettres), *44* (1958), 496–497.
14. O. Neugebauer: " Studies in Byzantine Astronomical Terminology ", *Trans. Am. Phil. Soc., 50* (1960), 3–45.
15. P. Tannery: *Mémoires scientifiques.* Paris, 1912.
16. E. S. Kennedy: " The Exact Sciences in Iran under the Seljuqs and Mongols ", *Cambridge History of Iran,* vol. 5, ed. J. A. Boyle. Cambridge, 1968.
17. J. Mogenet: "L'influence de l'astronomie arabe à Byzance du IXe au XIVe siècle", *Colloque d'Histoire des Sciences,* I (1972) et II (1973), 45–55. Université de Louvain, Recueil de Travaux d'Histoire et de Philologie, 6e série, fasc. 9. Louvain, 1976.
18. Kiyoshi Yabuuchi, " Indian and Arabian Astronomy in China", *Tōhō Gakuhō, 25* (part 1, Kyoto, 1954) (Silver Jubilee Volume on the Zinbun-Kagaku-Kenkyusyo, Kyoto University), 585–603.
19 *Ming Shih.* Peking: Chinese Press, 1974.
20. W. Hartner: " The Astronomical Instruments of Cha-ma-lu-ting, their Identification and their Relations to the Instruments of the Observatory of Marāghah ", *Isis, 41* (1950), 184–195; repr. *Oriens-Occidens* (Hildesheim, 1968).
21. P. A. Revilla: *Catálogo de los Códices Griegos de la Biblioteca de el Escorial.* Tomo 1. Madrid, 1936.
22. E. S. Kennedy: " A Survey of Islamic Astronomical Tables ", *Trans. Am. Phil. Soc., 46* (1956), 123–177.

23. J. A. Boyle: " The Longer Introduction to the ' Zij-i-Ilkhani ' of Nasir-ad-Din Tusi ", *J. Semitic Studies* (Manchester, 1963), 244–254.
24. L. P. E. A. Sédillot: *Prolégomènes des tables astronomiques d'Oloug-Beg.* 2 vols. Paris, 1847 and 1853.
25. L. Ideler: " Mémoire sur la chronologie de Khata et d'Igour", *J. Asiatique*, *15* (Paris, 1835), 305–348.
26. E. S. Kennedy: " The Chinese-Uighur Calendar as Described in the Islamic Sources ", *Isis*, *55* (1964), 435–443.
27. L. Bazin: *Les calendriers turcs anciens et médiévaux*, published as a microfiche by the École Nationale des Langues Orientales Vivantes. Série I ' 1876–1888, LOV 1, I à LOV 1, XX '.
28. P. Pelliot: *Notes on Marco Polo.* 3 vols. Paris 1959, 1963, 1973.
29. A. von Gabain: *Alttürkische Grammatik.* Leipzig, 1950.
30. A. Gaubil: " Traité de l'astronomie chinoise ", *Observations mathématiques.* 3 vols. Paris, 1732.
31. Sir Gerard Clauson: *Etymological Dictionary of pre-Thirteenth Century Turkish.* Oxford, 1972.
32. O. Neugebauer: *A History of Ancient Mathematical Astronomy.* Berlin, 1975.
33. *Yüan Shih.* Peking: Chinese Press, 1976.
34. Kiyoshi Yabuuchi: *Chūgoku no Tenmon Rekihō.* (Studies of Chinese Calendars and Astronomy, in Japanese.) Tokyo, 1969.
35 Kiyoshi Yabuuchi: " Astronomical Tables in China from the Wutai to the Ch'ing Dynasties ", *Japanese Studies in the History of Science*, no. 2 (Tokyo, 1963), 94–100.
36. Yeh-lü Ch'u-ts'ai: *Chan-jan Chü-shih Wên-chi.* Shanghai: Commercial Publishing House, 1939.
37. E. S. Kennedy: " The Astronomical Tables of Ibn al-A'lam ", *J. for the History of Arabic Science*, *1* (Aleppo, 1977), 13–23.
38. W. Hartner: " Naṣīr al-Dīn al-Ṭūsī's Lunar Theory ", *Physis*, *11* (1969), 287–304.
39. C. A. Nallino: *Al-Battānī, Opus astronomicum.* Milano, 1903–1907; repr. Frankfurt, 1969.
40. B. R. Goldstein: " A Medieval Table for Reckoning Time from Solar Altitude ", *Scripta Mathematica*, *27* (1964), 61–66.
41. O. Pedersen: *A Survey of the Almagest.* Odense, 1974.
42. Abū Rayḥān Muḥammad b. Aḥmad al-Bīrūnī: *Al-Qānūn'l-Mas'ūdī.* 3 vols. Hyderabad-Deccan: Osmania Oriental Publications Bureau, 1954–1956.
43. D. Pingree: " Daras to nun legomenon Taures ", *Bull. de l'Acad. Roy. de Belgique* (Classe de Lettres), *48* (1962), 323–326.

X

THE ASTRONOMICAL TABLES OF GEORGE GEMISTUS PLETHON

1. INTRODUCTION

Towards 1430 a set of astronomical tables were developed by George Gemistus Plethon, designed to implement a new calendar that he proposed.[1] The type of calendar that he intended was luni-solar, that is, one in which the year was tropical while the individual months were mean lunar months, running from one mean conjunction to another. At this time the only calendar of this type known in Byzantium was the Jewish calendar. The Christian calendars, both Greek and Latin, followed the Julian scheme, with months unrelated to the Moon, while the Islamic calendar bore no relation to the seasons. The Persian calendar continued the ancient Egyptian scheme with a simple sequence of 30 day months, unrelated to either Sun or Moon. The ancient Athenian calendar was also a true luni-solar arrangement, and indeed Plethon was not only well aware of its detailed structure, but we see details of it reproduced in his scheme, particularly in the way the days of the month were counted.

If he had simply wanted a luni-solar calendar he might have adopted the Jewish calendar, but this would not have satisfied his very evident wish to reconstruct one that was in line with that used in ancient Greece. This was after all a part of his effort to discover and resurrect the Hellenic heritage.

He proposed a calendar in which each year would begin at the mean lunation following the midnight after the winter solstice. This rule entails intercalation, so that some years will have 13 instead of 12 months. The calendar is outlined in an extant chapter of the *Laws*, and is implemented in full detail in the tables under consideration here.

His tables go well beyond the immediate needs of the calendar, in that he provides also what is needed to compute the latitude of the Moon, and also the positions of the planets.

2. THE ASTRONOMICAL CONTENT OF THE TABLES

2.1. *Mean Motions: Plethon*

Table 1 shows the opening and closing rows of one of Plethon's tables, that giving the various quantities at intervals of 19 years. The 19-year cycle was only a first approximation to the luni-solar parameters adopted by Plethon, but is used in his tabulation. The symbol ☌ represents the mean conjunction. The quantities tabulated

TABLE 1. Tables of mean motions in steps of 19 years, as far as 494, and also 497 years.

19-years	position ☊	time ☊	apogee of ♂, stars	*bathos* ☾	*platos* ☾	mean♄	mean♃	mean⌀	*bathos*♀	*bathos*☿
19	0;7,2	248;15,44	0;17,17	306;55,33	7;28,34	232;28,31	216;55,20	36;55,23	318;29,35	319;34,43
38	0;14,5	136;31,28	0;34,34	253;51,6	14;57,8	104;57,2	73;50,40	73;50,46	276;59,10	279;9,26
……										
494	3;3,2	334;49,10	7;29,22	60;4,18	194;22,44	284;21,26	239;58,40	239;59,58	0;49,10	29;2,38
497	0;0,0	202;20,20	7;32,6	295;16,23	252;14,13	320;57,35	330;48,43	92;37,18	314;27,17	183;31,54

TABLE 2. Table of mean motions in successive years 1 to 18.

simple years	position ☊	time ☊	apogee of ♂, stars	*bathos* ☾	*platos* ☾	mean♄	mean♃	mean⌀	*bathos*♀	*bathos*☿
1	18;23,15	323;10,50	0;0,52	335;37,2	20;19,47	12;51,38	31;54,26	201;11,44	236;40,50	112;40,54
2	7;40,16	92;20,50	0;1,43	285;25,3	39;5,44	24;43,55	61;21,37	26;54,14	95;9,16	133;37,5
3	26;3,22	58;31,40	0;2,35	261;2,5	59;25,31	37;35,34	93;16,3	228;5,39	331;50,7	246;17,59
4	15;20,13	190; 41,40	0;3,26	210;50,7	78;11,28	49;27,51	122,43,14	53;48,29	190;18,34	267;14,11

in columns 2–6 concern the Sun and Moon:

col. 2: position of Sun and Moon at conjunction,
col. 3: the *time* of the conjunction,
col. 4: the apogee of the Sun,
col. 5: *bathos*, the argument of the Moon, and
col. 6: *platos*, the longitude of the node.

Table 2 shows the opening rows of the table in which the same quantities are shown for successive years, 1 to 18. The line to line variations show that the successive differences are not constant, but vary in a way that shows that successive rows represent the motions accumulated in steps of either 12 or 13 months. From this table we can determine that the sequence is 13, 12, 13, 12, 12, 13, 12, 12, ...

The *time* is here indicated in degrees (*chronoi*), such that a full day is 360°. This is the fraction of the day, the number of whole days being omitted. The daily motions can be determined from the entries in this table. The length of the synodic month, denoted M, can be found from the column of *time*, once we appreciate that in 13 months there are 383 days + a fraction, so that we have

$$13M = 383 + 323;10,50/360, \text{ giving } M = 29.5305941 \text{ days.}$$

This in fact is just equal to the 'standard' value of the synodic month found in a major part of Babylonian material, in Ptolemy, and in many Islamic zijes.

The motion of the Sun is found from the first column by the equation

$$13M\sigma = 360 + 18;23,15$$

which we can solve, since we know M, to get σ = 0.985646636°/day. In a similar way we find all the daily motions that underlie the table.

In Table 1 he gives the motions at intervals of 19 years, going up to 494 years, but ending with 497. Clearly the motions are not bound exactly by the 19-year cycle, as is evident from the fact that successive rows of the luni-solar position are not multiples of a common value. Moreover the reference to 497 years in the last row suggests that 497 is meant as the true period, but we find in fact that not even this is exactly consistent with the motions implied by this or the previous table.

As the rates of motion were derived from Table 2, so they may be derived also from Table 1, but we find that these two sets are not consistent with each other! The daily motions implied by the Table 1 giving motions in steps of one year are very nearly those of the *Persian Syntaxis*, while in the Table 2 the daily motions are very close to those of al-Battānī.

So far we have determined only the daily rates of the motions. In order to fix the radices we turn to another table in which Plethon gives the values of each of the parameters for six different eras. These eras range from the 'Descent of the Heracleides', in the year –1102, down to the time of the accession of the Byzantine Emperor John VIII Palaeologus, 1426-48. John VIII was ruling at the time of Plethon's work on these tables, and he was also closely under the Emperor's patronage

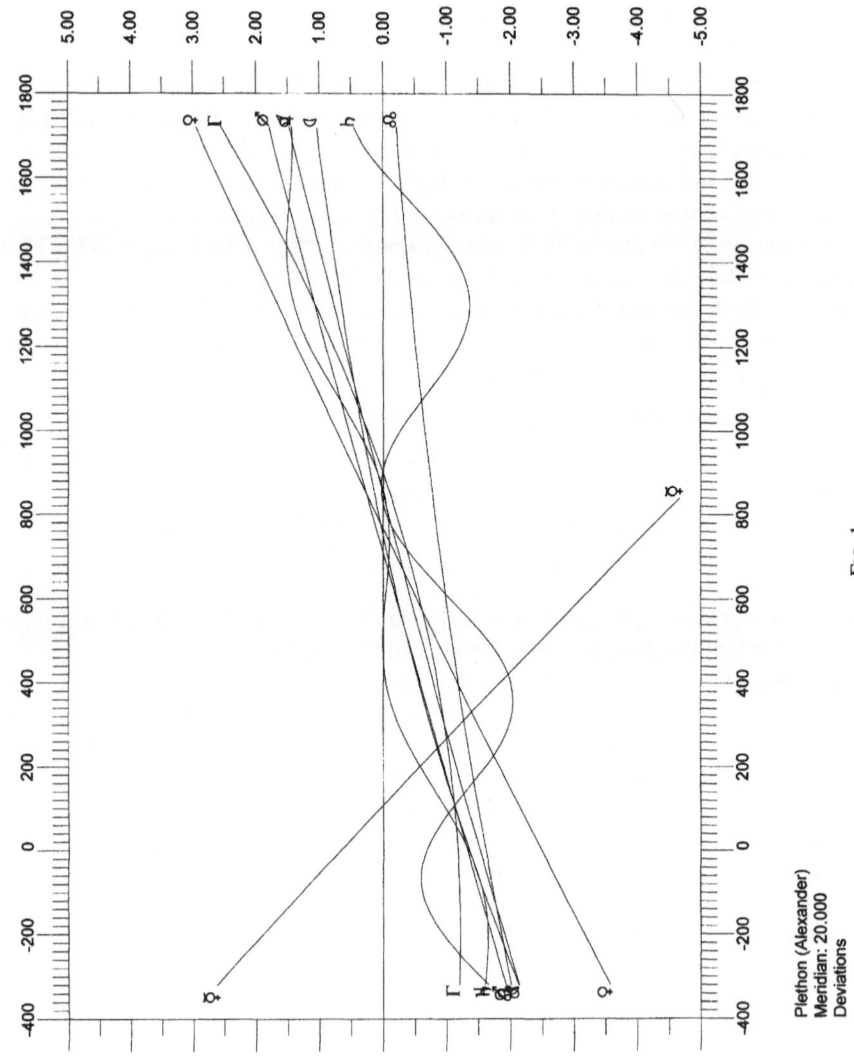
5.00
4.00
3.00
2.00
1.00
0.00
-1.00
-2.00
-3.00
-4.00
-5.00
-400
-200
0
200
400
600
800
1000
1200
1400
1600
1800
Plethon (Alexander)
Meridian: 20.000
Deviations

FIG. 1.

while living in Mistra. The reference to the 'Descent of the Heracleides', a great event in the mythology of early Greece, represents the beginning of the Kings of Peloponnese, and corresponds to some extent to the Dorian invasion. Plethon had a deep interest in the Peloponnese, and clearly sought out such references to its history as he could find.

Of these six eras three are commonly used, the eras of Nabonassar, Alexander, and the Arabs. However Plethon does not assume the normal dates for these. In every case he has chosen as the era the first day of the year, as defined by his calendar, which includes the era in question. For the 'Arabs', for example he uses 622 Jan 16 at midnight, plus the time interval 285;6 (that is 285;6/15 = 19^h;0,24), not 622 July 14. The era of Alexander is midnight, –330 Jan 12/13, evidently based on the battle of Gaugamela, for which the date –330 Oct 1 is established.

With a little experimenting one easily discovers the precise times intended for each of these eras. They are not, unfortunately, exactly consistent with one another, but only very nearly. We have only two manuscripts, neither of which is a true autograph; it is possible to isolate some errors as due to dittography, or simple arithmetical slips. Otherwise the inconsistencies must come down to deeper errors in calculation. Eventually we find by a careful study of all these radices that perhaps the radices referred to as 'Arabs' is closest to Plethon's source.

These radices show that Plethon always measures his longitudes from the head of Capricorn. In order to obtain radices referred to the head of Aries, we must add 270° to his tropical longitudes, but not of course to the *bathos* which is only the difference between tropical longitudes.

2.2. *Mean Motions: Proto-Plethon*

From the extant manuscripts relating to Plethon's work we found another set of tables which clearly amount to an early attempt at developing such material, a set which we call 'proto-Plethon'.[2] Here we have a similar set of tables, with tabulations at 1- and 19-year intervals. There is now complete internal consistency. Moreover the table for individual years is essentially the same as that found in the main version, indeed the entries in Table 2 shown here are simply a rounding down to two sexagesimal places from the entries in 'proto-Plethon', where he gives three places. There is now also a table of radices, but only for two eras, Nabonassar and Alexander.

In that way we arrive at an earlier set of parameters, 'proto-Plethon'.

2.3. *Deviation Curves*

It is helpful to take a synoptic view of the results by means of deviation curves.[3] In Figure 1 we see the curves for the mean longitudes according to Plethon's tables. These show about as much cogency as one finds for the mean longitudes in Islamic zijes. Broadly speaking, the deviations are nearest zero around the year A.D. 1000. Evidently we are not dealing with parameters fixed by observation in Plethon's own time.

TABLE 3. Dates in the Julian calendar of the first day of each month in a cycle of 19 years from 1433 Dec 13. The date is indicated by the month (where I = Jan), the day of the month, and the weekday (where A = Sunday). The calculations are according to mean lunations using the parameters of Plethon's tables.

	1	2	3	4	5	6	7	8	9	10	11	12	13
1	XII 13A	1434 I 11B	II 10D	III 11E	IV 10G	V 9A	VI 8C	VII 7D	VIII 6F	IX 4G	X 4B	XI 2C	XII 2E
2	XII 31F	1435 I 30A	II 28B	III 30D	IV 29F	V 28G	VI 27B	VII 26C	VIII 25E	IX 23F	X 23A	XI 21B	
3	XII 21D	1436 I 19E	II 18G	III 18A	IV 17C	V 16D	VI 15F	VII 14G	VIII 13B	IX 12D	X 11E	XI 10G	XII 9A
4	1437 I 8C	II 6D	III 8F	IV 6G	V 6B	VI 4C	VII 4E	VIII 2F	IX 1A	IX 30B	X 30D	XI 29F	
5	XII 28G	1438 I 27B	II 25C	III 27E	IV 25F	V 25A	VI 23B	VII 23D	VIII 21E	IX 20G	X 19A	XI 18C	
6	XII 17D	1439 I 16F	II 14G	III 16B	IV 15D	V 14E	VI 13G	VII 12A	VaIII 11C	IX 9D	X 9F	XI 7G	XII 7B
7	1440 I 5C	II 4E	III 4F	IV 3A	V 2B	VI 1D	VI 30E	VII 30G	VIII 29B	IX 27C	X 27E	XI 25F	
8	XII 25A	1441 I 23B	II 22D	III 23E	IV 22G	V 21A	VI 20C	VII 19D	VIII 18F	IX 16G	X 16B	XI 15D	
9	XII 14E	1442 I 13G	II 11A	III 13C	IV 11D	V 11F	VI 9G	VII 9B	VIII 7C	IX 6E	X 5F	XI 4A	XII 3B
10	1443 I 2D	1443 I 31E	III 2G	IV 1B	IV 30C	V 30E	VI 28F	VII 28A	VIII 26B	IX 25D	X 24E	XI 23G	
11	XII 22A	1444 I 21C	II 19D	III 20F	IV 18G	V 18B	VI 16C	VII 16E	VIII 15G	IX 13A	X 13C	XI 11D	XII 11F
12	1445 I 9G	II 8B	III 9C	IV 8E	V 7F	VI 6A	VII 5B	VIII 4D	IX 2E	X 2G	XI 1B	XI 30C	
13	XII 30E	1446 I 28F	II 27A	III 28B	IV 27D	V 26E	VI 25G	VII 24A	VIII 23C	IX 21D	X 21F	XI 19G	
14	XII 19B	1447 I 17C	II 16E	III 18G	IV 16A	V 16C	VI 14D	VII 14F	VIII 12G	IX 11B	X 10C	XI 9E	XII 8F
15	1448 I 7A	II 5B	III 6D	IV 4E	V 4G	VI 2A	VII 2C	VIII 1E	VIII 30F	IX 29A	X 28B	XI 27D	
16	XII 26E	1449 I 25G	II 23A	III 25C	IV 23D	V 23F	VI 21G	VII 21B	VIII 19C	IX 18E	X 18G	XI 16A	
17	XII 16C	1450 I 14D	II 13F	III 14G	IV 13B	V 12C	VI 11E	VII 10F	VIII 9A	IX 7B	X 7D	XI 5E	XII 5G
18	1451 I 3A	II 2C	III 4E	IV 2F	V 2A	V 31B	VI 30D	VII 29E	VIII 28G	IX 26A	X 26C	XI 24D	
19	XII 24F	1452 I 22G	II 21B	III 21C	IV 20E	V 19F	VI 18A	VII 18C	VIII 16D	IX 15F	X 14G	XI 13B	

2.4. *Equations of Sun, Moon and Planets*

The whole collection of the tables includes motions in months, fortnights and days, and then goes on to give the equations of Sun, Moon and planets. These tables of equations are lifted directly from Ptolemy's *Almagest* (it may be shown that this and not the *Handy tables* is the source). These are not satisfactorily related to the main tables, insofar as the apogees are not properly specified. Certainly Plethon did not adequately extend his work to cover the planets. The sole difference from Ptolemy's equations is found in an additional column in the equation of the Sun, a column headed 'al-Battānī'. Here however the entries are not exactly those from the zij of al-Battānī, although the maximum value 1;59,10 is in agreement. Elsewhere the entries were obtained by taking 5/6 of Ptolemy's value. This shows that Plethon did not actually have al-Battānī's tables before him; it is possible that this column was added by Bessarion, through whose hands our present copies passed.

2.5. *Calendar*

The calendar proposed by Plethon can be derived from his tabular parameters. It is a matter of determining the successive mean conjunctions, and then following his rule that the month begins with the midnight following the mean conjunction, while the year begins with the month that follows the winter solstice.

In Table 3 we have a table for the first 19 years of the calendar, beginning with the lunation of 1433 December 12. Not only is this a suitable year with which to start, since the lunation occurs just immediately after the solstice, but indeed he states expressly that this is the first year, according to a gloss on one of the tables.

It is easy for a modern student of this work to write a computer program which gives instant access to such a calendar. It is not known how much Plethon himself made calculations of dates, but two are given in the text. Nevertheless, the central problem was to find a conversion between his calendar and the ordinary Julian form used in Byzantium. Cardinal Bessarion questioned him about this in 1446, but in his reply Plethon not only avoided giving him a practical procedure, but even took the occasion to propose yet a third set of parameters.

2.6. *Summary*

We can summarise the position by saying that Plethon has presented a set of tables of Sun, Moon and planets based on luni-solar conjunctions. The unit of time is the mean synodic month, for which he uses the standard value which found in the *Almagest*, the value that continued to be used widely if not exclusively in Islamic astronomical circles, and which one is not surprised to find in this case. The periodicity is not precisely the 19-year cycle, because the solar motion is not geared to the synodic month and the 19-year cycle in the strictly consistent way that one finds, for example, in one version of the Jewish calendar.

TABLE 4. From Wing I of the *Six wings*.

Complete cycles from Creation of World	Moment of conjunction	Mean [motion] of luminaries	Argument of Sun	Argument of Moon	Argument of Node
269	2^d12;12	1,1;33	10,14;4	4,3;5	8,2;37
270	5^d 4;45	1,1;40	10,13;53	2,10;1	8,10;12
271	7^d 21;18	1,1;47	10,13;43	0,16;56	8,17;48
272	3^d 13;51	1,1;54	10,13;33	10,23;52	8,25;24
273	6^d 6;24	1,2;1	10,13;23	9,0;47	9,2;59
274	1^d 22;56	1,2;8	10,13;13	7,7;43	9,10;35
275	4^d 15;30	1,2;15	10,13;2	5.14;38	9,18;10
276	7^d 8;3	1,2;22	10,12;52	3,21;34	9,25;46
277	3^d 0;36	1,2;29	10,12;42	1,28;30	10,3;46
278	5^d 17;9	1,2;36	10,12;32	0,5;26	10,10;58
279	1^d 9;42	1,2;43	10,12;22	10,12;22	10,18;34
280	4^d 2;15	1,2;50	10,12;12	8,19;18	10,26;10

3. SOURCES

3.1. *Parameters*

From an initial inspection of the rates of motion we see that the parameters are actually very close to those of al-Battānī. Al-Battānī's zij was widely known in Islamic circles, and was also translated into Latin and Spanish. In spite of its evident popularity, in both and Eastern and Western Islamic areas, and perhaps especially in Spain, only one Arabic MS has been found, and no Latin version of the tables is known, only one made into Spanish under the aegis of Don Alfonso el Sabio *c.* 1265. There is no reason to suppose that a Greek version had been made.

When we compare Plethon's parameters with those of al-Battānī we find that they are very close, but not quite identical to them.

The parameters of 'proto-Plethon', on the other hand, follow those of the *Persian Syntaxis*, that is, the Greek version of the Zīj-i Īlkhānī of Naṣīr al-Dīn al-Ṭūsī.[4] This version was widely studied in the Byzantine world, and it is not surprising to find its parameters used here. However, while the rates of mean motions of the *Persian Syntaxis* are followed closely, the radices were badly calculated for the eras of 'proto-Plethon', with an error of about −1°.

3.2. *Tables of Conjunctions*

Plethon's tables follow a style that is commonly found for tables of conjunctions and oppositions, based apparently on Ptolemy's *Almagest*, Book VI, chap. 4. Ptolemy's tables are drawn up at interval of 309 synodic months, very nearly 25 Egyptian years. The tabulation includes essentially the same quantities that we find given by Plethon. There is no reference to the planets, and indeed no one before Plethon included the planets in a table of conjunctions. There is no such material in the *Handy tables*, although you might expect to find it there.

In al-Battānī's zij there are two tabulations of conjunctions and oppositions, in one case a continuation of Ptolemy's table in steps of very nearly 25 Egyptian years, and another based on the Julian year and the Era Dhū'l-Qarnayn, which al-Battānī uses elsewhere, giving conjunctions at intervals of 296 or 297 months. The quantities that he tabulates are essentially as in the *Almagest*: the day in the month Adhār and the *time*, the position of Sun and Moon, the lunar argument, and the Moon-node interval.

In order to find more material of this kind, particularly material related to Plethon, we have to look at Hebrew works. The Jewish calendar depends on the mean luni-solar conjunction, the *molad*. The sequence of molads is the absolute backbone of the whole calendrical system. Tabular aids for calendrical calculations are to found in various Hebrew tracts, beginning at least with that of Abraham bar Ḥiyya (1123). However, neither there, nor in that of his near successor, Maimonides, is there a tabulation of the kind we are looking for. Eventually in the very full treatise by Isaac Israeli, the *Yesōd 'Ōlam, Foundation of the world*, written in Toledo in 1320, we find separate tabular arrangements for the Sun, lunar anomaly, and the Moon-node interval, the tabulation being at intervals of one synodic month.

Nearer to Plethon's time we have a Hebrew work named the *Six wings*, by Emmanuel ben Jacob, known in Latin as Bonfils, written in Tarascon in Provence around 1350. In Table 4 we see one of the tables in the first 'Wing'. We have, for successive cycles of the Jewish calendar, numbered 269 to 280, that is for the years 1351 to 1560, a tabulation of the quantities: moment of conjunction, mean [motion] of luminaries, argument of Sun, argument of Moon, argument of node. The radices as such are not included in this table, but are given in another.

It is significant that Bonfils follows, and says that he follows, al-Battānī. This is not difficult to confirm from a determination of the luni-solar motions implied by his tables. Moreover there is a set of tables drawn up by Bonfils which represent al-Battānī in a Hebrew version.[5] That is to say, we have tables of Sun, Moon and planets, together with their equations, based closely on the zij of al-Battānī. This is not a direct transcription from the zij, or a translation of the Arabic text, but a recalculation for the meridian 33° west of Raqqa, effectively the meridian appropriate to Bonfils at Tarascon, and recalculated also for the Julian calendar, beginning 1341 Feb 28.

There was great interest in the *Six wings*, so much so that there were both Latin and Greek versions of the work.[6] It must be remembered that the *Six wings* was concerned only with the Sun and Moon, so that the planetary parameters of al-Battānī had to come by another route. Indeed even in the *Six wings* the luni-solar parameters are not given to the precision expressed in Plethon's tables.

3.3. *The Parameters Again*

Now from the point of view of finding the source for Plethon's parameters, which as we have noted are close to those of al-Battānī's zij, the important result is that the

TABLE 5.

	☌	*time*	apogee of ☌	☾−Γ☾	☊
Plethon	3;3,2	334;49,10	0;17,17	306;55,33	7;28,34
al-Battānī	3;3,3.5	334;50,36.8	0;17,16.4	306;55,33.4	7;27,46.7
al-Battānī (Hebrew)	3;2,59	334;50,57.6	0;17,16.9	306;55,33.2	7;28,33.8
	♄	♃	♂	♀−☌	☿ −☌
Plethon	232;28,31	216;55,20	36;55,23	318;29,35	319;34,43
al-Battānī	232;28,26.6	216;55,23.2	36;55,9.6	318;29,5.5	318;29,37.0
al-Battānī (Hebrew)	232;28,25.7	216;55,20.4	36;55,23.4	318;29,35.1	319;34,42.4

rates of motion derived from the Hebrew version are quite exactly in agreement. In the course of the recalculation from the zij of al-Battānī some minor variations slipped in, and the result is that one can distinguish between the two. The agreement with this source is best as regards the rates of motion, and not so impressive when the radices are included. This is evident from Table 5. This shows the comparison between entries in Table 1 and calculations made from al-Battānī, both his zij and the Hebrew version. For the ☌ and *time* the values are taken from the row '494 years', but the others are from the first row. This choice is made because for ☌ and *time* the entries in Table 1 are not simply multiples of the first row, so one obtains a better estimate by taking the longer interval.

While there is evidently an affinity with both versions of the parameters of al-Battānī, there can be no doubt that Plethon's parameters are based on the Hebrew version. The values

apogee of ☌, ☾−Γ☾, ☊, ♃, ♂, ♀−☌, ☿ −☌

obtained from al-Battānī (Hebrew) all agree exactly with Plethon's when rounded off to the nearest second, while the values of

apogee of ☌, ☊, ♃, ♂, ♀−☌, ☿ −☌

obtained from al-Battānī would disagree, in most cases by several seconds. This shows very clearly that the parameters are derived from al-Battānī (Hebrew). There remain very small disagreements in ☌, *time*, ♄.

3.4. *Related Manuscript Sources*

The question of historical links remains unsettled. We need to find the links between Plethon and both the Hebrew version of al-Battānī and the *Persian Syntaxis*. These links remain unknown, but it seems important to draw attention to further manuscript sources concerning the *Persian Syntaxis*.

A Venetian manuscript gives a Latin version of the *Persian Syntaxis*, and also has in it a set of tables drawn up for the argument of the Moon for years expressly given in the Jewish calendar in the cycle 272, that is, the years 1408–26.[7] We do not know exactly what role this might have played, but it is most interesting that the work was drawn up in Crete, where there was a substantial Jewish community intimately linked with the Romaniote world, and hence with Byzantium.

There was a Hebrew version of the *Persian Syntaxis*, made in Salonika around 1375, if one can judge by the dated examples given by its author Sharbīṭ ha-Zahab.[8] This is relevant to Plethon, because the parameters of 'proto-Plethon' depend on the parameters of the *Persian Syntaxis*. There is besides a detailed Hebrew commentary on the *Persian Syntaxis* by Mordekai Komtino in Constantinople, around 1425.[9]

REFERENCES

1. This is based on the edition and commentary of this material by Anne Tihon and myself: *Georges Gémiste Pléthon: Manuel d'astronomie*, by Anne Tihon and Raymond Mercier (Corpus des Astronomes Byzantins ix, Louvain-la-Neuve, 1998).
2. See Anne Tihon's article in this issue, "The astronomy of George Gemistus Plethon".
3. This convenient 'x-ray' of mean motions goes back to Roger Billard's study of Indian Canons, *L'Astronomie Indienne* (Paris, 1971).
4. Raymond Mercier, "The Greek 'Persian Syntaxis' and the Zīj-i Īlkhānī", *Archives internationales d'histoire des sciences*, xxxiv (1984), 35–60.
5. Munich Heb. 343, ff. 182v–200v; Munich Heb 386, ff. 15v seq.; New York, Jewish Theological Seminary of America, MS 2597, ff. 45a seq. I am indebted to Professor B. R. Goldstein for the latter two references. The attribution of the tables to Bonfils is missing from the first of these three.
6. There are two quite distinct but close Latin versions of the text, respectively from 1406, by an Italian Zacharias Lucae e Camerino (Florence conventi soppr. J.IV.20; Axel Björnbo, "Die mathematischen S. Marcohandschriften in Florenz", Part 3, *Bibliotheca mathematica*, xii (1912), 97–132, p. 113), and anonymously *c.* 1420 (MS Padua Escorial e.III.23). In the latter the literal Latin version is accompanied by another in more fluent Latin. There is also a Greek version of the tables, with instructions given in an original Greek text, including calculations for 1435, by Michael Chrysococces. It is interesting that the Latin versions were made before the Greek, and that the texts are close translations.
7. Venice Bibl. Marciana Latin Cl.VIII Cod 31; the lunar tables are in ff. 63r–64r.
8. MS Paris BN Heb 1042, ff. 34r–104v.
9. MSS Paris BN Heb 1084, 1085; Oxford Bodley Heb.d.5.

XI

THE DATE OF THE *MAHĀSIDDHANTA*

The work known as the *Mahāsiddhānta*, attributed to an author now generally referred to as Āryabhaṭa II, and edited by Sudhakara (1910), has recently attracted perhaps more attention than its intrinsic merits would justify. This is because there is a dispute about he date of its composition. Dikshit [Dikshit (1985), p. 97] as long ago as 1896 estimated the date of this work as approximately Śaka 900 (A.D. 978), and this was generally accepted. This date was however shown to be incorrect by Billard [Billard (1971), p. 157–161], who found the time of composition to be in the early sixteenth century. The *Mahāsiddhānta* includes in its second chapter another set of parameters which are attributed to 'Parāśara', and these as Billard shows point to the late fifteenth century. The conflict between these two very different findings can however be easily resolved, so that one can see why Dikshit arrived at the incorrect date.

Dikshit was the first to put the author's name in the form Āryabhaṭa II, and he is followed in this by the editor Sudhakara. In the text, in the opening *śloka* and elsewhere, the author refers to himself simply as Āryabhaṭa, acknowledge his dependence on 'Vṛddhāryabhaṭa' in *Pātādhikāra, śl.* 14, ed. p. 128.

1. THE METHOD

Dikshit and Billard appeal to the same basic procedure, which it would be as well to explain at the outset. The predictions according to the parameters in the text are compared with calculations which we make now according to the best modern parameters. Such a comparison is nothing unusual and is the resort of historians of astronomy for many purposes. Diskhit, like many others, made a rather casual and inexact use of it, without going into any detail whatever, and moreover considered only the Sun. Billard on the other hand has explained every step of his procedure, included a degree of statistical control, and also considered not merely the Sun, but the Moon and all the planets. This short paper is not the place to discuss the justification of the method. Apart from Billard's book one should consult Mercier (1987) and van der Waerden (1980).

The parameters of the work are available from the works of Billard and Dikshit, so that there is no need to list them all again. As usual the text prescribes a set of linear functions of time for the mean lingitudes, together with a set of procedures which lead to the true longitudes. To these sidereal positions must be added the precessional movement, in the form peculiar to Indian astronomy, the so-called *ayanāṃśa*, the movement of the equinoctial point along the ecliptic from a certain origin point which is siderially fixed. The resulting tropical longitude may then be compared with the tropical longitudes provided by modern astronomy. The formula for the *ayanāṃśa* according to the *Mahāsiddhānta* is

$$\sin^{-1}[\sin(0.504^{r} + 578159^{r}t') \cdot \sin 24°]$$

where ^{r} = revolutions

$$t' = t/1577917542000$$

Here t is the time measured in days from the beginning of the Kaliyuga at Sunrise, J.D. 588465.75 referred to the local meridian.

For Parāśara the formula is slightly different,

$$\sin^{-1}[\sin(0.5003^{r} + 581709^{r}t') \cdot \sin 24°]$$

$$t' = t/1577917570000$$

In practice, it is a question of plotting the 'deviation curves', that is the difference of mean lingitudes, in the sense of 'old-modern'. There are nine curves, for Sun, Moon, lunar argument, lunar node, and the five planets. This bundle of nine curves is generally found to converge closely at some period, indicating that the predictions from the text were most accurate at that time. We can then be sure that the work was based on observations taken close to that date, and that in all likelihood the work was itself composed at that time, certainly not before it. Dikshit and Billard agree on this approach. They nevertheless arrived at very different dates.

2. DIKSHIT

Dikshit gave his result as follows:

> The time when the *ayanāṃśa*, obtained from the second *Ārya Siddhānta*, would be equal to the Sun's tropical longitude at the true vernal equinox, comes to be about Śaka 900. If he had lived before this year, the date must have been only a few years earlier.
>
> Dikshit (1985), p. 97.

This rather compressed statement is unclear as it stands, and surely 'tropical' must be emended to read 'sidereal'. For, one looks for a date when the tropical longitude taken from the work equals that computed by modern parameters. At the vernal equinox the modern tropical longitude is zero, and if that predicted by the work is also zero, it means that sum of the sidereal longitude and the *ayanāṃśa* equals zero. Thus Dikshit must have found that the sidereal (not the tropical) longitude was equal in magnitude, but opposite in sign, to the *ayanāṃśa* in Śaka 900.

3. Deviation Curves*

The deviation curves for the *Mahāsiddhānta* are shown in Figs. 1 and 2, and those for Parāśara in Figs. 3 and 4. The curves are distinguished by the usual astronomical symbols, including Γ for the lunar apogee and ϒ for the *ayanāṃśa*. As the captions indicate, in one case the precession, the *ayanāṃśa*, is included, and in the other it is not. In the latter case, the *ayanāṃśa* is also plotted separately. Considering Fig. 1 then, which shows the deviations of sidereal longitudes against the modern tropical longitude, we see a convergence of the bundle towards the sixteenth century. At that time the deviations are approximately −17°, which is the required value of the *ayanāṃśa*. The *ayanāṃśa* itself is plotted, showing a portion of the oscillatory curve which varies between the limits ±24°. This curve, which vanishes on A.D. 603 Mar 19, is indeed of about the required value in the sixteenth century. In Fig. 2 the sidereal longitudes have all been increased by adding the *ayanāṃśa*. The vertical ordinate is plotted over a much smaller range, so that one sees less clearly the broad picture of the convergence of the bundle, but this allows one to observe clearly an important fact, namely that in the case of the Sun, the only part which Dikshit studied, the curve not only vanishes at A.D. 1595, but also near A.D. 1110. His own calculations could hardly have been carried out with the ease and precision permitted by modern computers, nor did he have modern parameters as accurate as ours. He therefore must have estimated by some technique of approximation the various values of the complicated expression for the *ayanāṃśa*. It is not surprising that he should get a year around A.D. 978, when the deviation is about −0;18, rather than the date when it is exactly zero. Having found a date in this way, he would not have looked for any other, especially one as late as the sixteenth century. Of course if he had looked at the other eight deviations, he would have been forced to a different conclusion. The great merit of Billard's approach is that its thoroughness and exact control leave no doubt about the matter.

In the case of Parāśara, the situation is very similar. Here the deviation curves

* The usual astronomical symbols used (vide Figures 1 to 4) are:

☉ for Sun	♃ for Jupiter
☾ for Moon	♀ for Venus
♂ for Mars	♄ for Saturn
☿ for Mercury	☊ for node

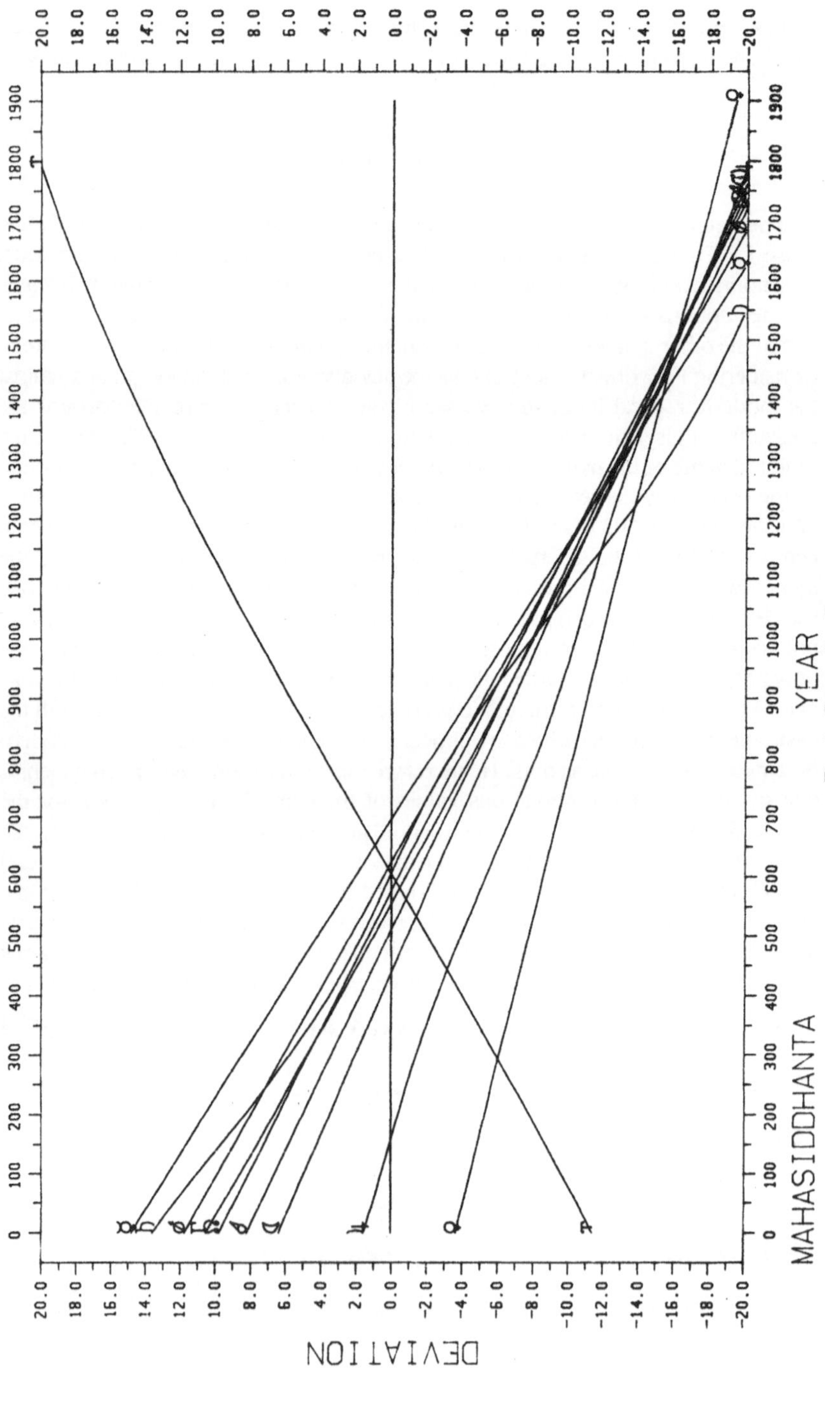

Fig. 1 Precession Separate

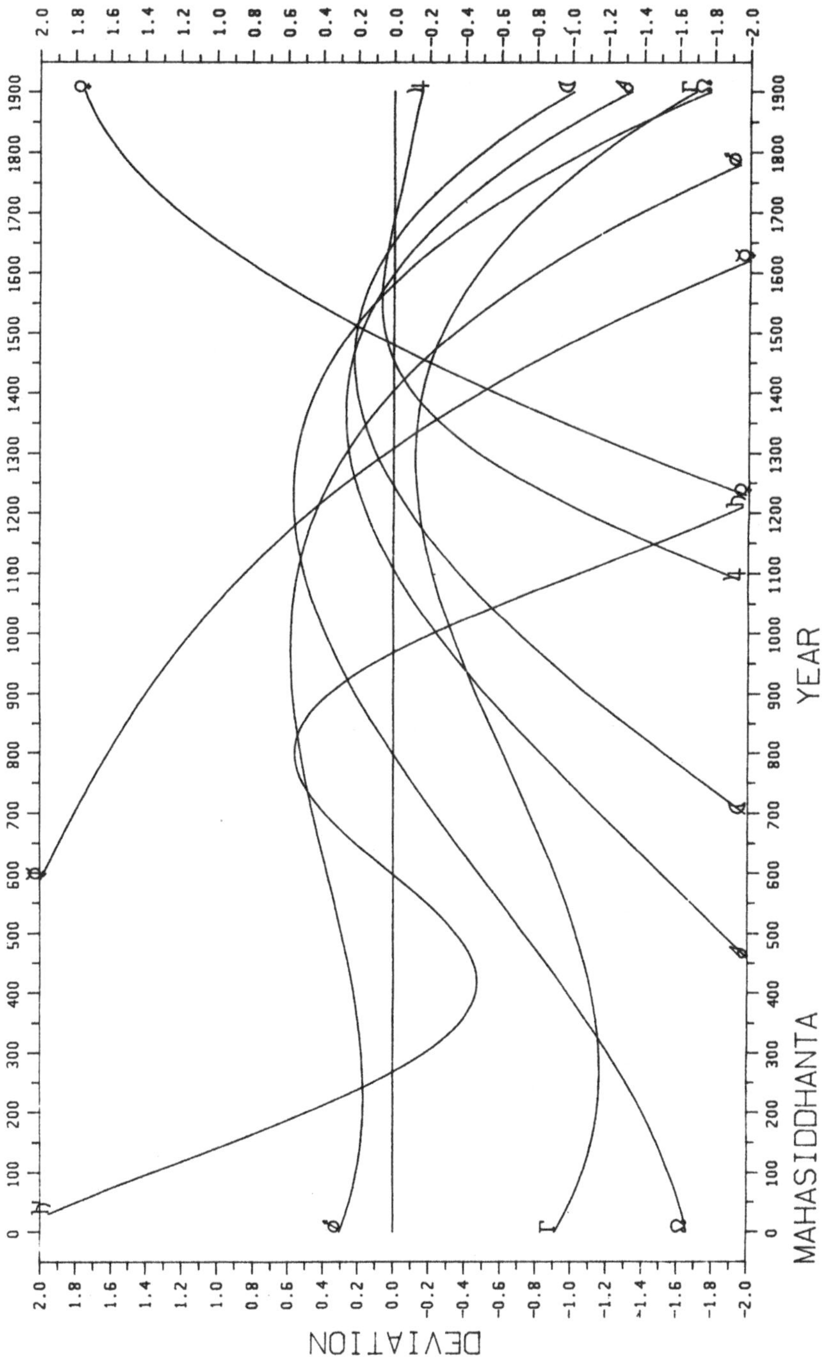

Fig. 2 Precession Included

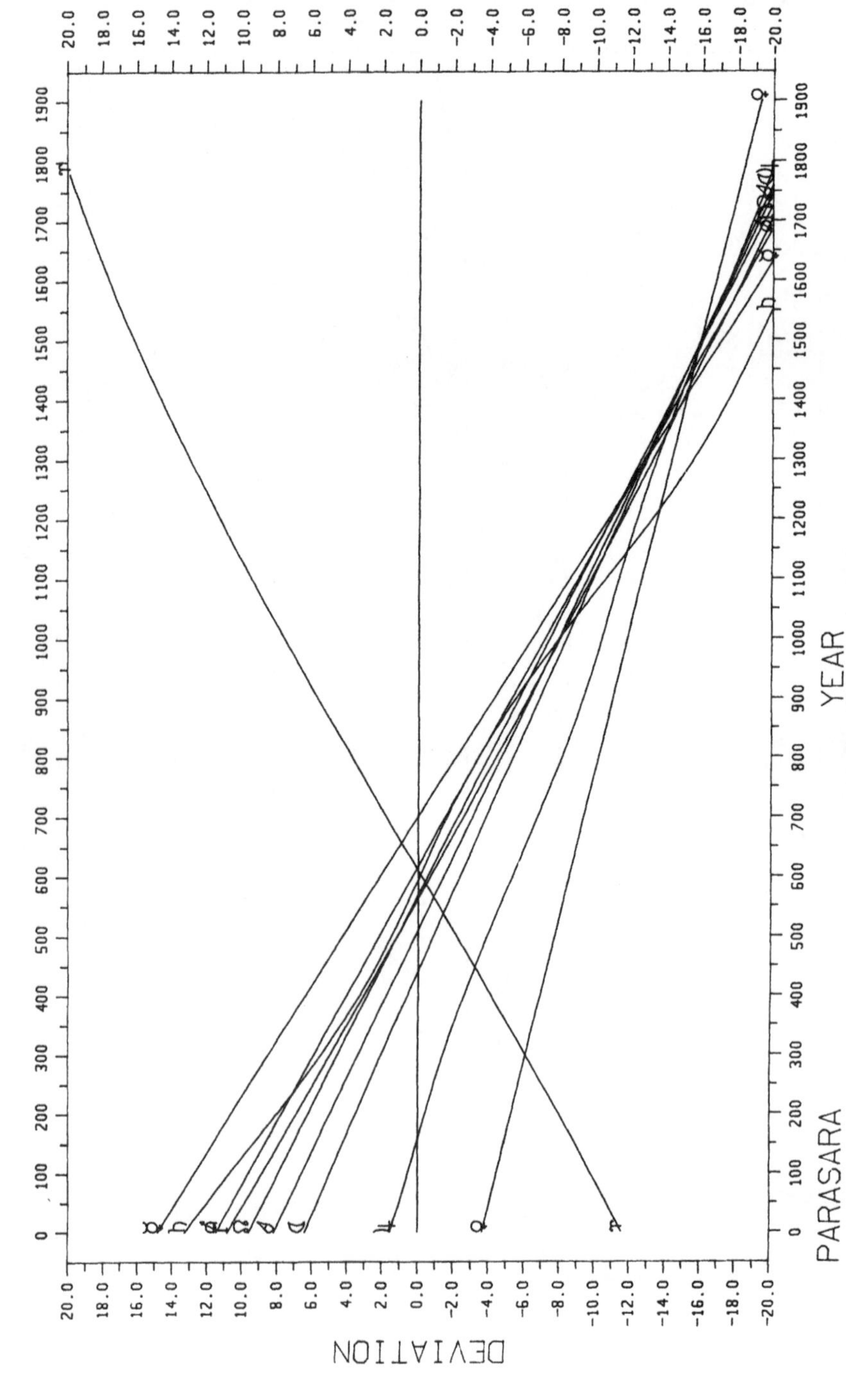

Fig. 3 Precession Separate

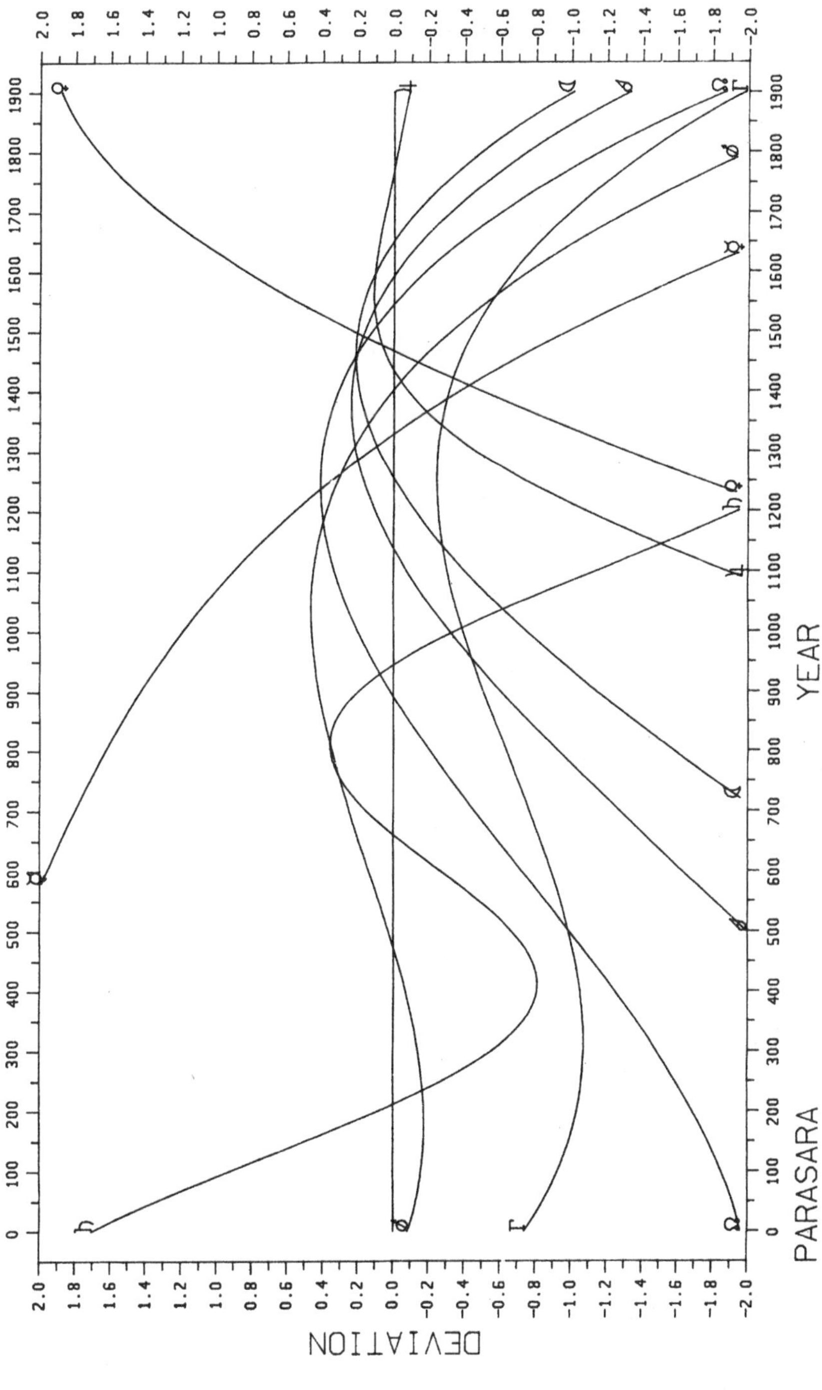

Fig. 4 Precession Included

present much the same pattern, and a detailed examination shows only that the convergence is to a slightly earlier date. The deviation for the tropical Sun vanished near A.D. 1140, 1590.

4. Āryabhaṭa and Bhāskara II

Dikshit (1985, pp. 33, 96) has brought forward another argument in order to show that the *Mahāsiddhānta* was composed before the *Siddhānta Śiromaṇi* of Bhāskara II, i.e. before A.D. 1150. In the *Siddhānta Śiromaṇi*: *Grahagaṇita*, Bhāskara writes, in his own commentary (Vāsanābhāṣya) on *Spaṣṭādhikāra*, śl. 65.

अत एवार्यभटादिभिः सूक्ष्मत्वार्थं दृकाणोदयाः पठिताः

ata evāryabhaṭādibhiḥ sūkṣmatvārthaṃ dṛkāṇodayāḥ paṭhitāḥ

That is, "for the sake of precision decans were used for the ascensions by Āryabhaṭa and others". Since in the *Āryabhaṭīya* of Āryabhaṭa the ascensions are reckoned in steps of 30°, not 10° (decans), while decans are used in the *Mahāsiddhānta* (*Tripraśnādhikāra*, śl. 38–41). Dikshit concluded that 'Āryabhaṭa' here must refer to the author of the *Mahāsiddhānta*, therefore putting the work earlier than the *Siddhānta Śiromaṇi*. The passage raises problems of some interest, but as will become clear during the following discussion, it does not suffice to prove the point as clearly as it might seem at first sight. Pingree in his anxiety to refute Billard, used 'Āryabhaṭa II' in his translation of this line, as he writes in the following astonishing paragraph,

> One of the secure facts regarding the date of Āryabhaṭa II—one that has been known to all serious historians of Indian astronomy since 1896, and to which I referred in my biography of Āryabhaṭa II in the *Dictionary of Scientific Biography*—is that Āryabhaṭa II is quoted by name by Bhāskara II in the commentary that he wrote on his own *Siddhānta Śromaṇi*. Since Bhāskara II composed the *Siddhānta Śiromaṇi* in A.D. 1150, Billard's marvellous method has led to a demonstrable error of more than 350 years. Pingree (1980), p. 61

Pingree's demonstrable error here is the unwarranted intrusion of 'II': only Āryabhaṭa, not Āryabhaṭa II, is named, so there is nothing secure here at all. *That* is clear to any serious historian.

In reviewing these problems, one must bear in mind the lost work on the midnight system of the early Āryabhaṭa. Shukla has published a series of intriguing extracts from a work referred to as *Āryabhaṭasiddhānta*, extracts quoted by Rāmakṛṣṇa Ārādhya (fl. A.D. 1472), in his commentary to the *Sūryasiddhānta*; Shukla (1967, 1977), Subbarayappa (1985), pp. 87–96. The extracts are concerned in many cases with instruments and their uses. According to Shukla this is the lost work by Āryabhaṭa on the midnight system. In a later article he gave extracts from this work as quoted in the commentaries by Mallikārjuna Sūri (A.D. 1178), and also the sixteenth century commentator Tamma Yajvā (A.D. 1599); Shukla (1977). It must be observed that in none of the extracts is there any positive indication that these are from a work on the midnight system. The word midnight-system, *ārdharātrika*, does not occur, nor does any numerical parameter which might link the text unambiguously with that source. The name *Āryabhaṭasiddhānta* alone seems to have been enough to convince Shukla, even though, as he observes, this is not likely to have been its original name. Bhāskara I had no name for it, calling it only *tantrāntara*, 'the other system'. The name '*Āryabhaṭīya*', similarly, is a later usage, not attested before Govindasvāmin (ca. 850); Bhāskara I simply called it *Āśmakatantra*, or *Bhaṭatantra*. The difficulty is that in the following we will see that the present work, which now goes under the name *Mahāsiddhānta*, was also referred to in the sixteenth century as *Āryabhaṭasiddhānta*. The mere use of this name is therefore not a proof of the connection that Shukla claims for it, and as he provided no other argument, it must be said that there appears to be no foundation for his assertion. Of course the *Mahāsiddhānta* is not the source of these texts.

In order to place references to Āryabhaṭa in a broader context, I have looked through the rest of the *Siddhānta Śiromaṇi* for references to Āryabhaṭa, and also through the commentaries by Nṛsiṃha and Munīśvara on the chapter on mean motions (Madhyamādhikāra) in the *Grahagaṇita*. Both commentators refer to the *Mahāsiddhānta* in a number of places. These commentaries may be dated as A.D. 1621, 1638, respectively. In the following the various forms of reference used by each commentator are given, together with a note on the passages quoted from the *Āryabtaṭīya* or the *Mahāsiddhānta*.

THE EARLY ĀRYABHAṬA

BHĀSKARA

'Āryabhaṭa'

Jhā 297. *bhūrekaiva kintu yat tvāryabhaṭādibhirācāryaiḥ satyapi niyāmake palāṃśadarśane 'nyathā' nyathā tatpramāṇamabhihitaṁ tatra*

ṣaṭsaptāṣṭayavamaṅgulaṁ kaniṣṭhikādibhedena śāstreṣūcyate

This is a reference to the fact that 'Ārabhaṭa and others' used varying divisions of the aṅgula (digit) into 6, 7 or 8 yavaṣ (barley corns), so resulting in varying values for the circumference of the earth.

Nṛsiṃha

'Vṛddhārya'

Jhā 113. *Āryabhaṭīya, Golādhyāya* śl. 9.
Jhā 118. *Āryabhaṭīya, Golādhāya* śl. 10.

Munīśvara

'Āryabhaṭa'

Jhā 43. *Āryabhaṭīya, Golādhyāya* śl. 10.

THE LATER ĀRYABHAṬA

Nṛsiṃha

'Āryabhaṭa'

Jhā 46. *Mahāsiddhānta, Parāśaramatā*, śl. 1a–3.

Munīśvara

'Āryabhaṭa'

Jhā 308. *Mahāsiddhānta, Madhyamādhyāya*, śl. 55 (ed. 39), on the circumference of the earth. In *Bhuvaṇakośapraśnottaram*, śl. 35 (ed. 191), the same circumference is given, expressed as tatāramā, 6625.

Jhā 309. 'Āryabhaṭa' assumes 6 yavas = 1 aṅgula. This is supported in śl. 55 just mentioned, where we have *tayavāṅgulamānena* . . . 'measuring 6 (=*ta*) yavas to an aṅgula . . .'

'*Aryabhaṭasiddhānta*'

Jha 51–2. *Madhyamādhikāra* śl 7, śl 3.

'Āryasiddhānta'

Jhā 243–4. On the size of orbit of the lunar nodes, *āryasiddhānte'pisvarbhānor vidatarkāṣṭadviśailākṣakhakuñjarāḥ*. Reading vedatarkā°, this means 'for Āryabhaṭa, ⟨orbit of⟩ Rāhu is 80572864'. The word-numerals used here are not used by either Āryabhaṭa, although they are common elsewhere. This is the orbit size given in the later *Sūryasiddhānta*, xii, 85. A number of versions of this number in the *Sūryasiddhānta* are quoted by Shukla in his edition of that work. In the printed text: *dasratatvābhiśailārthākāśakuñjarāhḥ* In the commentary of Rāmakṛṣṇa, *vedatarkāṣṭadviśailārthakhakuñjaṛāḥ*, in that of Yallaya, *vedatarkābdhiśailārthāśakuñjarān*. Only Rāmakṛṣṇa's version is clear to me. In the previous line, Munīśvara has '*vaśiṣṭhasiddhānte rāhorvedāṅganāgāśviśailabāṇakhakuñjarāḥ*, which again means 80582764. This is confirmed in the *Vṛddhavasiṣṭhasiddhānta* [Prasāda (1917), p. 73]. Elsewhere in the *vaśiṣṭhasiddhānte*; all the *bhagaṇa* values are those of the later Sūrya, and the expression attributed here to *Āryasiddhānta* evidently comes from the same tradition.

'Laghvāryabhaṭa'

Jhā 85. *Parāśaramatā*, śl. 1a–3a.

Jhā 121. On the mandocca of Saturn, he writes, *mūlaṃ tu laghvāryabhaṭoktaparāśaramatena* 54 iti. This is confirmed by the text, *Parāśaramatā* śl. 7, ed, 44.

Apart from these passages, I have also noticed that Munīśvara refers to the later Āryabhaṭa as 'Laghvāryabhaṭa' in his commentary to Bhāskara's *Golādhyāya: Golabandādhikara* śl. 17–9, a long discussion of the *ayanāṃśa* according to Bhāskara, in which the formulae given by the later Āryabhaṭa are referred to; Apte (1943), p. 217.

To summarise, the early Āryabhaṭa is referred to as 'Āryabhaṭa' by Bhāskara, Nṛsiṃha and Munīśvara, and Nṛsiṃha also refers to him as 'Vṛddhārya'. The later Āryabhaṭa is referred to as 'Āryabhaṭa' by Nṛsiṃha, and Munīśvara. The latter also calls him 'Laghvāryabhaṭa' and calls his work '*Āryabhaṭasiddhānta*'.

There is evidently room for confusion here, with this deplorable lack of precision and continuity in the names of the works. Such a tendency is hardly confined to the older period: why should the work by Bhāskara I which he himself called *Āryabhaṭakarmanibandha* now always be known as the *Mahābhāskarīya*?

The references used by the editors only increase the confusion. Sudhakara (p. 22) comments on the later author's remark (*Pātādhikāra*, śl. 14, ed. p. 128) that he has based his work on 'Vṛddhāryabhaṭa'. With unbelievable inconsistency, Sudhakara then refers to that source as Laghvāryabhaṭa, and to his work as *Laghvāryabhaṭīya*, asserting then that these are different from Āryabhaṭa and the *Āryabhaṭīya*! Jhā refers to either work indifferently as *Mahāryabhaṭasiddhānta*; on pp. 113, 118 this means the early work, the familiar *Āryabhaṭīya*, while on p. 52, this means the *Mahāsiddhānta*.

In any case, we now have a broader context in which to place the remark of Bhāskara with which this excursion began. We have another instance where he refers to Āryabhaṭa in a way that cannot be supported by the *Āryabhaṭīya*, i.e. Jhā 287, for there are no metrological references in the *Āryabhaṭīya*. We have seen from the habits of later authors that we cannot absolutely expect that when he wished to refer to separate works in the Āryabhaṭīya tradition, he would have used separate names for the authors. Should we not consider that he may have been referring to the work which Shukla has isolated under the title *Āryabhaṭasiddhānta*, whether or not that was the lost work on the midnight system.

In passing we may observe that references to Parāśara similarly show a distinction between the older and later authors of that name. The text included in the second chapter, *Parāśaramatādhyāyaḥ*, of the *Mahāsiddhānta*, is attributed to the 'Second Parāśara'. After a quotation of śl. 1–4 of Ch. 2, Munīśvara writes (Jhā (1917), p. 85) *iti dvitīyaparāśaramatādhikāre laghvārbhaṭoktamapi*. Elsewhere Munīśvara refers to Vṛddhaparāśara, e.g. Jhā (1917), p. 84. Nṛsiṃha does not mention the 'Second Parāśara', but only the older figure, linked with Brahma and Sūrya: *Pitāmahasauraparāśarādisiddhāntokta*.

5. Conclusion

The date of composition of anonymous *Mahāsiddhānta* is discussed in the light of the method of investigation established by R. Billard, who found it to be in the early 16th century. This leads to a re-examination of the earlier estimate of date *ca* A.D. 980 made by S.B. Dikshit a century ago, and the reason for Dikshit's erroneous date is explained. Dikshit introduced the style 'Āryabhaṭa II' for the author's name, since he calls himself Āryabhaṭa. A number of references to the work are found in commentaries written by Nṛsiṃha and Munīśvara in the early 17th century, and these are compared with references to the early Āryabhaṭa. When an attempt was made to distinguish them, they were referred to as Vṛddhāryabhaṭa and Laghvāryabaṭa (Old-, Young-Āryabhaṭa). Bhāskara (in A.D. 1150) made more than one reference to 'Āryabhaṭa' which cannot be supported by the extant *Āryabhaṭīya*, and it is suggested that this may have been a work named *Āryabhaṭasiddhānta* which was known at that time. This is a work of which a few fragments are extant, which Shukla believed was the lost work of Āryabhaṭa on the midnight system, although that claim does not seem to be justified.

REFERENCES

Mahādeva Cimaṇāji Āpṭe, *Vāsanābhāṣya-Marīcinkābhyām sahitaḥ Śribhāskarācārya-viracitaḥ Golādhyāyaḥ, pūrvārdharūpaḥ prathamo bhāgaḥ*, Ānandāśrama Sanskrit Series 122, Poona, 1943.

Roger Billard, *L'Astronomie Indienne, Investigation des textes Sanskrits et des données numériques*, École française d'Extrême-Orient, Maisonneuve, Paris, 1971.

Śankar Bālakṛṣṇa Dīkshit, *History of Indian Astronomy*, Part II: History of Astronomy during the Siddhantic and Modern periods, Government of India Press, 1985. Translated by R.V. Vaidya from *Bhāratīya Jyotiṣa Śāstrā cā Itihās āṇi Paricaya* (in Marathi), 1896.

Pandit Muralīdhara Jhā, *Siddhānta Śiromaṇe Grahagaṇitasya Madhyamādhikārāntaḥ Vāsanābhāshyavāsanavārtikamarīcibhiḥ sahitaḥ*, A Treatise on Astronomy by Bhāskarācārya, Vol I: *Madhyamādhikāra*, with *Vāsanābhāshya, Vāsanavārtika* and *Marīci*, Lazarus & Co., Medical Hall Press, Benares, 1917.

Raymond Mercier, "The meridians of reference of Indian astronomical canons", in G. Swarup, A.K. Bag and K.S. Shukla (Ed.), *History of Oriental Astronomy*, 1987, pp. 97–107.

David Pingree, Reply to van der Waeden (1980), *J. for History of Astronomy*, **11** (1980), 58–62.

V. Prasāda (Ed.), *Jyautiṣasiddhantasaṁgraha*, Benares Sanskrit Series No. 154, Benares, 1917.

K.S. Shukla (Ed.), *The Sūrya-Siddhānta, with the commentry of Parameśvara*, Lucknow, 1957.

K.S. Shukla, "Āryabhaṭa I's astronomy with midnight day-reckoning", *Gaṇita*, **18** (1967), 83–105.

________, "Glimpses from the *Āryabhaṭa-siddhānta*", *Indian J. of History of Science*, **12** (1977), 181–6.

Mahāmahopādhyāya Sudhākara Dvivedi (Ed.), *Mahāsiddhānta: A treatise on astronomy by Āryabhaṭa*, Benares, 1910.

B.V. Subbarayappa and K.V. Sarma (Comp.), *Indian Astronomy: A Source-Book* (based primarily on Sanskrit texts), Nehru Centre, Bombay, 1985.

G. Swarup, A.K. Bag and K.S. Shukla (Eds.), *History of Oriental Astronomy*, Proceedings of IAU Colloquium No. 91 (New Delhi, 13–16 Nov., 1985), Cambridge University Press, 1987.

B.L. van der Waerden, "Two treatises on Indian astronomy", *J. for History of Astronomy*, **11** (1980), 50–58.

ADDENDA AND CORRIGENDA

IIa

p.202 *For* 2s 18;50 *read* 2^s 17;50.

It is assumed without discussion that the text *De motu octave spere* is due to Thābit ibn Qurra. The disproof of this common assumption was given in article III.

Further, in article III it is also shown that the insistence on the precise reading of the Latin text regarding the constraint on the ecliptic was rather overdone in articles IIa, IIb, and that consideration of the variability of the obliquity shows that the simpler, if not literally exact, interpretation of the Latin text is to be preferred.

IIb

p.35 Line 6 from the end, *for* equator *read* ecliptic.

p.53 *For* Davis *read* Davies.

p.58 Line 2, *For* vol.2, p.124 *read* vol.1, p.42.

p.58 Line 2 from end, it is better to put $a_2 = \arcsin(\sin 9 \sin \theta)$.

p.70 *For* Ptolomaeus *read* Ptolemeus.

III

p.324 n.29. The reference to Blachère (1935) was omitted: *Ṣâ'id al-Andalusī, Kitâb Ṭabaḳât Al-Umam (Livre des Catégories des Nations)*, Traduction avec Notes et Indices précédées d'une Introduction par Régis Blachère, Paris : Larose 1935.

pp.331 seq. The references to the text relating to Ibn Kammād are to the Manuscript Madrid Biblioteca Nacional 10023.

IV

p.452 n.7 *For* Payne Smith (1910), *read* Robert Payne Smith, *Thesaurus Syriacus*, 1868-1901; col. (not page) 3089.

p.457 As well as jīng, there is also bīan 編 weave, compile.

p.460 Rizvi (1965) : the full name of the author is Sayyid Samad Husain Rizvi.

p.460 West (1892) : the full name of the author is Edward William West.

V

p.99 Add :
r = number of the parameters to be fixed: r = 3 for t and φ, r = 2 for φ or t only.

p.100. Billard had assumed incorrectly that only Gaussian statistics were required in the calculation of confidence bounds. That would be correct only if the number of degrees of freedom is large, but in this case it is at most 6, so that one must use the Student Distribution.

In finding the optimum meridian I used the older formula for ΔT, due to Spencer Jones. This results in meridian positions near to 76, whereas with the more recent formulae for ΔT, due to Morrison and Stephenson the result is nearer 80. In fact the uncertainty in the value of ΔT greatly outweighs that in the calculation of the geocentric longitude. There is therefore no urgent need to replace the older procedures based on the formulae of Newcomb, Brown, and Leverrier, with those of Bretagnon.

VI

p.22 *For* Selden supra 17 *read* Selden supra 16.
p.27 *For* Muhya *read* Muḥyī

VII

p.112 Central column of table, *For* 134;36 *read* 134;43.

IX

p.37 *For* Sanjarīa nd *read* Sanjarī and.
p.43 In the line T.2.2.10 :

For
Equation of days and nights to be subtracted from Sun
read
Equation of time given for true Sun and subtracted from true Moon

p.46 *For* Sintaxis *read* Syntaxis.
p.55 *For* ☌-♄ +6;0 *read* ☌-♄ +7;0.
For ☌-♂-12 ;0 *read* ☌-♂+12 ;0.

X

p.127 note 6, *For* MS Padua Escorial e.III.23 *read* MS Escorial e.III.23.

INDEX OF MANUSCRIPTS

INDEX OF AUTHORS